An Introduction to Electrical Measurements

Text for ECE 3041

Fifth Edition

Thomas E. Brewer
Georgia Institute of Technology

Kendall Hunt
publishing company

www.kendallhunt.com
Send all inquiries to:
4050 Westmark Drive
Dubuque, IA 52004-1840

ISBN 978-0-7575-7917-2

Printed in the United States of America
10 9 8 7 6 5 4 3

Contents

Chapter 1

Orientation

1.1 Objective

The object of this experiment is to obtain familiarity with basic laboratory instrumentation by performing an elementary experiment.

1.2 Equipment

1.2.1 Agilent Technologies 3630A Triple DC Power Supply

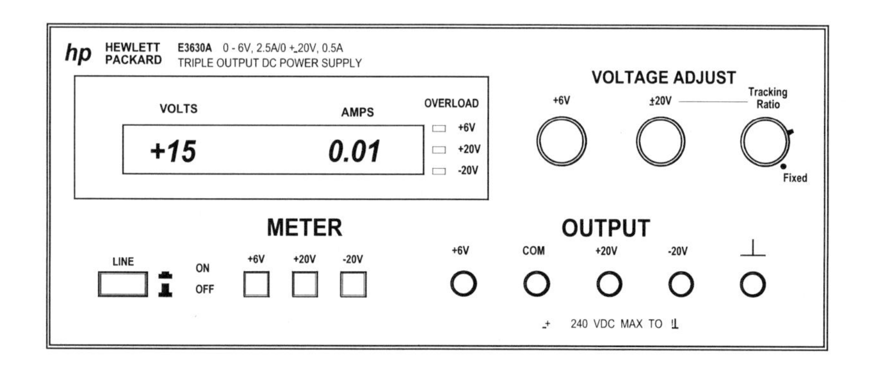

Figure 1.1: Agilent Technologies 3630A triple output dc power supply.

The dc energy source for this experiment will be the **Agilent Technologies 3630A** dc power supply shown in Fig. 1.1. This is a triple dc power supply which means that there are three separate dc supplies in the same instrument chassis. Each power supply is variable. One can be varied from 0 to $+ 6\ V$, another can be varied from 0 to $20\ V$, and the last from 0 to $-20\ V$.

It has five output connectors known as female binding posts. Each binding post accepts a male banana plug type connector. The binding post on the far right has the symbol $\perp$ which means that it is connected to the ground wire in the ac power cord for the instrument; it will not be used in this or subsequent experiments. The binding post labeled COM is known as the common because it is common to all 3 variable dc supplies and is the reference for $0\ V$. The binding post labeled $+6\ V$ can be varied from 0 to $+6\ V$ and has a maximum current capacity of $2.5\ A$. The binding post labeled $+20\ V$ can be varied from 0 to $20\ V$ and has a maximum current capacity of 0.5 A. The binding post labeled -20 V can be varied from 0 to -20 V and has a maximum current capacity of 0.5 A.

An internal digital voltmeter and ammeter is used to monitor the output of one of the power supplies. The user selects which power supply is being monitored by pressing the buttons below the display. If only a single dc power supply is desired, the $+20$ V supply can be used. The $+20$ V meter button is pushed which then means that the meters will monitor this supply. The $+20$ V knob is then varied until the desired dc voltage is shown. The COM terminal is then the minus voltage on the dc supply and the $+20$ V is the plus terminal.

This type of dc supply is known as a regulated dc supply. For this type of supply the terminal voltage does not change as the load current changes as long as the current is less than a value known as the current limit. An internal feedback circuit is used to maintain this constant terminal voltage. The current limit is 2.5 A for the 6 V supply and 0.5 A for the other two. If the current limit is reached, a yellow LED lights to the right of the display; this usually indicates that the user has made a grievous error.

The tracking ratio knob sets the ratio of the magnitude of the -20 V to the $+20$ V output. When the tracking ratio knob is set to its fully clockwise position the magnitude of these two variable outputs are equal to within 1 %. Turning this knob counter clockwise makes the magnitude of the -20 V a fraction of the $+20$ V output. Normally, this knob will be set to its fully clockwise position.

A normal configuration for this triple output dc power supply would be to set the variable $+6$ V output to $+5$ V and the magnitude of the two variable 20 V supplies to $+15$ V. The ±15 V are normally used to bias op amps and the $+5$ V to bias TTL digital ICs.

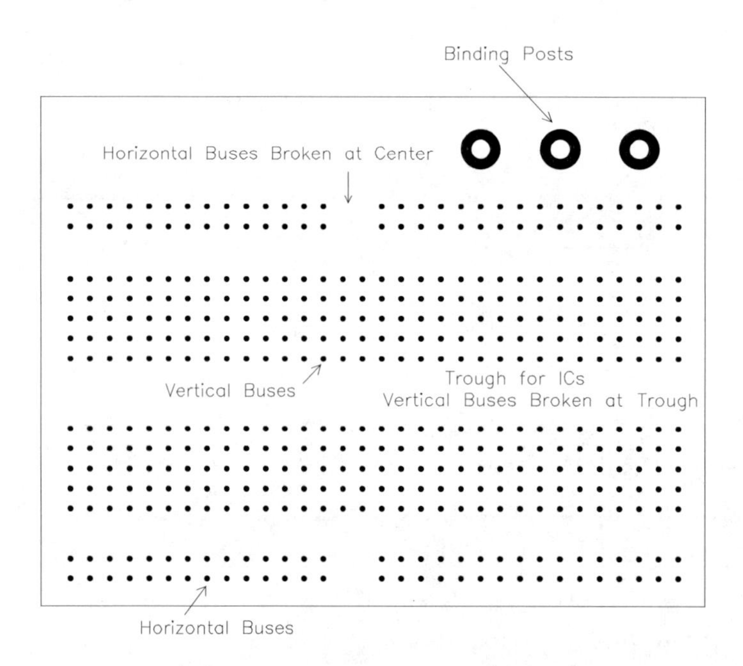

Figure 1.2: Solderless breadboard.

1.2.2 Breadboard or Protoboard

In the design of an electronic circuit it is often desirable to construct a prototype circuit to evaluate the design. The circuit construction technique that is used to interconnect the components should provide a certain degree of mechanical rigidity for the interconnections and still offer the flexibility for the circuit constructor to easily alter the circuit until the design is finalized. Solderless breadboards or protoboards are devices on which these prototype electronic circuits can be constructed with the above properties.

Most solderless breadboards consist of arrays of holes that are formed in blocks of plastic (Fig. 1.2) to accommodate No. 22 gauge single stranded hook-up wire so that components such as low-power and -voltage resistors, capacitors, diodes, transistors, integrated circuits, etc., can be inserted onto one of these boards and interconnected. Such boards have rows and columns of holes for component insertion which are connected together under the board with jumpers for ease of circuit construction. Holes which are electrically common form a bus. A circuit constructor would choose separate buses for each node of the circuit. Normally either the rows and columns form a bus with many breadboards offering a combination of the two. It is imperative that the circuit constructor understand the topology of the breadboard being used or many hours of circuit construction may be wasted.

The blocks of plastic are usually placed on metal plates with one or more binding posts mounted on them. These binding posts (a type of electrical connector) can easily be connected to power supplies, function generators, etc. with leads which have banana jack terminals. The binding post have holes cut in them so that they may be electrically connected to buses on the breadboard. Normally the binding post that is colored black is electrically connected to the metal plate on which the plastic blocks rest and should be used as the ground plane in an electronic circuit.

The term breadboard may seem to be a bizarre label for an object on which electrical circuits are constructed. This term dates from the early days of electrical engineering when radios were constructed on highly varnished wooden boards which looked like the boards on which bread dough was kneaded. The term has become part of the lexicon of electrical engineering and is used to describe any structure on which electrical or electronic circuits are constructed in a prototype version. Some engineering publications use the word breadboard as a verb as in "to breadboard a circuit". Such usage is undesirable and the term is properly used only as a noun.

Figure 1.3: Alligator clip and banana plug.

1.2.3 Alligator Clips or Connectors

An alligator clip is an electrical connector that attaches to a banana jack lead and has metal prongs (resembling the jaws of a small metal alligator Fig. 1.3) that are used to make a pressure connection to another lead.

1.2.4 Resistors

The resistors that will be used in this laboratory are 1/4 watt carbon film resistors. The quarter watt is the maximum power rating for the resistor; powers larger than this will destroy the resistor (turn into smoke).

These resistors have a color code place on them to indicate the nominal value of the resistor (Fig. 1.4). The actual value of the resistor can differ from the nominal value by the tolerance which is usually given as a percentage of the nominal value of the resistor.

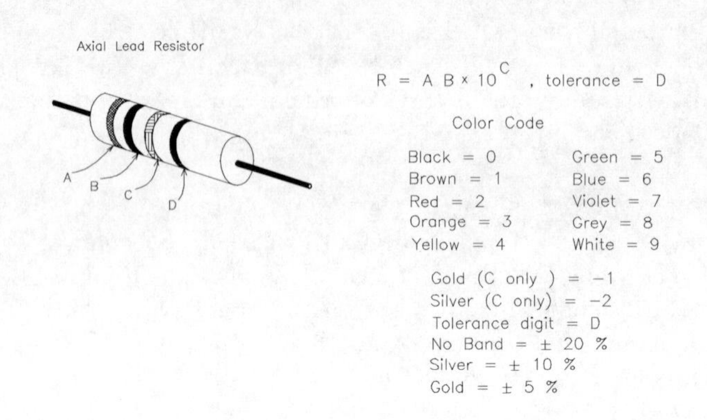

Figure 1.4: Color code for carbon film 5% resistors.

Figure 1.5: Agilent 34401A Digital Multimeter.

1.2.5 Agilent Technologies 33301A Digital Multimeter (DMM)

This is a digital instrument that displays the electrical parameter being measured (dc or ac voltage or current or electrical resistance) to six and one half digits [Fig. 1.5]. The designation one-half digit means that the most significant digit cannot take on all ten of the decimals digits. The number of displayed digits can be varied from 4 to 6.

When this instrument is turned on (white button located in lower left) it goes through a initial check-out procedure and illuminates the entire blue fluorescent display for a few seconds. The instrument then is automatically set to measure a dc voltage in the AUTO ranging mode (which simply means that the instrument selects an appropriate range voltage instead requiring that the user make the selection). Each time a measurement is made an asterisk in the lower left of the display flashes.

The input terminals for this instrument are located on the right; there are five female banana jack type connectors. The two terminals indicated as "Ω $4W$ $Sense/Ratio$ Ref" are used to make 4 wire resistance measurements (separate leads for the voltmeter and ammeter inputs to the DMM) which are used to measure very small resistances and will never be used in this laboratory. The inputs are normally used are the three female banana jacks on the far right. The input selector button should always be in the $FRONT$ position (button out).

To measure a dc voltage the button labeled "$DC\ V$" should be pressed and the inputs located in the upper right labeled HI (red) and LO (black) used. (The two connectors below the symbols $Input\ V\ \Omega\ \rightarrow$.) To measure the rms value of an ac voltage the button labeled "$AC\ V$ " should be pressed and the same input terminals should be used. The red terminal is the high or plus input and the black is the low or negative input. If the terminals are reversed when measuring a dc voltage, the voltage will be displayed with a sign change. Obviously there is no sign change with an ac voltage.

To measure electrical resistance the button labeled "Ω $2W$" should be pressed. The input terminals are the same as for voltage measurement. A continuity checker position is available for which an audible beep is emitted by the instrument when a small resistance is measured. A pn junction measure position is available so that when the "resistance" of a pn junction is measured a sufficient voltage is provided to break down the junction.

To measure a dc current the blue button labeled "$Shift$" should be pressed and then "$DC\ I$" which shares the same button with "$DC\ V$". This instrument is functional similar to a HP calculator in that when the blue "$Shift$" button is pressed the blue functions above the buttons are active. The input terminals used to measure current are the red "$3A$" and the black "LO". The "$3A$" means that an internal 3 Ampere fuse is placed in series with the input. Thus if a current larger than 3 $Amps$ were placed through the instrument this fuse would blow. For ac currents, the button "$AC\ I$" should be pressed.

For periodic functions of time this instrument can be used to measure frequencies from 3 Hz to 300 kHz. The corresponding period can also be measured.

For sinusoidal waveforms, level changes in decibels can be measured. A reference of dB or dBm can be used.

The status of this instrument is constantly displayed. The instrument will indicate whether it is set to measure dc, ac voltage or current or electrical resistance or frequency.

1.2.6 Hook-Up Wire

Electrical connection are made between buses on the protoboard using hoop-up wire. This is single stranded (one piece of wire) wire with a diameter known as No. 22 gauge. The wire must be cut into short lengths that make for easy connections and the insulation must be stripped off each end (about 3/8 inch).

1.2.7 Hewlett-Packard 3311A Function Generator

The instrument shown in Fig. 1.6 is known as a function generator because it produces voltages which are functions of time. It can produce, among other things, a sinusoidal waveform with a dc level added to the sine wave. Namely, it can produce the waveform

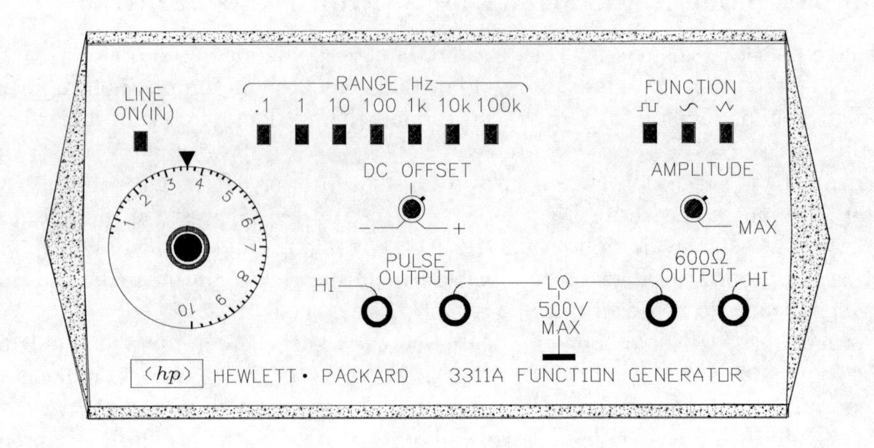

Figure 1.6: Hewlett Packard 3311A function generator.

$$v(t) = V_{os} + V_p \sin(\omega t) \tag{1.1}$$

where the voltage V_{os} is set with the dc level or offset control which is a rotary knob, the voltage V_p is set with the amplitude control which is also a rotary knob and the frequency f is set with a range selector and vernier knob. To produce a sine wave, i.e. a sinusoidal function of time, the function selector is set to sine. The output is taken from the terminals labeled 600 Ω which simply means that this function generator is equivalent to an ideal voltage source in series with an internal resistance of 600 Ω. It should be borne in mind that the 600 Ω resistor is an internal resistor contained inside the function generator.

1.2.8 Simpson Meter Model 260-7

This is a nonelectronic multimeter which uses a d'Arsonval meter movement as its readout device. It is shown in Fig. 1.7. The input terminals that are used are in the lower left corner of the plastic case and are labeled "+" and "*COMMON*". To measure a dc current or voltage the function switch should be set to "+*DC*" and the terminal labeled "+" should be higher in potential than the "*COMMON*" terminal for the pointer to deflect upscale. The range should be selected prior to connecting the instrument to a circuit and should be set to a larger value than the voltage being measured; if the voltage being measured is not even approximately known, the largest range voltage should be used and the range decreased until an angular deflection is obtained that can be easily read. For ac voltage and current measurement the function is switched to ac and the terminal connections are immaterial. When the pointer deflects, the value of the circuit parameter is read off the appropriate scale, e.g. if the 10 *Volt* range is selected that goes from 0 to 10 is used.

The **Simpson** meter has a dc sensitivity of $S = 20\ k\Omega/Volt$ and an ac sensitivity of $S = 5\ k\Omega/Volt$. Thus the input impedance when used as a dc voltmeter is $R_{in} = SV_{fs}$ where V_{fs} is the range voltage where $S = 20\ k\Omega/Volt$. When the function switch is changed to ac the voltmeter will measured the rms value of a sinusoidal voltage with a dc level of zero (it is not a true rms voltmeter and will not measure the correct rms value of any other type of voltage).

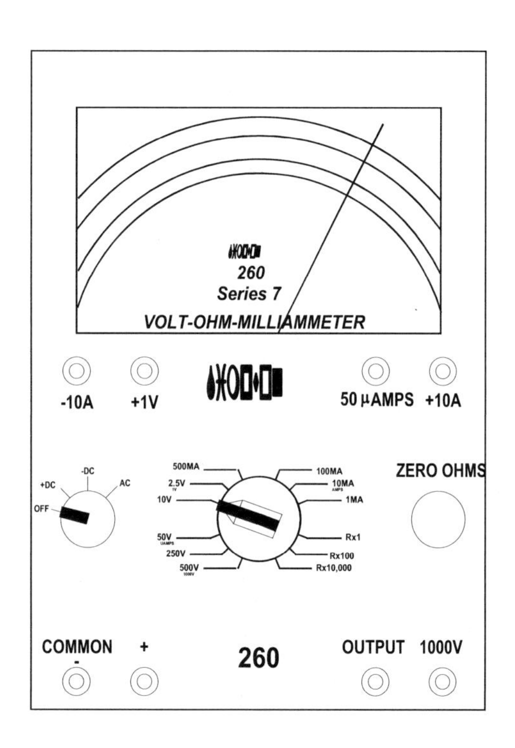

Figure 1.7: Simpson Meter Model 260-7 analog multimeter.

1.2.9 Capacitors

The capacitors that will be used in this laboratory are ceramic disks. These will either have the capacitance printed directly on them in microfarads or will have a capacitor code. A typical capacitor code would be "104" which means 10×10^4 picofarads $= 0.1\ \mu F$. Although not universally adhered to, the code that is normally used is that if the number printed on the capacitor is less than one then this is the capacitance in microfarads and if the number is greater than 1 then this is the capacitance in picofarads where the last number indicates the power of ten that the first two numbers are to be multiplied by to yield the nominal capacitance.

1.2.10 ELVIS

ELVIS is an acronym that stands for Educational Laboratory Virtual Instrumentation Suite. It consists of a PCI board known as a DAQ (Data Acquisition), a Breadboard station, and a connecting cable. It contains a multimeter, an oscilloscope, an LCR meter, a spectrum analyzer, and other instruments. A software package known as LabVIEW is used to provide a Soft Front Panel for these instruments. The interface with the analog circuits and signals is the breadboard station. The version of **ELVIS** currently used is **ELVIS II+**.

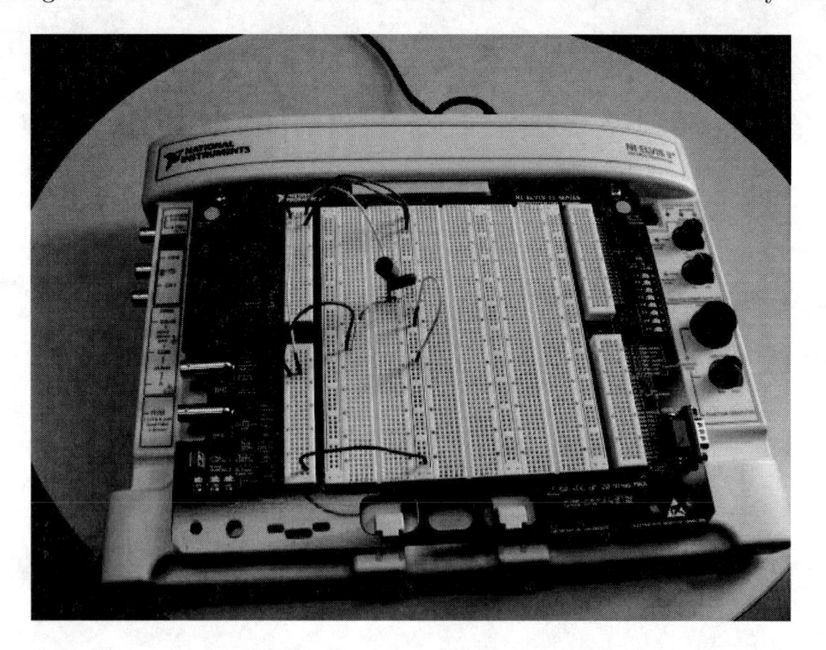

Figure 1.8: ELVIS II+.

1.3 Procedure

1.3.1 Resistance Measurement

Three Ohmmeters will be used to measure resistance: the **Agilent 34401A DMM**, **ELVIS,** and the **Simpson Meter Model 260-7**. The first is a digital multimeter, the second a digital integrated pc measurement system, and the last is an analog multimeter. The digital meters provide more accuracy and most do not have to be "zeroed" or calibrated for zero Ohms.

The resistors that will be used in this experiment are 1/4 Watt, 5% resistors which means that they have a maximum power dissipation of 1/4 Watt and have an actual resistance that lies within 5% of the nominal value which is specified by a color code.

Chapter 2

Signal Characteristics

2.1 Objective

The object of this experiment is to measure the gross parameter or characteristics of signals with meters.

2.2 Theory

A signal is function of time containing information about physical process. For simplicity it will be assumed that the signal is a real single valued function. Examples of signals are the instantaneous values of voltages, currents, temperatures, pressures, etc. This experiment will be restricted to voltage and current signals. If the signal is a periodic function of time it is also termed a waveform or wave.

A complete description of a signal requires a complete plot of the signal as a function of time or frequency. An oscilloscope or spectrum analyzer can be used such plots. Oftentimes gross parameters of the signal such as the dc level or rms value is all that is required. These gross parameters can be obtained with instruments know as meters.

2.2.1 dc Level

The dc level or value of a signal is simply the time average value of the signal. For a voltage, $v(t)$, the dc level, V_{dc}, is given by

$$V_{dc} = \lim_{T\to\infty} \frac{1}{T} \int_{-T/2}^{T/2} v(t)dt \tag{2.1}$$

Physically the dc level represents the component of $v(t)$ which would cause a dc current to flow through a resistance. The dc level is also know as the average value, steady value, or dc component.

If the voltage is a periodic function of time with period T the dc level is given by

$$V_{dc} = \frac{1}{T} \int_{\alpha}^{\alpha+T} v(t)dt \tag{2.2}$$

where α is an real number. A function such as $v(t)$ is a periodic function of time with period T if

$$v(t) = v(t \pm nT) \tag{2.3}$$

where n is any integer and T is the smallest value of time for which Eqn. 2.3 is true.

2.2.2 rms Value

The rms value of a voltage or current is the value of an equivalent dc voltage or current which would produce the same heating effect in a resistor. For a voltage $v(t)$ the rms value of $v(t)$ is defined as

$$V_{rms} = \sqrt{\lim_{T\to\infty} \frac{1}{T} \int_{-T/2}^{T/2} [v(t)]^2 dt} \tag{2.4}$$

which is the square root of the time average value of the voltage squared. If $v(t)$ is periodic with period T

$$V_{rms} = \sqrt{\frac{1}{T} \int_{\alpha}^{\alpha+T} [v(t)]^2 dt} \tag{2.5}$$

where α is any real number. The rms value of an ac voltage is often just called the ac value of the voltage.

2.2.3 Crest Factor, cf

The crest factor of a signal is simply the ratio of the peak value to the rms value. This gross parameter cannot be determined solely with a meter because the peak value of the signal must be determined.

2.3 Examples

1. Sine Wave.

$$v(t) = A \sin \omega t \tag{2.6}$$

where A is the peak value in volts, $\omega = 2\pi f$ is the angular frequency in radians per second, f is the frequency in Hertz, and $f = 1/T$ where T is the period in seconds. For this signal $V_{dc} = 0$, $V_{rms} = A/\sqrt{2}$, $cf = \sqrt{2}$.

2. Square Wave.

$$v(t) = \begin{cases} -A \text{ for } -T/2 < t < 0 \\ A \text{ for } 0 < t < T/2 \\ 0 \text{ for } t = 0 \\ v(t \pm nT) \end{cases} \tag{2.7}$$

where A is the peak value of the square wave in volts, T is the period in seconds, and n is any integer. For this signal $V_{dc} = 0$, $V_{rms} = A$, and $cf = 1$.

3. Triangular Wave.

$$v(t) = \begin{cases} -\frac{4t}{T}A - 2A \text{ for } -T/2 < t < -T/4 \\ 4t/T \text{ for } -T/4 < t < T/4 \\ -\frac{4t}{T}A + 2A \text{ for } T/4 < t < T/2 \\ v(t \pm nT) \end{cases} \tag{2.8}$$

where A is the peak value of the square wave in volts, T is the period in seconds, and n is any integer. For this signal $V_{dc} = 0$, $V_{rms} = A/\sqrt{3}$, and $cf = \sqrt{3}$.

2.4 Procedure

2.4.1 Measurement of Function Generator Signals

Sine Wave

Turn on the **HP 3311A Function Generator** and set it for a sine wave with a frequency of 100 Hz. Do this by setting the range to 10 and the vernier dial to 10. Turn on the **Agilent 34401A Digital Multimeter**

(DMM) and set it to measure dc voltage. Connect the output labeled "600 Ω Output" of the function generator to the DMM with two banana plug leads.

Vary the dc level or dc offset control on the function generator until the dc voltage measured by the DMM is nulled or zero. It cannot be set exactly to zero but try to adjust it until the dc voltage is less than 10 mV. Whether positive or negative is not the goal; the magnitude of the dc voltage should be less than 10 mV. It may be easiest just to watch the sign on the DMM and determine the setting where the voltage switches from positive to negative.

Set the DMM to measure ac voltage. Vary the amplitude control on the function generator until the DMM indicates an ac voltage of approximately 1 V. Record the measured value of the dc and ac values of the signal.

Without changing the setting on the function generator connect the output of the function generator to the **Simpson** meter with two banana plug leads. Measure and record the dc and ac voltages with this meter. Use the red ac voltage scale when measuring ac voltages.

Launch **ELVIS** by double clicking on the **NI ELVISmx Instrument Launcher** icon on the desktop of the pc or through the START programs sequence. Connect the output of the HP function generator to the voltage banana jack (female banana jacks) on the upper left side of **ELVIS** with two banana plug leads. Connect the HI output of the hp function generator to the V input banana jack on **ELVIS** and the LO output of the hp function generator to the COM input banana jack on **ELVIS**. Double Click on the Digital Multimeter (DMM) on the INSTRUMENT LAUNCHER on **ELVIS.** Measure dc and ac voltage with **ELVIS** by clicking on dc and ac voltage on the panel and then the Run arrow. If the reading is constantly changing change the Acquisition Mode from Run Continuously to Run One. Click the Run arrow and record the value. Click on Run 4 more times and average the 5 values. Record the measured values.

Square Wave

Change the function on the hp function generator from sine to square. Do not change the dc offset or amplitude settings on the hp function generator. Measure and record the ac and dc voltages indicated by the **Agilent DMM**, the Simpson meter, and **ELVIS**.

Triangular Wave

Change the function on the hp function generator from square to triangular. Do not change the dc offset or amplitude setting on the hp function generator. Measure and record the ac and dc voltages indicated by the **Agilent DMM**, the **Simpson** meter, and **ELVIS**. Close the **ELVIS** digital multimeter soft front panel.

2.4.2 Observation of Function Generator Signals

Set the function on the HP function generator back to sine. Do not change the setting of either the dc offset or amplitude on the HP function generator. Use a banana plug to BNC connector to connect the output of the HP function generator to CH 0 BNC input connector on **ELVIS** (lower right). On the **ELVIS** Instrument Launcher Double Click on Oscilloscope (Scope). Click the Run arrow. Change the VERTICAL Scale to 500 mV/div. Change the TIMEBASE to 2 ms/div. Change the Trigger Type to Edge.

The oscilloscope screen display will now be plotted. Click Stop on the oscilloscope soft front panel; this stops the continuous acquisition of data and permits saving the data points to a file. Click LOG. (It is very important that Stop and LOG only be clicked once. If it is clicked more than once the array of data will be appended with each click.) Save the file somewhere on the pc. Start EXCEL. Click File Open on Excel and change the file type to text. Import the file saved from **ELVIS**. This should start the text import wizard. Select import starting at row 6. Leave the File Type as Delimited. Click Next. Step 2 of the text import wizard should now appear. Select Space, Tab, Comma, and Other as delimiters and enter a colon in the Other box. Click Next. Click Finish. There should now be 3 columns in the spreadsheet. Delete the first two. There should now be only one column which is the values of the voltage measured. Insert a column at the beginning of the spreadsheet. Open the data file and enter 1, 2, 3 times the value shown for delta t in the first 3 rows of the blank column. Highlight these first 3 rows of column 1. Place the pointer at the lower right of the 3 highlight spreadsheet entries until it turns into a plus sign. Double click it; this should fill in values for the time in column A. Select the entire two columns of data and copy the selection. Minimize or close Excel.

Start KaleidaGraph by clicking on the desktop icon or the Start Programs sequence. Click on the clipboard icon on the toolbar. This should fill in the two columns of time and voltage. Click on Gallery and then Linear and Line. Select the X axis as column A and the Y axis as column B. Select New Plot. A plot of a sine wave should now appear. Double click on the title and change it to Sine Wave. Double Click on the titles on the horizontal and vertical axes and change the names to time and voltage. Click on the text box ("T" in upper left of screen). Enter your name and GTID number, the date, time, and lab section. Print the graph and include it in the lab report.

Delete the file saved from **ELVIS**. Repeat the above procedure for the square and triangular waves. Turn off the hp function generator and remove all leads attached to it.

2.4.3 Arbitrary Waveform

In this procedure Mathcad will be used to specify an arbitrary function which will then be imported into DAC0 (Digital to Analog Converter 0) of **ELVIS**. The dc and rms value of the waveform will then be measured with meters. And the waveform will be observed on the **ELVIS** oscilloscope. The arbitrary waveform will be specified by the laboratory instructor. The following is a Mathcad sheet (shown as Figure 2-1) which illustrates how to generate a parabola Use the same procedure for the waveform specified by the laboratory instructor.

Start Mathcad from either the desktop icon or the Start Programs sequence.

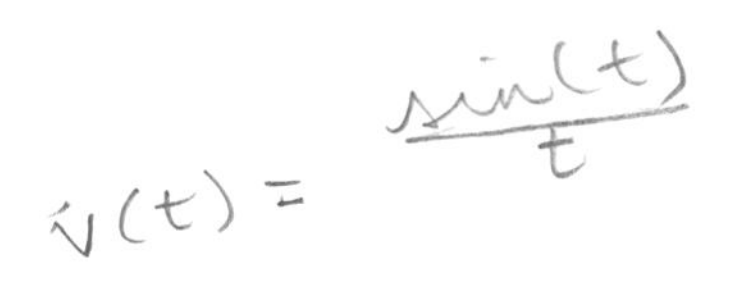

Mathcad File for Parabola (This Portion of the Worksheet is Text)

$A := 1$ Specfies Amplitude of 1 V. Keyboard Sequence A : 1

$T := 1$ Specifies Period of 1 s. Mathcad is case sensitive

$v(t) := A\cdot\left(\frac{t}{T}\right)^2$ Specifies one cycle for v(t). Keyboard sequence v(t):A*t/T space ^2

Next v(t) must be digitized for export to arbitrary waveform generator with N points $N := 100$

$i := 0..N-1$ Index i from 0 to N-1 Keyboard sequency i:0;N-1

$d_i := v\left(\frac{i}{N-1}\right)$ Specifies a data vector with N points by sampling v(t)

The i on d is not a subscript but an array index. Keyboard sequency d[i:v(i/N-1 space space)

Next the data must be normalized from -1 to +1

$Max := max(d)$ $Min := min(d)$

$d_i := \left(d_i - \frac{Max + Min}{2}\right)\cdot\frac{2}{Max - Min}$ Next the first 15 entries will be displayed by typing d= on the keyboard (no colon)

d =

	0
0	-1
1	-1
2	-0.999
3	-0.998
4	-0.997
5	-0.995
6	-0.993
7	-0.99
8	-0.987
9	-0.983
10	-0.98
11	-0.975
12	-0.971
13	-0.966
14	-0.96
15	...

Place the pointer inside the vector on the left, right click and select Select All. This should highlight the entire vectorl. Right Click again and select Copy Selection which copies it to the Windows Clipboard. Minimize Mathcad, start Windows Notepad, and paste the data in.Save the file with the .txt extension. It will be imported into the ELVIS Arbitrary Waveform Generator.

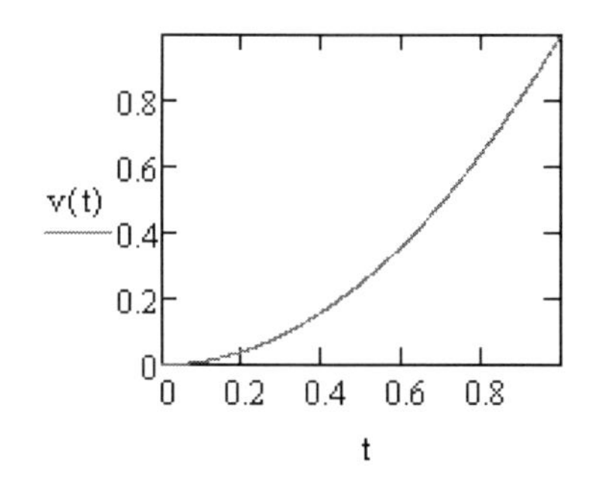

Mathcad Code for Arb

Now the file which was saved as a text file can be imported into **ELVIS**. Use the instrument launcher on **ELVIS** to start the Arbitrary Waveform Generator (ARB); double click it. Click on Waveform Editor on the Soft Front Panel of the Arbitrary Waveform Generator. Click Edit, Import From File, and then import the text file produced by Mathcad. Click ok on the Open Text File Wizard. Place the pointer on "2: (10 sec)" in the box in the left hand corner and click so that it is highlighted. This is the blank second segment. Delete it. There should now only be one cycle of the parabola displayed by the Waveform Editor. Click on segment 1 and change the duration to 1 second. Save the sampling rate to 1,000 and, therefore, the number of samples to 1,000. Save the file. For the file type choose the default with extension .wdt. Choose the default number of samples and sample rate. Close or minimize the Waveform Editor.

On the Arbitrary Waveform Generator Soft Front Panel click on the choice for AO 0 (Analog Output Zero) and select the file that was generated by the Waveform Editor and saved in the previous step. Click Enabled. One cycle of the parabola should now be displayed by the Arbitrary Waveform Generator. The frequency of the waveform supplied by AO 0 is given by the frequency of the waveform stored time the update rate divided by the number of points. The frequency will now be set to 100 Hz by changing the update rate to 1 MS/s. Press the Run arrow.

Use a jumper wire to connect AO 0 on the **ELVIS** breadboard to the AI 7+ (Analog Input 7). Use another jumper wire to connect the AI 7 - to the ground bus.

Launch the ELVIS oscilloscope, switch the input to AI 7, and use the same procedure that was used for the sine, square, and triangular wave to plot the display. Close the oscilloscope.

Use a BNC to banana plug lead to connect the output of BNC 1 on the ELVIS to the input of the **Simpson** meter. Connect AO 0 to BNC 1+ with a jumper wire and BNC 1- to ground. Measure and record the dc and ac voltage indicated. Repeat for the **Agilent 34401A DMM**. Repeat for the **ELVIS** DMM.

2.5 Laboratory Report

Compare the measured and expected values of the dc and ac components for the waveforms examined. Which instrument was the most accurate? Which was the least?

Turn in all plots and data taken.

2.6 References

1. Paton, B, *Hand-On NI ELVIS*, National Instruments, Part No. 323777A-01, 2004.
2. National Instruments, *NI Education Laboratory Virtual Instrumentation Suite (NI ELVIS)*, Part No.323363A-01, 2003.

Chapter 3

Meter Movements

3.1 Objective

The objective of this experiment is to examine an elementary analog meter movement. The electrical and electromechanical properties of the meter movement will be obtained and used to design an rudimentary analog multimeter.

3.2 Theory

A meter movement is an instrument that can be used to accurately measure dc current. It is an electromechanical device that converts an electrical current into a mechanical torque that moves the position of a pointer or indicator along a calibrated scale. The position of the pointer on the calibrated scale provides the measure of the dc current.

Meter movements are used as the readout device in a wide variety of electronic and nonelectronic instruments in the measurement of current, voltage, and power. Such instruments are analog instruments because the physical parameter that is being measured is determined by an analog quantity, the angular displacement or position of a pointer. The user must read the position of the pointer and convert this reading into the numerical value of the parameter being measured.

Analog instruments have been replaced in many applications by digital instruments which directly display the numerical value of the parameter being measured to a fixed number of decimal places. Digital instruments are superior to analog instruments in that they are more accurate, portable, rugged, reliable, easier to read, and can be directly interfaced to computers. Nevertheless, there are many applications for which analog meters are preferable and they will have a permanent niche in electrical measurements.

There are many types of meter movements. The subject of this experiment will be the d'Arsonval meter movement which is also known as the permanent magnet moving coil meter movement. This instrument uses the force produced on a current carrying conductor immersed in a uniform magnetic field produced by a permanent magnet and a restoring spring to cause a deflection of a pointer that is linearly proportional to the dc current flowing through it. The properties of the d'Arsonval meter movement as well as the use of this instrument in nonelectronic ammeters, voltmeters, and ohmmeters will be investigated.

3.2.1 Properties of the d'Arsonval Meter Movement

In 1820 Hans Oersted discovered that when a loop of wire was immersed in a uniform magnetic field (Fig. 3.1) between the pole faces of a permanent magnet that a torque was exerted on the loop that tended to cause it to rotate so that the magnetic flux passing through the loop was maximized. If the current is perpendicular to the magnetic flux lines, then the force F [newtons] on each side of the loop that is perpendicular to the lines of magnetic flux is given by the product of the magnetic flux density, B [webers/m^2]; the length of the

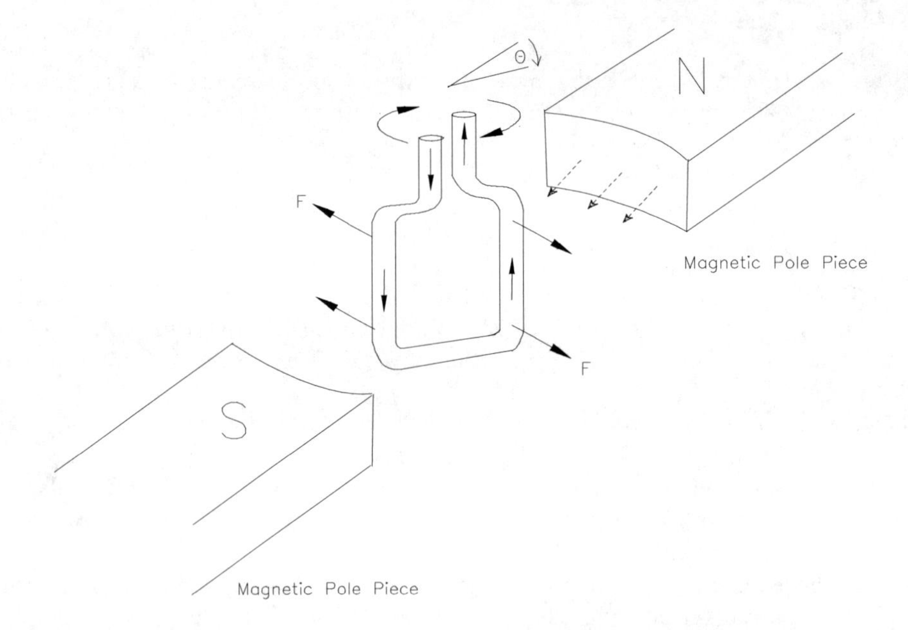

Figure 3.1: Interaction of electric current with magnetic field.

side of the loop of wire that is perpendicular to the magnetic field, ℓ [meters]; and the dc current, I [amperes] or

$$F = B\ell I \tag{3.1}$$

where the units given above are in the mks system. If the radius of the loop of wire is r [meters], then the torque applied to the loop is

$$T = 2rB\ell I \cos\theta \tag{3.2}$$

where θ is the angle that the loop makes with respect to the magnetic flux lines. If θ is sufficiently small, $T \simeq 2rB\ell I$. The units of T are newton-meters.

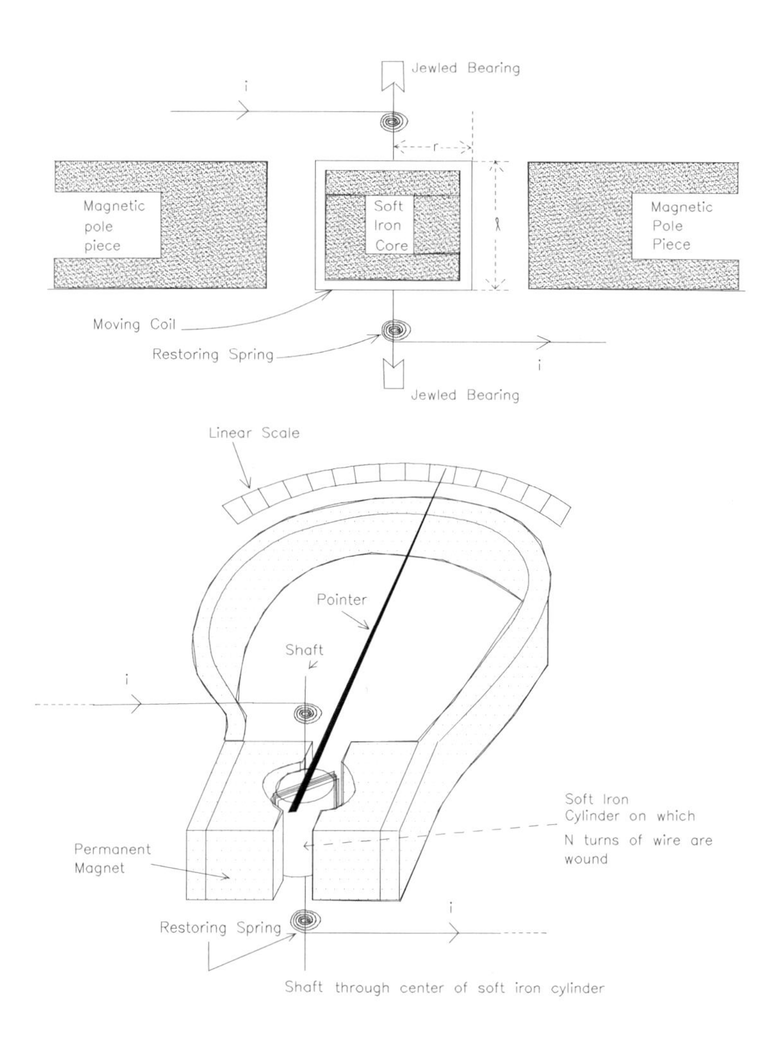

Figure 3.2: d'Arsonval meter movement.

Oersted's discovery of the interaction between electricity and magnetism was used by Jacques d'Arsonval in 1881 to construct an instrument to accurately measure dc currents. The basic mechanism is shown in Fig. 3.2. A soft iron or aluminum cylinder on which N turns of wire have been wound is placed between the pole faces of a permanent magnet. A thin conducting metal shaft supports the cylinder and is held in place by jeweled bearings at each end of the shaft. Mechanical restoring springs are placed at the top and bottom of the shaft which apply a torque which opposes the electromagnetic torque. [Some d'Arsonval meter movements use a taunt metal band or ribbon instead of the metal shaft and restoring spring to add mechanical rigidity to the mechanism.]

Current enters the instrument through one of the external terminals, passes through the restoring spring connected to this external terminal, through the shaft connected to that restoring spring, through the N turns of wire, through the other end of the shaft and attached restoring spring, and finally out through the other external terminal. If current has been applied with the correct polarity, the pointer will deflect upscale. As the pointer deflects upscale three torques are applied to the cylinder: the electromechanical torque applied by the N turns of wire through which the current flows, the mechanical restoring torque applied by the restoring springs, and a damping torque produced by eddy current induced in the metal cylinder as the cylinder rotates through the magnetic field. The restoring torque and damping torque both oppose the electromechanical torque. The restoring torque is linearly proportional to the angular deflection of the pointer while the damping torque is linearly proportional to the angular velocity of the cylinder.

In order to the determine the angular position of the pointer $\theta(t)$ as a current $i(t)$ [a general time varying current not restricted to dc currents] flows through the meter movement, the torques will be summed about the axis of rotation and equated to the product of the moment of inertia with respect to this axis times the second derivative of $\theta(t)$ with respect to t [Newton's Second Law of Motion]. The system equation is then

$$J\frac{d^2\theta(t)}{dt^2} = T(t) - D\frac{d\theta(t)}{dt} - K_s\theta(t) \tag{3.3}$$

where J is the moment of inertia of the cylinder with respect to the axis of rotation [newton-meters/radian/sec^2], D is the rotational damping constant [newton-meters/radian/sec], K_s is the rotational spring constant [newton-meters/radian], and $T(t)$ is the electromagnetic torque

$$T(t) = 2r\ell BNi(t) \tag{3.4}$$

[newton-meters]. (The symbol J is used for moment of inertia instead of I because I is a sacrosanct symbol reserved for electrical current in electrical engineering literature.)

The meter movement can be regarded as a system in which the current is the input (excitation) and the angular deflection of the pointer is the output (response). The response, $\theta(t)$, will be determined for three types of excitation: dc current, step function current, and ac sinusoidal currents. These three excitations provide insight into the operating characteristics and limitations of the d'Arsonval meter movement.

If the external circuit that supplies electrical current to the meter movement consists of a voltage source $e(t)$ in series with a resistor R then the current flowing through the terminals of the meter movement $i(t)$ is related to this voltage by

$$e(t) = Ri(t) + L\frac{di(t)}{dt} + e_g(t) \tag{3.5}$$

where L is the inductance of the meter coils and $e_g(t)$ is the voltage generated by the coil moving through the magnetic field. This is given by

$$e_g(t) = 2r\ell BN\frac{d\theta(t)}{dt} \tag{3.6}$$

where $d\theta(t)/dt$ is the angular velocity of the coil in rad/s. The conservation of energy requires that

$$e_g(t)i(t) = T(t)\frac{d\theta(t)}{dt} \tag{3.7}$$

This simply states that the electrical power developed by the coil rotating in the magnetic field is equal to the mechanical power required to rotate it. If the inductance of the coil is neglected, Eqs. 3.3 through 3.7 may be combined to yield

$$J\frac{d^2\theta(t)}{dt^2} + \left[D + \frac{4r^2\ell^2B^2N^2}{R}\right]\frac{d\theta(t)}{dt} + K_s\theta(t) = \frac{2r\ell BN}{R}e(t) \tag{3.8}$$

as the system equation for the meter movement. This equation illustrates that the meter is a second order system which can be under damped, critically damped, or over damped. The damping, which is the coefficient of $d\theta(t)/dt$, is produced by a combination mechanical damping and electromagnetic damping. If the meter movement is being excited by a current source, then, of course, $R = \infty$ in Eq. 3.8.

It is the damping that makes the meter most useful for certain types of measurements. If a current is being measured which consists of a dc component and a small-amplitude, high-frequency, time-varying component, the damping of the analog meter movement prevents the meter from responding to this time-varying component. In general, a digital meter has a much more rapid response to the time-varying component which can make the measurement of the dc component tedious.

3.2.2 DC Current

A dc current is one that is equal to a constant as time varies from $-\infty$ to ∞. Functionally, $i(t) = I_m$ for all t, $i(t) = I_m \forall t$, [the upper case symbol I is used since the current is a dc current and the subscript "m" indicates that this current flows through the meter movement]. Since the current is a constant, thc solution for $\theta(t)$ can be obtained from the system equation by equating all the derivatives with respect to time to zero since there is no time variation of any parameter is a purely dc circuit.

The solution for $\theta(t)$ is then

$$\theta(t) = \frac{2r\ell BN}{K_s}I_m \tag{3.9}$$

which indicates that the angular deflection of the pointer is linearly proportional to the current flowing through the meter movement. Thus if the dc current, I_m, is zero the pointer is positioned to $\theta = 0°$ by the restoring spring. For larger value of current, the pointer deflects until the restoring torque equals the electromagnetic torque. The value of dc current that causes the pointer to deflect all the way across the calibrated scale is called the full scale deflection current, I_{mfs}, and the corresponding angular deflection, θ_{fs}, is called the full scale deflection angle. The linear relationship between θ and I_m means that

$$\frac{I_m}{I_{mfs}} = \frac{\theta}{\theta_{fs}} \tag{3.10}$$

which means that to measure an unknown current the percentage angular deflection of the pointer is observed on the calibrated scale and multiplied by the full scale deflection current I_{mfs}.

Two parameters are required to characterize the d'Arsonval meter movement for use in dc circuits: the full scale deflection current, I_{mfs}, and the internal resistance of the meter movement, R_m. The resistor R_m is just the resistance of the N turns of wire wound on the cylinder. Fig. 3.3 shows the circuit symbol that will be used for the d'Arsonval meter movement in dc circuits. The circle with an arrow in it is a mnemonic symbol known as an ideal indicator to emphasize that a meter movement is located in this branch of the circuit. The two small circles on the ends of the circuit are the external terminals to which electrical connections are made. To an external dc circuit a d'Arsonval meter movement appears to be just a resistor of R_m ohms.

3.2.3 Step Function Current

A step function is a function of time that is zero prior to a certain time and has a constant value for values of time greater than this "turn on" time. Functionally,

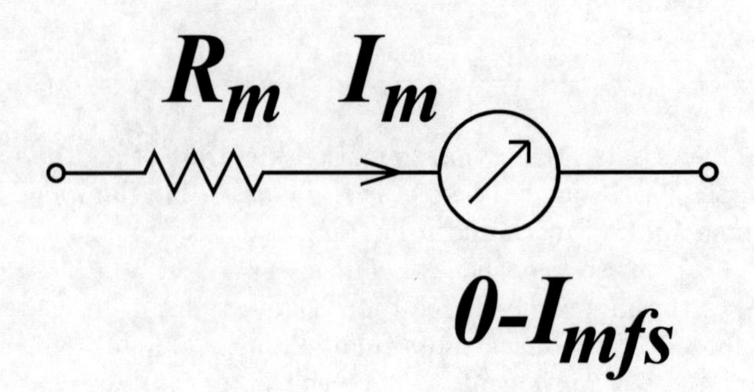

Figure 3.3: DC circuit symbol for d'Arsonval meter movement.

$$i(t) = I_o u(t) \tag{3.11}$$

is a step function of current applied at time $t = 0$. The symbol $u(t)$ is known as a unit step function which is zero prior to $t = 0$ and 1 for t greater than zero. This type of current is more realistic than the dc currents previously considered. Namely, $t = 0$ would correspond to the instant of time at which the d'Arsonval meter movement was connected to the dc circuit or the instant of time at which a constant voltage or current source is "turned on" in the circuit. Prior to $t = 0$ no current is flowing through the meter movement and after $t = 0$ the current is I_o amperes. The solution for the angle $\theta(t)$ requires that the system differential equation be solved for a step function excitation.

The system differential equation for a step function excitation is

$$J\frac{d^2\theta(t)}{dt^2} + D\frac{d\theta(t)}{dt} + K_s\theta(t) = T_o u(t) \tag{3.12}$$

where $T_o = 2r\ell BNI_o$ is the amplitude of the step torque. If both sides of the system equation are divided by J, the differential equation becomes

$$\frac{d^2\theta(t)}{dt^2} + 2\zeta\omega_o\frac{d\theta(t)}{dt} + \omega_o^2\theta(t) = \frac{T_o}{J}u(t) \tag{3.13}$$

where

$$\zeta = \frac{D}{2\sqrt{JK_s}} \text{ [dimensionless]} \tag{3.14}$$

is known as the damping factor or ratio of the meter movement and

$$\omega_o = \sqrt{K_s/J}\ [radians/sec] \tag{3.15}$$

is known as the natural frequency of the meter movement. The form of the solution depends on ζ.

If $\zeta < 1$ the meter movement is said to be underdamped and the solution is

$$\theta(t) = \theta_o\left[1 - \frac{e^{-\zeta\omega_o t}}{\sqrt{1-\zeta^2}}\sin(\omega_o\sqrt{1-\zeta^2}t + \cos^{-1}\zeta)\right]u(t) \tag{3.16}$$

where θ_o is the final angular position of the pointer given by $\theta_o = T_o/K_s$. Thus the pointer begins at $\theta = 0$ at $t = 0$ [$\theta(t)$ must be a continuous function of time since the cylinder has inertia], exhibits decayed oscillations about θ_o, and approaches $\theta(t) = \theta_o$ as t tends to infinity. If $\zeta = 0$ [no rotational damping] the pointer would oscillate from 0 to $2\theta_o$ and never begin to settle to θ_o. Thus some damping is both desirable and mandatory. As the damping is increased, the amount of time required for the pointer to reach the final value is increased.

If $\zeta = 1$ the meter movement is said to be critically damped. For this case the solution to the system differential equation is

$$\theta(t) = \theta_o[1 - (1 + \omega_o t)e^{-\omega_o t}]u(t) \tag{3.17}$$

where $\theta_o = T_o/K_s$. There is no oscillation about the final value of $\theta(t) = \theta_o$; the pointer rises exponentially from $\theta = 0$ to the final value of $\theta = \theta_o$.

Finally, if $\zeta > 1$ the meter movement is said to be overdamped. The solution to the system differential equation is

$$\theta(t) = \theta_o \left[1 - \frac{e^{-\zeta\omega_o t}}{\sqrt{\zeta^2 - 1}} \sinh(\omega_o \sqrt{\zeta^2 - 1}t + \cosh^{-1}\zeta)\right] u(t) \tag{3.18}$$

where $\theta_o = T_o/K_s$. This is similar to the solution for critical damping in that it is nonoscillatory. As the damping is increased, the amount of time required to reach a particular percentage of the final angular deflection increases.

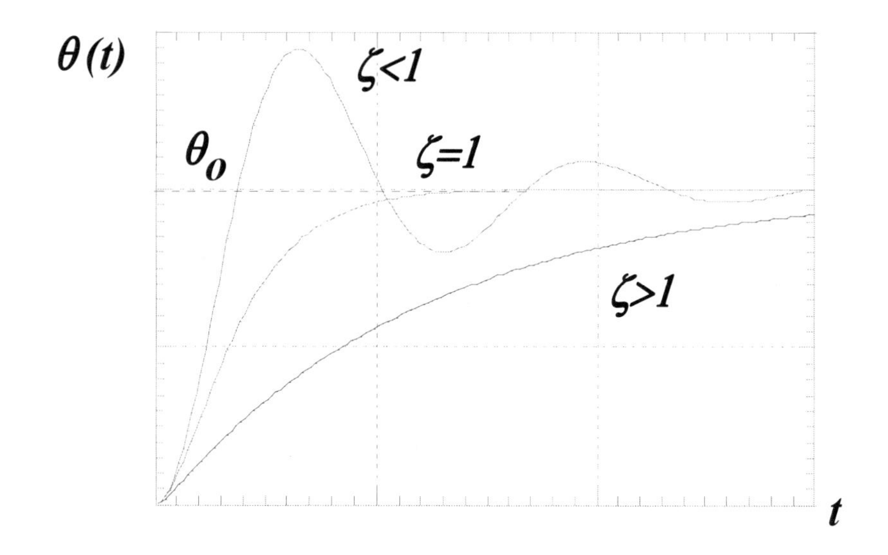

Figure 3.4: Step function response of d'Arsonval meter movement.

Shown in Fig. 3.4 is a plot of $\theta(t)$ versus t for step function excitation of the d'Arsonval meter movement for various value of ζ. As this plot illustrates, the most desirable value of ζ for meter movements is close to critical damping. Small values of ζ cause the pointer to take an undesirably long time to settle to the final value of the angular deflection, θ_o. Large value of ζ require an undesirably long time for the pointer to deflect to within a small percentage of the final value of the angular deflection. Most meter movements are designed to have a ζ that is slightly less than one.

3.2.4 AC Sinusoidal Current

If the current passing through the meter movement is

$$i(t) = I_{mfs}/2 + I_1 \sin(\omega t) \tag{3.19}$$

the dc component of this current, $I_{mfs}/2$, will position the pointer to the half scale reading, $\theta_{fs}/2$, and the sinusoidal component, $I_1 \sin(\omega t)$ will cause the pointer to fluctuate about this mid-scale reading. The system equation for this current is given by

$$\frac{d^2\theta(t)}{dt^2} + 2\zeta\omega_o \frac{d\theta(t)}{dt} + \omega_o^2\theta(t) = [T_{mfs}/2 + T_1 \sin(\omega t)]/J \tag{3.20}$$

where $T_{mfs} = 2r\ell BNI_{mfs}$ and $T_1 = 2r\ell BNI_1$. The solution to this differential equation is

$$\theta(t) = \theta_{fs}/2 + A(\omega)\sin(\omega t + \phi) \tag{3.21}$$

where

$$A(\omega) = \frac{T_1/J}{\sqrt{[\omega_o^2 - \omega^2]^2 + [2\zeta\omega_o\omega]^2}} \tag{3.22}$$

and

$$\phi = -\tan^{-1}\left[\frac{2\zeta\omega\omega_o}{\omega_o^2 - \omega^2}\right] \tag{3.23}$$

are the amplitude and phase of the sinusoidal component of the response, $\theta(t)$. If $A(\omega)$ is normalized by $A(0)$, Eq. 3.22 becomes

$$\frac{A(\omega)}{A(0)} = \frac{A(f)}{A(0)} = \frac{1}{\sqrt{\left(1 - \left[\frac{f}{f_o}\right]^2\right)^2 + \left(2\zeta\frac{f}{f_o}\right)^2}} \tag{3.24}$$

where $A(f)$ simply means the amplitude expressed as a function of f and not, of course, that f equals ω. The natural frequency expressed in Hertz, f_o, is equal to $\omega_o/(2\pi)$ where ω_o is the natural frequency expressed in radians per second.

The amplitude of the sinusoidal component of the response is plotted in Fig. 3.5. For sufficient large values of f the amplitude, $A(f)$, is negligible. It is this property of the d'Arsonval meter movement that makes it possible to use it as a readout device in an ac ammeter. Namely, that the meter movement will respond only to the dc component of the current flowing through it if the time varying component has a frequency $f >> f_o$.

The damping factor for a d'Arsonval meter movement may be increased by placing an external resistor in parallel with the meter movement. If a resistor, R_c, is placed in parallel with the meter movement, the new damping factor is approximately

$$\zeta = \frac{D + \dfrac{4\,(NlrB)^2}{R_c}}{2\sqrt{K_s J}} \tag{3.25}$$

if the inductance of the meter movement is negligible. Meter movements often are shipped with a short across the terminals which maximizes the damping factor and therefore minimizes movement of the pointer during transit (Lenz's Law).

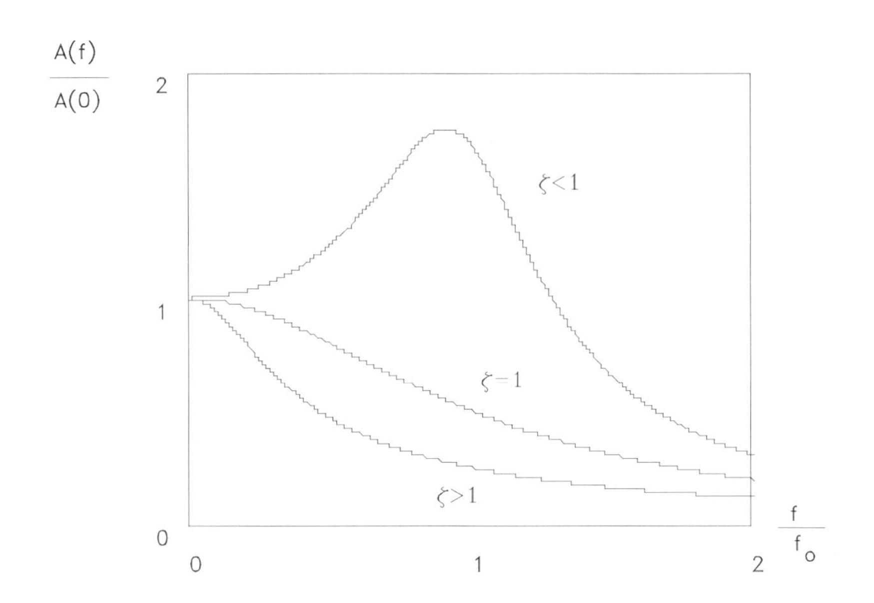

Figure 3.5: Sinusoidal response of d'Arsonval meter movement.

3.2.5 Least Squares Estimation

If experimental data of $\theta(t)$ versus t or $A(f)$ versus f is taken, the mechanical parameters ζ, the damping factor, and f_o, the natural frequency expressed in Hertz, could be estimated. These two mechanical parameters would be varied until the plot of the theoretical curve matched, in some sense, the experimental curve. If the criteria for being matched is that the sum of the squared error between the experimental and theoretical curves at the data points is minimized, then this is known as least squares estimation.

In this experiment, data is taken of $A(f)$ versus f for a particular d'Arsonval meter movement and from this data a least squared estimate of ζ and f_o is made. The parameters ζ and f_o will be varied in Eq. 3.20 until the squared error between this theoretical equation and the data is minimized.

A total of N data points will be taken. This data will be of the form

$$(f_i, d_i) \; i = 1, ..., N \tag{3.26}$$

where f_i are the frequencies at which data is taken and $d_i = A(f_i)/A(0)$ is the amplitude data taken at frequency f_i normalized by $A(0)$. This data is to be fitted to Eq. 3.24. It's easier to match this data to the denominator squared of Eq. 3.24, i.e. to match the denominator to (x_i, y_i) where $x_i = f_i^2$ and $y_i = 1/d_i^2$. The least squares estimation problem is to then determine estimates of ζ and f_o that minimize

$$\epsilon = \sum_{i=1}^{N} [\hat{y} - y_i]^2 \tag{3.27}$$

where

$$\hat{y} = 1 + \alpha x + \beta x^2 \tag{3.28}$$

$$\alpha = \frac{4\zeta^2 - 2}{f_o^2} \tag{3.29}$$

and

$$\beta = 1/f_o^4 \tag{3.30}$$

$x = f^2$, and ϵ is the squared error. The least squares estimation problem is to then find values of α and β that minimize ϵ. This can be done by taking partial derivatives of ϵ with respect to each of these parameters and setting these partial derivatives equal to zero which assures a local minimum.

Taking partial derivatives of ϵ with respect to α and β and setting these equal to zero yields

$$\frac{\partial\, \epsilon}{\partial\, \alpha} = \sum_{i=1}^{N} 2[1 + \alpha x_i + \beta x_i^2 - y_i]x_i = 0 \tag{3.31}$$

$$\frac{\partial\, \epsilon}{\partial\, \beta} = \sum_{i=1}^{N} 2[1 + \alpha x_i + \beta x_i^2 - y_i]x_i^2 = 0 \tag{3.32}$$

which can be solve simultaneously for α and β.

Solving Eqs. 3.31 and 3.32 simultaneously yields

$$\alpha = \frac{\left[\sum_{i=1}^N f_i^8\right]\left[\sum_{i=1}^N f_i^2\left(\frac{1}{d_i^2} - 1\right)\right] - \left[\sum_{i=1}^N f_i^6\right]\left[\sum_{i=1}^N f_i^4\left(\frac{1}{d_i^2} - 1\right)\right]}{\left[\sum_{i=1}^N f_i^8\right]\left[\sum_{i=1}^N f_i^4\right] - \left[\sum_{i=1}^N f_i^6\right]^2} \tag{3.33}$$

and

$$\beta = \frac{\left[\sum_{i=1}^N f_i^2\left(\frac{1}{d_i^2} - 1\right)\right] - \alpha\left[\sum_{i=1}^N f_i^4\right]}{\left[\sum_{i=1}^N f_i^6\right]} \tag{3.34}$$

from which the least squares estimate of the damping factor and natural frequency of the meter movement are

$$\hat{f}_o = \left[\frac{1}{\beta}\right]^{(1/4)} \tag{3.35}$$

and

$$\hat{\zeta} = \sqrt{\frac{\alpha f_o^2 + 2}{4}} \tag{3.36}$$

where the "^" superscript simply means that these are the least squares estimates of these mechanical parameters.

3.2.6 Meter Movement Applications

Nonelectronic DC Ammeter

The d'Arsonval meter movement is basically a dc ammeter. It can measure dc currents from zero to I_{mfs}. The maximum current that any ammeter can measure is called the range current of the ammeter. Thus the d'Arsonval meter movement's range current would be I_{mfs}. If it desired to increase the range current of such an ammeter, I_{mfs} would have to be increased by changing either the restoring spring or the flux produced by the magnet. It is much easier to simply place a resistor in parallel with the meter movement to increase the range current.

A nonelectronic dc ammeter is shown in Fig. 3.6 which uses a simple shunt resistor in parallel with the meter movement to increase the range current of the instrument. The term nonelectronic is used because this instrument does not employ an electronic amplifier. If the range current of this ammeter is to be I_{fs} then as

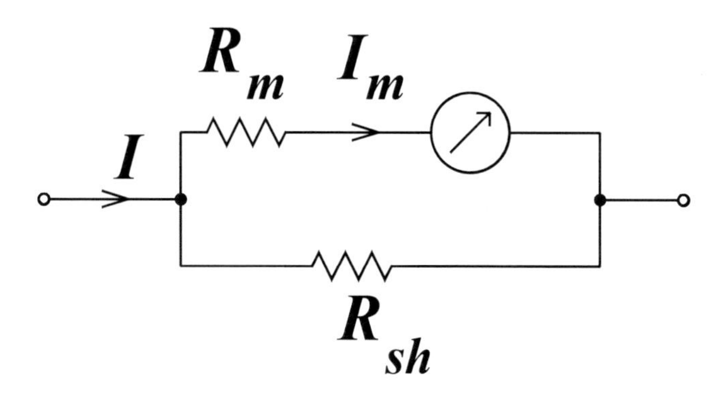

Figure 3.6: Nonelectronic dc ammeter.

the current I varies from 0 to I_{fs} then the current flowing through the meter movement must correspondingly vary from 0 to I_{mfs}. Thus if the dc parameters of the meter movement are known (R_m, I_{mfs}) and it is desired that the range current of the ammeter be I_{fs}, then

$$R_{sh} = \frac{R_m I_{mfs}}{I_{fs} - I_{mfs}} = \frac{R_m}{\left[\dfrac{I_{fs}}{I_{mfs}}\right] - 1} \tag{3.37}$$

must be the value of the shunt resistor. The scale on this instrument would have linear tic marks from 0 to I_{mfs}. Thus an unknown current would be measured by measuring the percentage angular deflection of the pointer and multiplying this percentage by the range current of the ammeter, I_{fs}.

An important parameter of any ammeter is its internal resistance which determines what effect the ammeter will have on the circuit in which the current is to be measured. This is the resistance that would be measured between the terminals of the instrument with an ideal ohmmeter. For the simple nonelectronic dc ammeter shown above the internal resistance, R_i, is $R_i = R_m \| R_{sh}$ because the shunt resistor is in parallel with the meter movement.

Nonelectronic DC Voltmeter

To measure a dc voltage with a d'Arsonval meter movement it is necessary to employ a transducer that converts the voltage to be measured to a current that flows through the meter movement. The transducer is obviously a resistor. A nonelectronic dc voltmeter is shown in Fig. 3.7. Since the relationships between the meter movement current and the voltage is linear, to measure an unknown voltage the percentage angular deflection of the pointer is observed and multiplied by the voltage that produces full scale deflection, V_{fs}. If it is desired to measure dc voltage, V, from 0 to V_{fs}, i. e. the range voltage of the voltmeter is to be V_{fs}, then the resistor R_s must be selected so that

$$V_{fs} = (R_s + R_m) I_{mfs} \tag{3.38}$$

so that a voltage V_{fs} causes full scale deflection of the pointer on the meter movement. Thus if the parameters of the meter movement are specified, the design of the voltmeter simply involves determining the value of R_s for a specified range voltage, V_{fs}.

A most important parameter of a voltmeter is its input impedance. This is the resistance seen looking into the terminals. The input impedance determines what effect the voltmeter has on the circuit in which the voltage is to be measured.

For the simple nonelectronic dc voltmeter considered above, the input impedance is $R_{in} = R_s + R_m$. For multirange nonelectronic dc voltmeters that employ the same meter movement as the readout device,

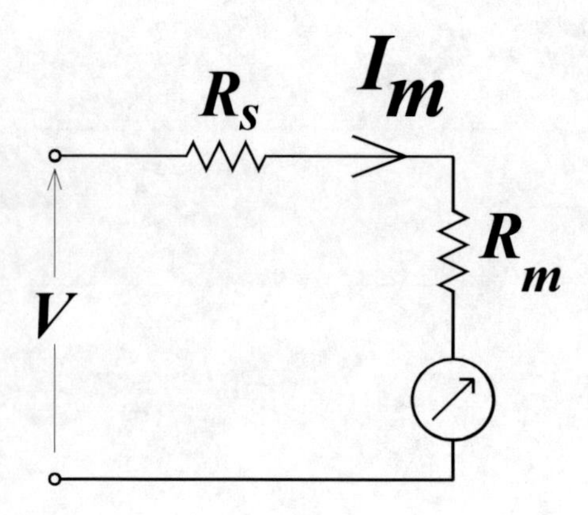

Figure 3.7: Nonelectronic dc voltmeter.

the input impedance changes when the range is changed, i. e., for different ranges different values of R_s must be used. For such instruments the sensitivity of the meter movement, S, is usually specified where $S = 1/I_{mfs}$ [Ohms/Volt]. The input impedance of the voltmeter is then the product of the range setting of the voltmeter and the sensitivity of the meter movement which is also known as the sensitivity of the voltmeter, $R_{in} = SV_{fs}$.

Nonelectronic Ohmmeter

An ohmmeter is an instrument that measures electrical resistance. To use a d'Arsonval meter movement as the readout device requires that a transducer be employed to convert resistance to a current that can be used to actuate the meter movement. An nonelectronic ohmmeter that is to be used to measure unknown resistances, R_x, is shown in Fig. 3.8.

For the ohmmeter shown in Fig. 3.8, the current flowing through the meter movement is

$$I_m = \frac{E_o}{R_x + R_s + R_m} \tag{3.39}$$

which discloses that the relationship between the current, I_m, and the resistance, R_x, is nonlinear. The tic marks on the resistance scale would, therefore, not be linearly spaced.

To make maximum use of the scale on the meter movement the parameters E_o and R_s are selected so that $R_x = 0$ produces full scale deflection, i. e., $I_m = I_{mfs}$ when $R_x = 0$. The value of the resistor R_x that makes the angular deflection of the pointer 50% of the full scale deflection is known as the mid scale resistance reading and for this ohmmeter is given by $R_{ms} = R_s + R_m$. If the resistance $R_x = \infty$ (open circuit), then the angular deflection of the pointer is 0.

Nonelectronic AC Voltmeter

For sinusoidal voltages, $v(t) = A \sin(\omega t)$, the parameter of interest is the rms value of $v(t)$, $V_{rms} = A/\sqrt{2}$, which is the value of an equivalent dc voltage source which would produce the same heating effect in a resistor. To measure the rms value of a sinusoidal voltage with a d'Arsonval meter movement a transducer must be used that will convert the rms value of the ac voltage to a dc current that will be used to actuate the meter movement. Such a transducer is a resistor in series with a diode. Such a voltmeter is shown in Fig. 3.9.

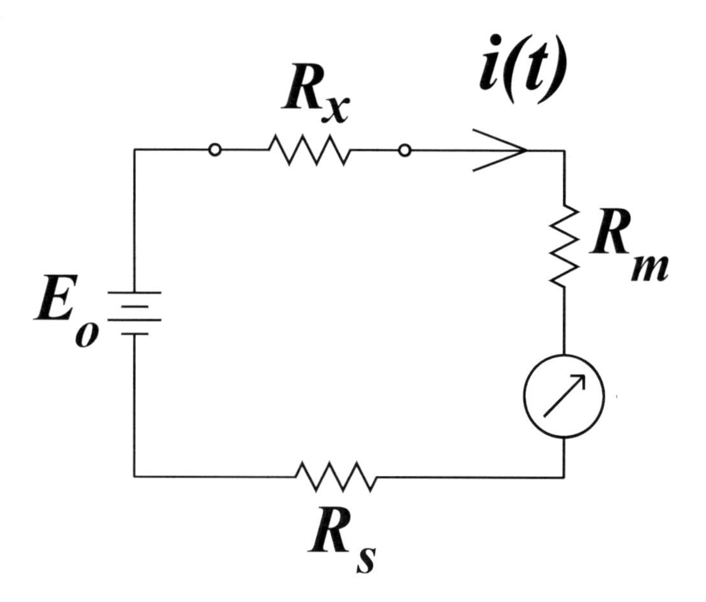

Figure 3.8: Nonelectronic ohmmeter.

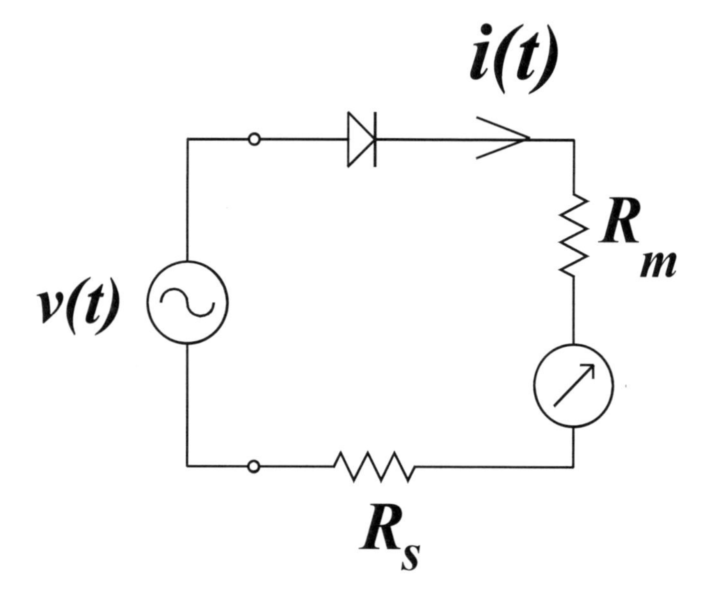

Figure 3.9: Nonelectronic ac voltmeter.

A diode can be considered to be a circuit component which is a short circuit when current flows through it in one direction and an open circuit when current flows through it in the opposite direction [this is a very simplistic model of a diode but adequate for the present discussion]. The diode in Fig. 3.9 is the circuit element with the symbol consisting of a triangle touching a vertical line. It permits current to flow in the clockwise direction but not in the counterclockwise direction.

When the voltage $v(t)$ in Fig. 3.9 is positive, the current flowing through the circuit is

$$i_m(t) = \frac{v(t)}{R_s + R_m} \tag{3.40}$$

and zero when $v(t)$ is negative. This current can be resolved into a dc component

$$I_m = V_{rms} \frac{\sqrt{2}}{\pi(R_s + R_m)} \tag{3.41}$$

plus a time varying component. If the frequency of $v(t)$ is large enough, then the meter movement will respond to only the dc component, I_m. The rms value of $v(t)$ is therefore

$$V_{rms} = (R_s + R_m) I_{mfs} \frac{\theta}{\theta_{fs}} \frac{\pi}{\sqrt{2}} \tag{3.42}$$

which means that the percentage angular deflection of the pointer (θ/θ_{fs}) must be multiplied by the above factor.

An instrument which can measure dc and ac current and voltage and electrical resistance is known as a multimeter. The principles discussed above are used an electrical instrument known as a nonelectronic analog multimeter which is widely used in electrical measurements.

3.3 Equipment

3.3.1 The d'Arsonval Meter Movement.

This has two wires colored red and black to which connections are made to an external circuit. The red wire is known as the high or plus lead and the black wire is known as the minus or low lead. When current enters the red wire and leaves the black wire the pointer on the meter movement deflects upscale. It is protected by a half amp fuse in series with the meter movement and two diodes in parallel with the meter movement with opposite orientations to protect the meter movement from both excessive currents and voltages.

3.4 Procedure

3.4.1 Measurement of R_m

Measure and record the internal resistance, R_m, of the d'Arsonval meter movement with the **Agilent 34401A** digital multimeter. Connect the red lead of the meter movement shown in Fig. 3.10 to the *HI Input* on the DMM and the black lead to the *LO* input on the DMM. Press "Ω 2*W*" on the **Agilent 34401A** DMM.

If the pointer on the meter movement does not deflect, the fuse on the back of the meter movement is probably blown. Have a skilled individual, such as the laboratory instructor, check the fuse if this is the case.

R_m = ______________________________

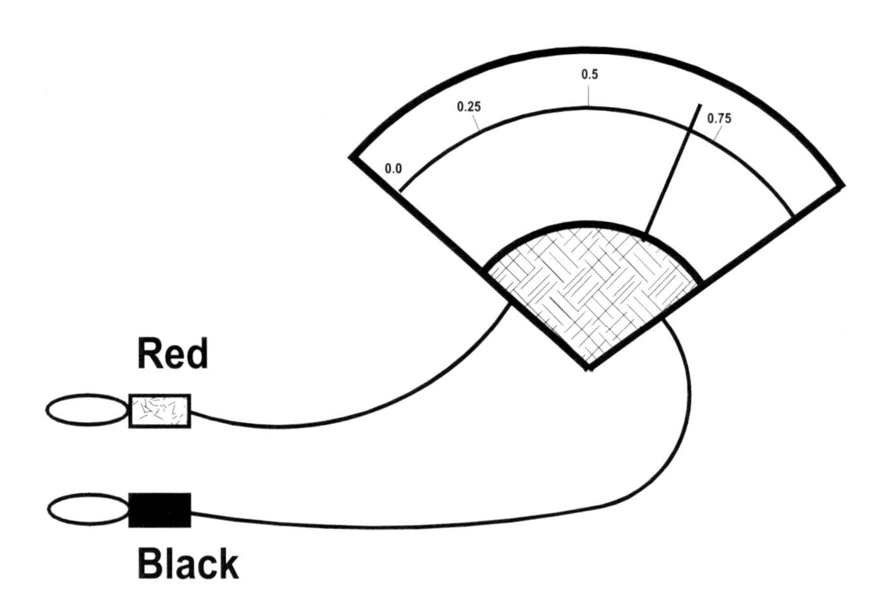

Figure 3.10: Meter movement.

3.4.2 Measurement of I_{mfs}

Assemble the circuit shown in Fig. 3.11 with $R_s = 10\ k\Omega$ [brown, black, orange] and $R_{sh} = \infty$ (open circuit, *i.e.* nothing there) with the voltage adjust knob set to the complete counter clockwise position. The **Agilent 34401A** DMM is to be used as the ammeter standard [AM] which requires that the blue button labeled "*Shift*" be pressed followed by the blue "*DC I*" and the input jack labeled "*3A*", the plus input, and the "*LO*", the minus input, be used. The dc voltage source is the **Agilent 3630A** dc power supply. The "+20 *V*" binding post on the **Agilent 3630A** dc power supply is the plus terminal and the "*COM*" is the minus. Press the button to monitor the "+20 *V*" supply.

Connect alligator clips on the banana plug leads from the d'Arsonval meter movement under study. Then connect a short length of No. 22 gauge hook-up wire to the alligator clips and insert the other end of the hoop-up wire into the appropriate hole on the protoboard.

Turn on the power supply and turn the voltage adjust knob on the power supply clockwise until the pointer on the d'Arsonval meter movement under study deflects full scale. Record this current (measured on the **Agilent 34401A**) as I_{mfs}. Turn the voltage adjust knob on the power supply to the completely counterclockwise position.

I_{mfs} =____

3.4.3 Nonelectronic DC Ammeter

Change the value of R_s in Fig. 3.11 to 2 $k\Omega$ [red, black, red]. Change the value of R_{sh} to 100 Ω [brown, black, brown].

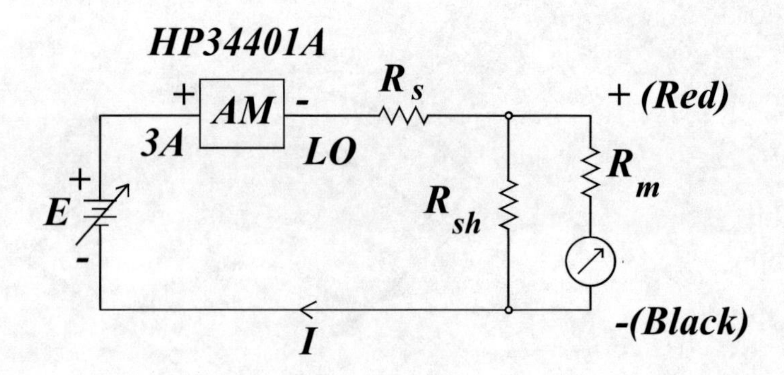

Figure 3.11: Nonelectronic dc ammeter.

Turn the voltage adjust knob on the power supply clockwise until the pointer on the meter movement deflects full scale and measure and record the current I on the **Agilent 34401A** as I_{fs} for this value of R_{sh}. Turn the voltage adjust knob completely counterclockwise.

Change R_{sh} to 20 Ω [red, black, black] and repeat the above.

Change R_{sh} to 10 Ω [brown, black, black] and repeat the above.

R_{sh}	I_{fs} (mA)
100 Ω	
20 Ω	
10 Ω	

3.4.4 Nonelectronic Voltmeter

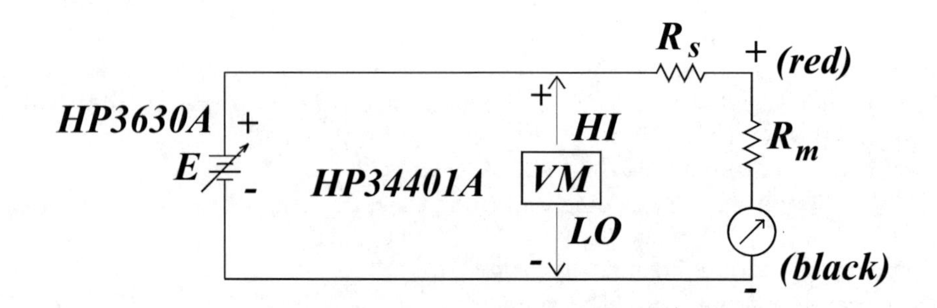

Figure 3.12: Nonelectronic dc voltmeter.

Assemble the circuit shown in Fig. 3.12. Use the **Agilent 34401A** as the voltmeter standard (VM) which requires that the button labeled "*DC V*" be pressed and the jacks labeled "*Input HI*" and "*LO*" be used as the + and − inputs respectively. Turn the voltage adjust knob on the power supply to the completely counterclockwise position.

Use $R_s = 10\ k\Omega$ [brown, black, orange]. Turn the voltage adjust knob clockwise until the pointer deflects full scale. Record the voltage indicated on the **Agilent 34401A** as V_{fs}.

Change $R_s = 18\ k\Omega$ [brown, gray, orange] and repeat the above.

R_s	V_{fs} $(Volts)$
$10\ k\Omega$	
$18\ k\Omega$	

3.4.5 Dynamics of d'Arsonval Meter Movement

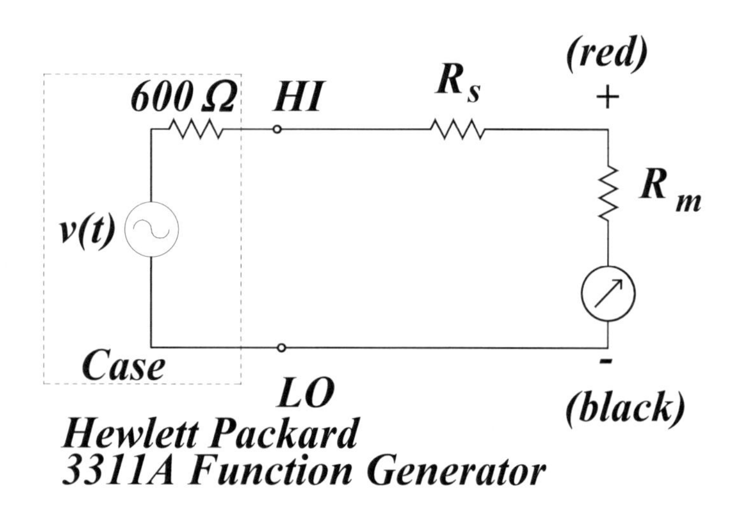

Figure 3.13: Dynamic response of d'Arsonval meter movement.

Assemble the circuit shown in Fig. 3.13. Use a value of $R_s = 2\ k\Omega$ [red, black, red].

The voltage source that will be used is the **Hewlett-Packard 3311A** Function Generator. Set the frequency to 400 Hz by setting the frequency vernier to 4 and the range multiplier to 100. Set the function to sine. The output terminals are labeled 600 Ω output.

Turn on the function generator and adjust the dc offset control until the pointer on the meter movement deflects half-scale. Turn the Amplitude adjust knob completely counterclockwise. Reduce the frequency to 1 Hz (range multiplier set to 1 and the frequency vernier set to 1). Increase the amplitude knob until the pointer swings from 0.2 to 0.8 on the scale of the meter movement. This corresponds to an amplitude of 0.3 $(0.8 - 0.5 = 0.5 - 0.2)$. [The amplitude is the maximum deviation of the pointer from the center position.]

Record the frequency and amplitude swing, $A(f)$, as the frequency varies from 0.1 Hz to 7 Hz. Increase the frequency until pointer movement can no longer be detected. Record this frequency.

f	$A(f)$ [$percentage\ of\ full\ scale$]
0.1 Hz	
0.5 Hz	
1.0 Hz	
1.5 Hz	
2.0 Hz	
2.5 Hz	
3.0 Hz	
3.5 Hz	
4.0 Hz	
5.0 Hz	
6.0 Hz	
7.0 Hz	

Maximum frequency for which perceptible movement of the pointer could be observed

$f_{\max}$ =____

3.4.6 Nonelectronic AC Voltmeter

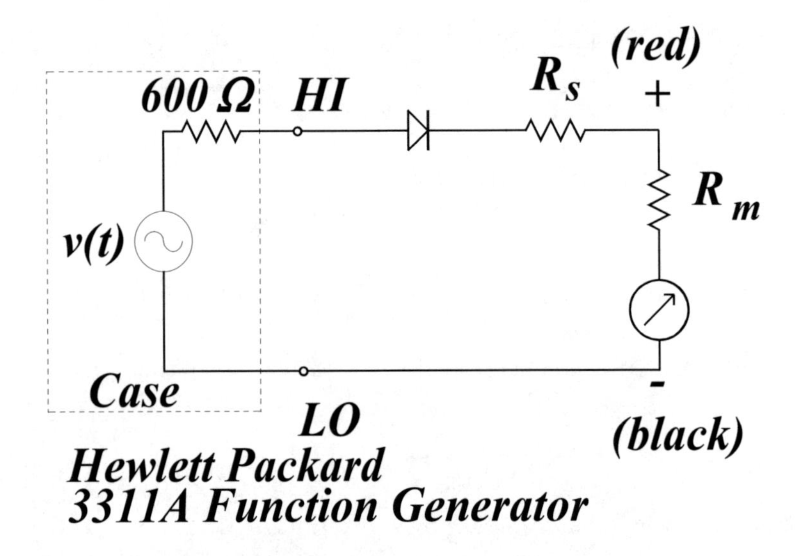

Figure 3.14: Nonelectronic ac voltmeter.

Assemble the circuit shown in Fig. 3.14. Use $R_s = 2\ k\Omega$ [red, black, red].

Before connecting $v(t)$ to the circuit set the frequency of the **Hewlett Packard 3311A** function generator to 400 Hz and connect the output directly to the input of the **Agilent 34401A**. Vary the dc offset control on the function generator until the dc voltage is as close to zero as it can be adjusted. (The knob labeled "*DC V*"should be pressed and the inputs labeled HI and LO used.) Do not connect the DMM to the circuit.

The diode has a band on it that corresponds to the straight line on the circuit symbol. The band should be on the end nearest the + on the meter movement in the circuit. The diode will have printed on it 1N4148 [the 1N means that the device is a diode (2N would be a transistor) and the 4148 means that approximately 4,147 diodes were catalogued before this one].

Increase the amplitude knob on the function generator until the pointer deflects full scale on the meter movement. Disconnect the voltage source (without changing any knob settings) from the circuit and measure the ac rms value of $v(t)$ with the **Agilent 34401A** DMM. Press the button labeled "*AC V*" and record the value as the range voltage of the nonelectronic ac voltmeter.

$V_{rms,fs}$ =____

3.4.7 Ammeter Design

Design an ammeter with a range specified by the laboratory instructor. Verify the design. Demonstrate the design to the laboratory instructor.

I_{fs} (mA) Design	I_{fs} (mA) Experimental	R_{sh} (Ω) Calculated	R_{sh} (Ω) Actual

3.5 Laboratory Report

All data that was taken as well as all calculations that were performed should be included in the laboratory report.

For step 3.4.3 of the procedure compare the actual value of the range current of the nonelectronic ammeter with the value that would be theoretically expected for an ammeter with the measured values of I_{mfs}, R_{sh}, and R_m.

For step 3.4.4 of the procedure compare the actual value of the range voltage of the nonelectronic voltmeter with the value that would be expected using the measured values of R_s, R_m, and I_{mfs}.

For step 3.4.5 of the procedure plot the experiment data for $A(f)$ versus f for the d'Arsonval meter movement. The plot may be made manually or by using a spreadsheet. From this plot estimate whether the meter movement is under, critically, or over damped.

For step 3.4.6 of the procedure compare the experimentally obtained value of the range voltage of the nonelectronic ac voltmeter with the value that would be expected using a voltmeter with the measured values of R_m, I_{mfs}, and R_s. The diode used acts like an idealized diode in series with a dc voltage of about 0.6 Volt. Therefore the theoretically expected range voltage should be given by

$$V_{rms,fs} = [R_{tot}I_{mfs} + 0.6]\pi/\sqrt{2}$$

where R_{tot} is the total resistance of the circuit ($600\ \Omega + R_s + R_m$).

If the laboratory experiment and/or report cannot be completed by the end of the three hour lab session, turn in what has been completed to the laboratory instructor.

3.6 References

1. Brewer, T. E., "d'Arsonval's Meter Movement Will Be Moving into the Twenty-First Century", *Annual Conference Proceedings of ASEE*, 1993.
2. Cooper, W. D., *Electronic Instrumentation and Measurements Techniques*, 2nd edition, Prentice-Hall, 1978.
3. Jones, L., and Chin, A. F., *Electronic Instruments and Measurements*, John Wiley, 1983.

4. Krenz, J. H., *Introduction to Electrical Circuits and Electronic Devices: A Laboratory Approach*, Prentice-Hall, 1987.
5. Kantrowitz, P., et. al., *Electronic Measurements*, Prentice-Hall, 1979.
6. Oliver, B. M., and Cage, J. M., *Electronic Measurements and Instrumentation*, McGraw-Hill, New York, 1971.
7. Riaz, M., *Electrical Engineering Laboratory Manual*, McGraw-Hill, 1965.
8. Wolf, S., *Guide to Electronic Measurements and Laboratory Practices*, Prentice-Hall, 1973.
9. Wolf, S., *Guide to Electronic Measurements and Laboratory Practices*, 2nd edition, Prentice-Hall, 1983.
10. Wolf, S., and Smith, R. F. M., *Student Reference Manual for Electronic Instrumentation Laboratories*, Prentice-Hall, 1990.

Chapter 4

Voltage and Current Measurement with Meters

4.1 Objective

The object of this experiment is to examine the proper use and performance limitations of voltmeters and ammeters in the measurement of voltage and current in an electrical circuit. Both analog and digital meters are examined for measurements in dc and ac circuits. The primary performance limitation that is examined is meter loading. An elementary examination of energy sources for electrical circuits is made. Since most physical parameters such as pressure, temperature, position, velocity, acceleration, optical intensity, flow, etc. can be converted into a voltage or current by a transducer or sensor, a knowledge of the proper use and limitations of voltmeters and ammeters is crucial to all engineering disciplines and branches of science as well as seemingly unrelated fields such as medicine.

4.2 Theory

4.2.1 DC Voltage Sources

Historically, the first man/woman made dc voltage source was the battery. Batteries function by releasing electrochemical energy and converting it to dc electrical energy. This is accomplished by inserting two different types of metals, known as electrodes, into a liquid or semi liquid known as an electrolyte.

An electrolyte is a liquid or semiliquid solution in which most of its molecules break up into positive and negative ions. One of the electrodes of the battery attracts the positive ions and acquires a positive charge while the other attracts the negative ions and acquires a negative charge. This charge builds up on the two electrodes until they acquire a sufficiently large voltage to repel the remaining ions. If a resistor is connected between the positive and negative electrodes, a current can flow which reduces the charge built up on the electrodes and permits additional ions to interact with the electrodes to sustain the electric current being drawn by the external circuit.

Some common battery types are: zinc-carbon, lead storage, mercury, and manganese-alkaline. All utilize the same basic electrochemical principles to generate a DC voltage. They differ only in the types of electrolyte and electrodes and the elementary cell voltage that is produced. Some can be recharged while others suffer an irreversible loss of their ability to generate a DC voltage.

Shown in Fig. 4.1 is the zinc-carbon battery that was originally developed by Georges Le Clanche in 1868. It is known as a dry cell because the electrolyte is a semiliquid paste made of ammonium chloride and zinc chloride. The electrodes are made of carbon and zinc. Each combination of two electrodes and an electrolyte is known as a cell. When freshly prepared this cell has a terminal voltage of about 1.55 V;

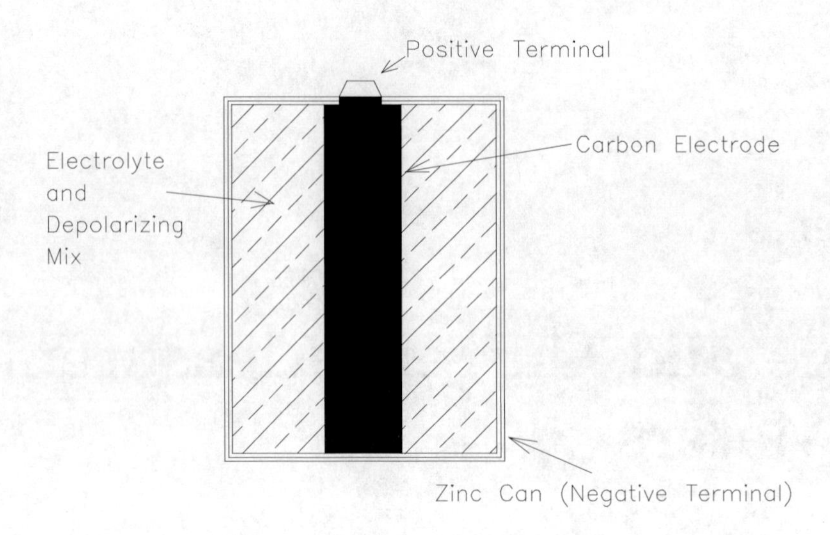

Figure 4.1: Zinc carbon battery.

larger voltages are obtained by connecting a sufficient number of cells in series (the term battery comes from "battery of cells").

The carbon electrode of the zinc-carbon cell becomes surrounded by hydrogen gas when the electrolyte interacts with it. For this reason a depolarizer consisting of manganese dioxide is added to remove this gas. The hydrogen gas must be removed or the resistance to the flow of current will become so large that the battery will be rendered useless; also, if the gas pressure becomes large enough, the case of the battery can be ruptured which would result in the spilling of the corrosive materials.

After a sufficient amount of use, the zinc electrode is degraded to the point that this type of battery can no longer be used. Even if no current is drawn the zinc electrode will eventually decompose. This type of battery has a shelf life of no more than two years. This type of battery is also known as a primary dry cell which means that it cannot be recharged.

Shown in Fig. 4.2 is a lead storage battery which is the type that is used in automobiles. It is made of secondary wet cells. The term wet cells is used because the electrolyte is a liquid (sulfuric acid diluted with distilled water). The term "secondary" is used because this type of battery can be recharged. The two electrodes are lead (negative) and lead peroxide (positive). When fully charged, each cell in this battery has a charge of 2.2 V and a total of six cells are used which yields an open circuit voltage of 13.2 V. When a car is starting this type of battery can supply a current of hundreds of amps.

The lead storage battery is a secondary battery which means that it can be recharged. This means that the battery can be restored to its original state by passing a current into the positive terminal and out of the negative terminal. In a car, this is done automatically by the alternator.

A circuit model for a battery is shown in Fig. 4.3a. It consists of an ideal voltage source, E_{oc}, in series with a resistor, R_s, which is the source resistance of the battery. Every battery has a source resistance because current must flow through the electrolyte and electrodes and these have nonzero resistance. The source resistance for a fresh zinc-carbon battery might be as small as 0.05 Ω while a the same battery after considerable use may have a source resistance of several hundred ohms.

Batteries are examples of unregulated dc voltage sources in that the terminal voltage depends on the current drawn by the load. Once a load resistor, R_L, is connected to the battery, the terminal voltage drops to

$$V_t = E_{oc}\frac{R_L}{R_S + R_L} \tag{4.1}$$

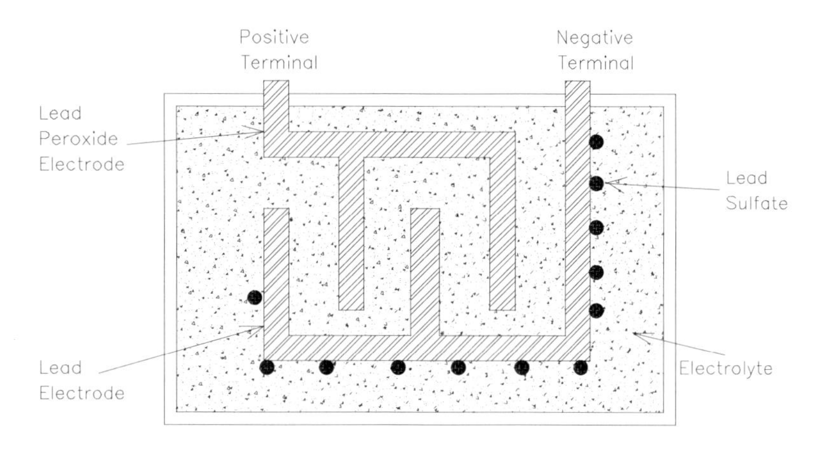

Figure 4.2: Lead storage battery.

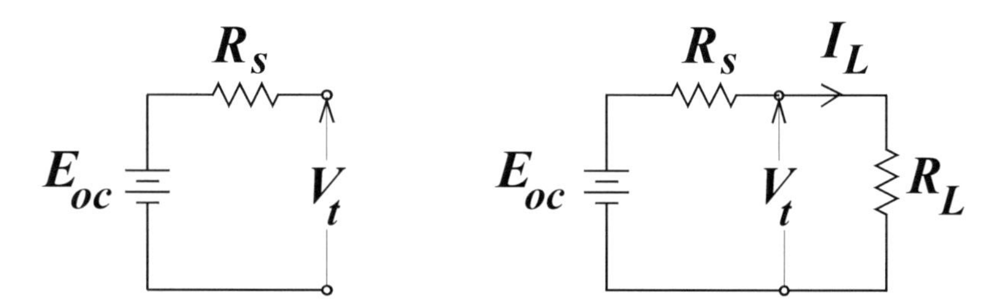

Figure 4.3: Circuit model for a battery; (a) open circuit model, (b) battery with a load.

which means that the terminal voltage is a function of the load.

Batteries are normally only used when it is inconvenient or impractical to obtain dc indirectly from the ac available from utility outlets. Batteries are bulky, contain corrosive chemicals, are unregulated (the terminal voltage depends on the load), and need to be frequently replaced or recharged.

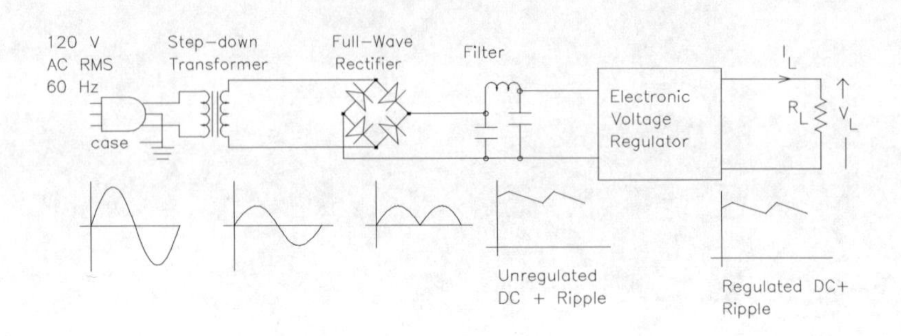

Figure 4.4: Laboratory dc power supply.

Shown in Fig. 4.4 is a block diagram representation of a laboratory dc power supply. Such as instrument has a power cord that plugs into an ac utility outlet that has 120 V ac rms. This ac voltage is then stepped-down to a smaller (or sometimes larger) ac voltage by a transformer. The ac voltage is then supplied as the input to a combination of four diodes known as a full-wave rectifier which yields the absolute value of the input which is a time-varying voltage with both a dc and an alternating component. The output of the full-wave rectifier is then supplied as the input to a filter which removes most of the time-varying component leaving only a dc component and a small time-varying component known as the ripple voltage. The output of the filter is unregulated in that the dc component of this voltage depends on both the load current and variations in the line voltage. The output of the filter is then placed through an electronic circuit known as a voltage regulator which makes the dc component of the output essentially independent of variations in both the line voltage and the load voltage.

The case or chassis of this dc power supply is connected to the earth ground through the center prong or ground wire on the ac power cord. This is done for safety reasons. An external ground connection is usually provided on a dc power supply if the user wishes to connect either the plus or minus (but not both) terminals to the earth ground.

Laboratory dc power supplies may contain more than one dc power supply. A common arrangement is a triple power supply in which one is a fixed voltage supply (such as 5 V) and the other two are adjustable supplies. The adjustable supplies will provide a constant voltage between their output terminals until the load current reaches the current limit value. The current limit can be adjusted from zero to a value selected by the user to an upper limit set by the particular dc supply (500 mA is typical for the maximum value for the current limit).

4.2.2 AC Voltage Sources

The most commonly used ac voltage source in electrical engineering laboratories is the function generator. A function generator is an instrument which can produce several periodic functions such as sine, square, triangular, ramp, pulse, etc., waves. The term "wave" is used in electrical engineering literature to indicate a periodic function of time. Both the amplitude, frequency, and dc offset of these waves can be set by the user. Many function generators can also be amplitude and frequency modulated. Function generators are preferred to the ac line voltage because the latter can produce only sine waves at the frequency 60 Hz.

Function generators require dc voltages for the electronic circuits required for the generation of the periodic waveforms. Rarely, are they battery operated and they, therefore, must have a dc power supply section. The case or chassis of every function generator is connected to the earth ground via the ground wire on the power cord for safety considerations. Some function generators have one of their output connectors also internally connected to the earth ground and these types of function generators are known as grounded function generators. If neither of the output connectors of a function generator are internally connected to the earth ground, this type of function generator is known as a "floating" function generator.

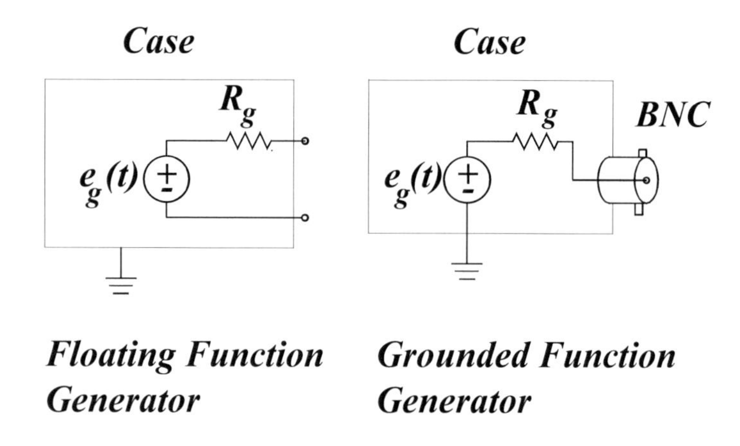

Figure 4.5: Function generator; (a) floating, (b) grounded.

Functional block diagrams of a floating and a grounded function generator are shown in Fig. 4.5. The floating function generator has a two wire output connector while the grounded function generator has a cylindrical output connector known as a BNC connector for which the outer conductor, known as the shield, is connected internally to the earth ground. In each case, the function generator consists of an ideal voltage source, $e_g(t)$, in series with an internal resistance, R_g. The resistance R_g is usually precisely known.

4.2.3 Voltage and Current Measurement

Instruments that are designed to measure the voltage between two points in an electrical or electronic circuit are appropriately denominated voltmeters. Voltmeters are always connected in parallel with the circuit element across which the voltage is to be measured. The types of voltages that can be measured depends on the instrument. There are specialized voltmeters than can measure the peak value of voltages (peak voltmeters) and the magnitude and phase of an ac voltage (vector voltmeters) but most voltmeters than are part of an instrument known as a multimeter can measure only the dc component of a voltage and the rms value of this voltage.

Voltmeters are dichotomized as electronic and non-electronic voltmeters. Non-electronic voltmeters use a dropping resistor to set the range (maximum voltage to be measured) and employ a d'Arsonval meter movement as the readout device; this type of voltmeter was examined in a previous experiment. Electronic voltmeters employ one or more electronic amplifiers to enhance the measurement capability of this instrument. Electronic voltmeters can be further dichotomized as analog or digital depending on the type of readout device used.

The instruments that are designed to measure the current flowing in a branch of a circuit are known as ammeters. They can measure either dc or ac currents. Ammeters are always connected in series with the circuit element through which the current to be measured is flowing.

When set to dc the meter measures the average or dc component of the voltage (current). For voltages, this is given by

$$V_{dc} = \lim_{T\to\infty} \frac{1}{T} \int_0^T v(t)dt \tag{4.2}$$

which is the dc component or average value of $v(t)$. If $v(t)$ is a dc voltage, i.e. $v(t) = V \forall t$, then Eq. 4.2 states the obvious $V_{dc} = V$. If $v(t)$ is a periodic function with period T, i.e. $v(t+T) = v(t)$, then Eq. 4.2 becomes

$$V_{dc} = \frac{1}{T} \int_0^T v(t)dt \tag{4.3}$$

which is the average value of $v(t)$ over one cycle.

When set to ac, a voltmeter will measure the rms value of $v(t)$ where the rms value of $v(t)$ is given by

$$V_{rms} = \sqrt{\lim_{T\to\infty} \frac{1}{T} \int_0^T v^2(t)dt} \tag{4.4}$$

which is the average value of $v^2(t)$. If $v(t)$ is periodic with period T, Eq. 4.4 becomes

$$V_{rms} = \sqrt{\frac{1}{T} \int_0^T v^2(t)dt} \tag{4.5}$$

which is just the average value of $v^2(t)$ averaged over a cycle. Another name for the rms value of a waveform is the effective value. The rms value of a waveform is just the value of an equivalent dc voltage source which would produce the same heating effect in a resistor.

All properly calibrated ac voltmeters will correctly measure the rms value of a sine wave with a dc level of zero ($A/\sqrt{2}$ where A is the peak value). However, only meters known as true rms meters will measure the rms value of other types of waveforms.

4.2.4 Accuracy, Resolution, Range, and Precision

One of the prime concerns in any measurement system is how accurate is the measurement. Namely, how much does the actual value of the parameter being measured differ from the value indicated by the instrument. For voltmeters and ammeters this is usually expressed as a percentage of the reading plus a percentage of the full scale.

A concept close related to accuracy is resolution. The resolution of an instrument is the smallest change in the parameter being measured that the instrument can measure. Obviously, an instrument cannot be any more accurate than its resolution. The resolution of a digital instrument is set by the amount of the parameter represented by the least significant digit. Although analog instruments theoretically have infinite resolution, the actual resolution is the by the smallest change in the reading that the experimenter can resolve which is rarely more than three significant figures.

The range of an instrument is the difference between the maximum and minimum value of the parameter that can be measured. The larger the range the smaller the resolution becomes. Therefore, it is always desirable to use the range that is just larger than the value of the parameter that is being measured.

A precise instrument is one than indicates the same value for a parameter each time the measurement is made. The measurement may be erroneous but if the same value is obtained each time a measurement is made the instrument is precise. For non precise instruments a number of readings may be taken and the results averaged if the errors are randomly distributed about the actual value. For instruments with systematic or catastrophic errors, averaging the readings will not yield an accurate value for the parameter being measured.

4.2.5 Meter Loading

One of the hoarier laws of nature is that any attempt to directly observe a system requires an interaction between the observer and the system that alters the state of the system. Some energy must be supplied to the observer by the system in order for the observer to make such a measurement of a physical parameter. Therefore, any direct measurement of a physical parameter alters, to some extent, the parameter being measured. It is simply not possible to directly measure with infinite precision any physical parameter.

Electrical measurements of voltage or current are made with instruments known as voltmeters and ammeters. These instruments become circuit elements when connected to an electrical circuit to measure the appropriate circuit variable. When a meter is connected to a circuit all the voltages and currents in branches not containing a voltage or current source are altered by the insertion of the meter. This alteration of the voltages and currents in a circuit by the insertion of a meter to measure a voltage or current is known as meter loading.

4.2.6 Voltmeter Loading

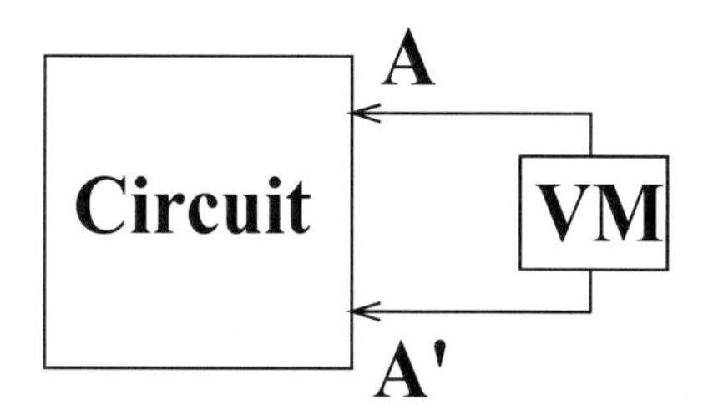

Figure 4.6: Voltage measurement with a voltmeter.

A voltmeter is being used to measure the voltage between the nodes A and $A\prime$ in the circuit in Fig. 4.6. Initially it will be assumed that the voltage being measured is dc. Prior to the connection of the voltmeter, the voltage between these two points is V volts. After the connection of the voltmeter the voltage between these two points is V'. The difference between V and V' determines the usefulness of this voltmeter in measuring these voltages. If the difference is small, then this instrument can be used to obtain useful data.

To obtain a measure of the usefulness of measurements made by a voltmeter, a measure known as the percentage error due to voltmeter loading will be introduced. The percentage error due to voltmeter loading in the measurement of dc voltages is defined as

$$\%\, error = 100 \times \frac{V' - V}{V} \tag{4.6}$$

which simply expresses the percentage by which the measured voltage differs from the voltage that existed between those two points in the circuit prior to the connection of the voltmeter to measure the voltage. This measure is valid unless the voltage being measured is zero; in this case it is the numerator that is of interest. If the voltage being measured is zero the percentage error will also be zero since the application of a voltmeter across terminals A and A' will not alter, in any way, the voltage.

The determination of the percentage error is facilitated by replacing the circuit with its Thévenin equivalent as shown in Fig. 4.7. The voltage that is present across terminals AA' before the voltmeter is applied is V and after the voltmeter is applied the voltage across these terminals is V'. From the Thévenin equivalent circuit it can be easily seen that

$$V = E_{th} \text{ and } V' = \frac{R_{in}}{R_{th} + R_{in}} E_{th} \tag{4.7}$$

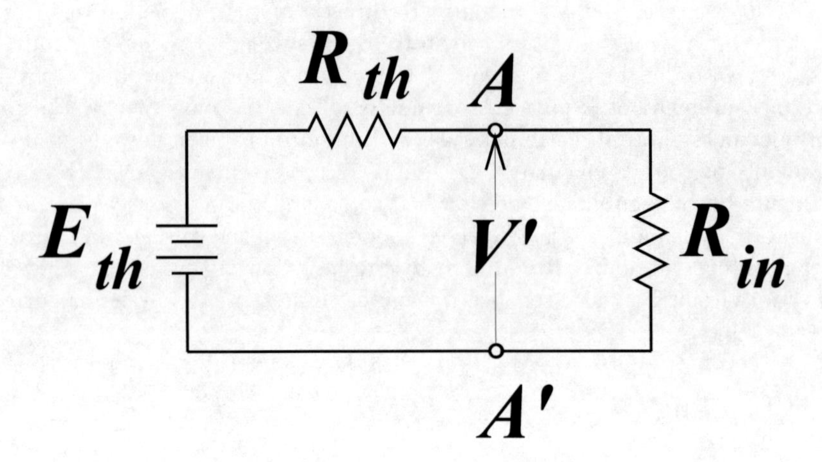

Figure 4.7: Thévenin equivalent for dc circuits.

which illustrates that $V' < V$ which is the reason why the term loading is used. Namely, the voltmeter loads the circuit down and reduces all the voltages and currents not set by a voltage or current source.

The percentage error due to voltmeter loading is then given by

$$\%\ error = -100 \times \frac{R_{th}}{R_{th} + R_{in}} \tag{4.8}$$

which illustrates that the percentage error due to voltmeter loading is not a function of the Thévenin equivalent voltage but is a function of the input impedance of the voltmeter and the Thévenin equivalent impedance of the circuit seen looking back into the circuit from the terminals at which the voltmeter has been connected. A voltmeter with an infinite input impedance would produce no error due to voltmeter loading but it also couldn't measure a voltage since it could not draw any current from the circuit. In order for accurate voltage measurements to be made with a voltmeter the relationship $R_{in} >> R_{th}$ must be satisfied.

The analysis of loading effects for voltage measurements with voltmeters in ac circuits parallels the analysis for dc voltage measurements. The ac Thévenin equivalent circuit is shown in Fig. 4.8 where the Thévenin equivalent voltage is expressed as a rms complex phasor. The ac voltage across the terminals AA' before the voltmeter is applied is $\bar{V} = \bar{E}_{th}$ and after the voltmeter is connected to the circuit the voltage is

$$\overline{V}' = \frac{\overline{Z}_{in}}{\overline{Z}_{in} + \overline{Z}_{th}} \overline{E}_{th} \tag{4.9}$$

which differs from $\bar{V}$ in both magnitude and phase. Since most ac voltmeters measure only the rms value of a time varying voltage, the quantity of interest is the percentage error due to voltmeter loading in the measurement of the rms magnitude of the voltage which would be given by

$$\%\ error = 100 \times \frac{|\,\overline{V}'\,| - |\,\overline{V}\,|}{|\,\overline{V}\,|} \tag{4.10}$$

where the vertical bars indicate the magnitude of the complex phasor voltages. In terms of the circuit and voltmeter impedances the expression for the percentage error due to voltmeter loading in the measurement of the rms magnitude of an ac voltage becomes

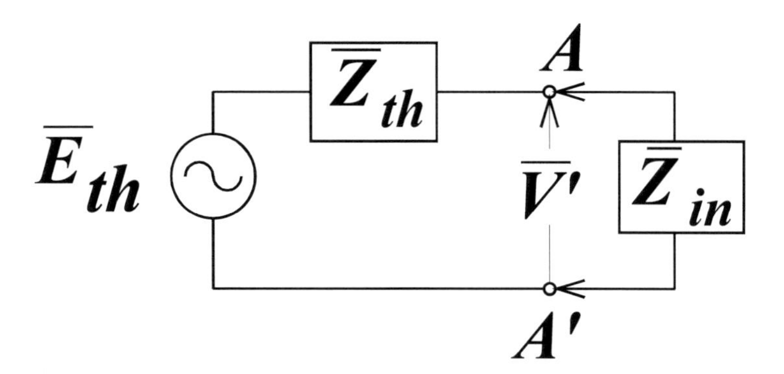

Figure 4.8: Thévenin equivalent for ac circuits.

$$\%\ error = 100 \times \left[\frac{1}{\left| 1 + \frac{\bar{Z}_{th}}{\bar{Z}_{in}} \right|} - 1 \right] \tag{4.11}$$

which cannot be placed into as compact a form as the expression for voltmeter loading in dc circuits because the impedances are complex. Accurate measurements of the rms value of ac voltages, therefore, requires that $|\bar{Z}_{in}| >> |\bar{Z}_{th}|$.

4.2.7 Ammeter Loading

The ammeter loading problem is the dual of the voltmeter loading problem. Namely, the insertion of an ammeter to measure a current in a branch of a circuit changes the current from I to I' (dc circuit). The percentage error due to ammeter loading may be defined as

$$\%\ error = 100 \times \frac{I' - I}{I} \tag{4.12}$$

which in terms of the circuit and instrument impedances becomes

$$\%\ error = -100 \frac{R_{in}}{R_{in} + R_{th}} \tag{4.13}$$

where R_{in} is the internal resistance of the ammeter (the resistance that would be measured between the terminals of the ammeter) and R_{th} is the Thévenin equivalent impedance of the circuit as seen from the terminals to which the ammeter has been connected. This expression discloses that accurate measurements can be made by an ammeter as long as $R_{th} >> R_{in}$. There would be no meter loading if $R_{in} = 0$ but the ammeter could not draw any energy from the circuit and no measurement could be made. For ac current measurements, the expression for the percentage error due to ammeter loading becomes

$$\%\ error = 100 \times \left[\frac{1}{\left| 1 + \frac{\bar{Z}_{in}}{\bar{Z}_{th}} \right|} - 1 \right] \tag{4.14}$$

which is the dual of the expression for the percentage error due to voltmeter loading.

4.2.8 Electronic Multimeters

An instrument that is used to measure voltage, current, and resistance and employs an electronic amplifier to enhance its measurement capabilities is known as an electronic multimeter. Electronic multimeters may have either analog or digital readouts with the digital format becoming dominant. Some instruments use both analog and digital readout devices.

Some electronic multimeters can measure the rms value of time varying voltages and currents as well as dc voltages and currents. Not all multimeters that have ac voltage measurement capability can measure the rms value of an arbitrary voltage but will measure the rms value of only a sinusoidal with a dc level of zero volts. Multimeters that measure the correct rms value for arbitrary voltages are known as "true rms" multimeters.

The loading effect produced by an electronic voltmeter is usually less than that produced by a nonelectronic voltmeter because it usually has a much higher input impedance. Often the electronic amplifier is used to simply increase the input impedance of the voltmeter.

4.2.9 Electronic Analog Voltmeter

When vacuum tube devices were developed in the early part of the twentieth century one of the first applications found for them were in electronic amplifiers that were used to enhance the measurement capabilities of voltmeters. Such voltmeters were known as VTVMs (vacuum tube voltmeters). Today almost all electronic amplifiers use solid state devices as the amplifying elements or active devices because these devices are more rugged and consume less power than vacuum tubes. Electronic voltmeters that employ solid state devices as the active elements as usually designated with the acronym EVM (electronic voltmeter) or if a field effect transistor is the active device the acronym is FET VOM (field effect transistor volt ohm meter).

An amplifier is a circuit which has two ports (a port is a set of two wires) known as the input and output ports for which the energy at the output port can be controlled by a considerably smaller energy at the input port. The purpose of the amplifier in a voltmeter is to reduce the amount of energy that must be withdrawn from the circuit to measure a voltage and, therefore, to minimize voltmeter loading. Thus the amplifier acts a buffering device which effectively isolates the circuit in which the voltage is to be measured from the readout device, such as a d'Arsonval meter movement, which is used to obtain a measure of this voltage.

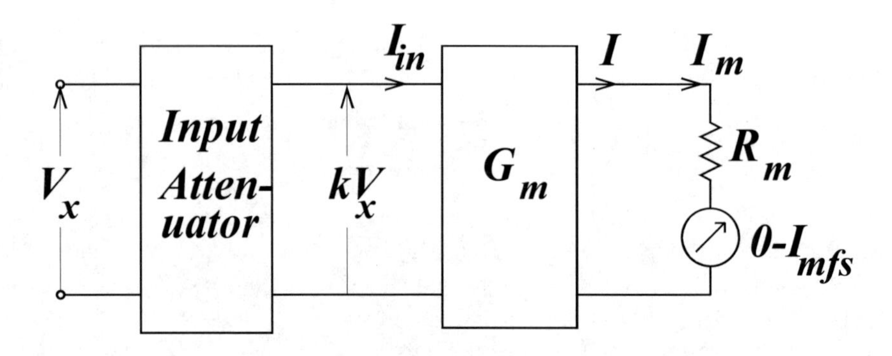

Figure 4.9: Electronic voltmeter with d'Arsonval meter movement readout.

Shown in Fig. 4.9 is a block diagram of an electronic voltmeter with a analog readout device which is a simple d'Arsonval meter movement. The voltage to be measured in the circuit, V_x, is attenuated to a normalized range, kV_x, which is placed across the input terminals or port of an electronic amplifier. The attenuation factor k is known and is selected so that as V_x varies across a certain range the attenuated voltage kV_x varies across a normalized range such as 0 to 1 Volt.

The amplifier shown in Fig. 4.9 is known as a transconductance amplifier. The current I_{in} flowing into the input port is essentially zero since the active devices used to implement the amplifier have a large input impedance (typically FET, field-effect transistors, which have input impedances on the order of 10^{14} Ω). The relationship between the output current and the input voltage is given by

$$I = G_m k V_x \tag{4.15}$$

where G_m is a constant known as the transconductance of the amplifier (the parameter G_m is known as the transconductance of the amplifier because when multiplied by a voltage a current results and the prefix trans is used because the current is at the output port of the amplifier and the voltage is at the input port). Thus this amplifier has a negligible input power and an output current that is a linear function of the input voltage. The amplifier's internal circuitry is designed so that the transconductance is such that as the input voltage to the amplifier varies over a normalized range such as $0 - 1\ Volt$, the output current ranges over $0 - I_{mfs}$.

Were it not for the input attenuator this type of voltmeter would have an input impedance set by that of the active device. The attenuator is necessary if it is desired to have a voltmeter than can measure voltages in several different ranges using an amplifier with a specified transconductance. It is much easier to change the attenuation factor of the input attenuator than to change the transconductance of the amplifier.

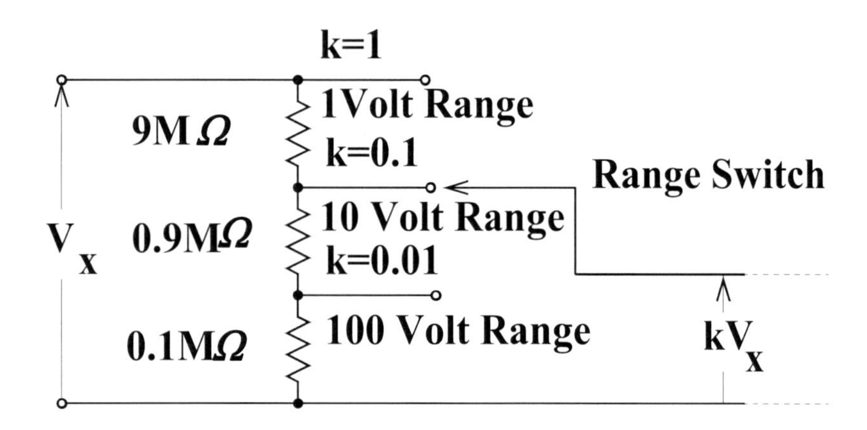

Figure 4.10: Input attenuator for digital voltmeter.

A typical input attenuator is shown in Fig. 4.10. It consists of the series combination of three resistors having values of 9 $M\Omega$, 0.9 $M\Omega$, and 0.1 $M\Omega$. These values are chosen so that the sum of the resistance values is 10 $M\Omega$. Since the input impedance of the amplifier is immense, this means that the input impedance of this voltmeter will be 10 $M\Omega$ which is a typical value for electronic voltmeters.

There are three range settings for this voltmeter: 1 Volt, 10 Volts, and 100 Volts. The selection of a range involves setting the switch to the settings shown (the voltages shown to the right of the circles are the position to which the switch is to be set for the range setting for the voltmeter and certainly not the voltage at those points). The most sensitive or smallest range for this voltmeter is 1 Volt, i.e. when the range switch is in the top position the attenuation factor is 1 and the voltages V_x and kV_x are identical. For the 10Volt range the switch is set to the center position for which $k = (0.9\ M\Omega + 0.1\ M\Omega)/(9\ M\Omega + 0.9\ M\Omega + 0.1\ M\Omega) = 0.1$ which means that the voltage at the input to the amplifier is one tenth of the voltage at the input to the voltmeter. For the 100 Volt range, the switch is set to the lower of the three settings for which

$k = 0.1\ M\Omega/(9\ M\Omega + 0.9\ M\Omega + 0.1\ M\Omega) = 0.01$ which means that the voltage to the input of the amplifier is one hundredth of the voltage at the input to the voltmeter.

For any setting of the input attenuator the input impedance of this voltmeter is 10 $M\Omega$. This is the inherent superiority of the electronic voltmeter over the non-electronic voltmeter. The input impedance of the electronic voltmeter is constant and large compared to the non-electronic voltmeter. Therefore, less meter loading is normally obtained with an electronic voltmeter.

4.2.10 Electronic Analog Ammeter

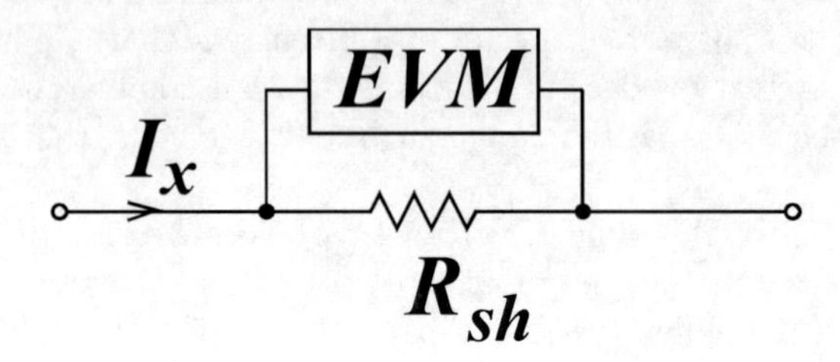

Figure 4.11: Electronic ammeter.

An electronic analog ammeter which uses a d'Arsonval meter movement as the readout device is shown in Fig. 4.11. The current to be measured is passed through the parallel combination of a known shunt resistor, R_{sh}, and the electronic voltmeter with analog readout considered in the previous section. If $R_{sh} << R_{in}$ where R_{in} is the input impedance of the electronic voltmeter [usually 10 $M\Omega$], then essentially all of the current I_x passes through the shunt resistor and the current is given by $I_x = V/R_{sh}$ where V is the voltage measured with the electronic voltmeter. The internal resistance of this ammeter is essentially R_{sh}.

4.2.11 Electronic Analog Ohmmeter

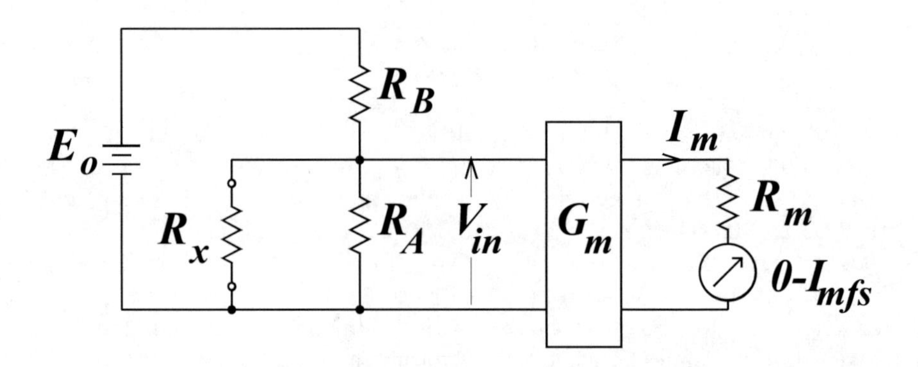

Figure 4.12: Electronic analog ohmmeter.

A block diagram of an electronic analog ohmmeter that employs a d'Arsonval meter movement as the analog readout device is shown in Fig. 4.12. The resistor that is to be measured is R_x. The resistors R_B and the parallel combination of R_x and R_A form a voltage divider that sets the voltage V_{in} at the input port of the transconductance amplifier as

$$V_{in} = \frac{R_A \parallel R_x}{R_A \parallel R_x + R_B} E_o \tag{4.16}$$

where E_o is a fixed dc voltage. Since the current flowing through the meter movement is given by $I = G_m V_{in}$ the current flowing through the meter movement is a function of R_x, albeit nonlinear, and a resistance scale can be marked off on the scale of the meter movement.

The current flowing through the meter movement is given by

$$I = G_m \left[\frac{R_x}{R_x + R_A \parallel R_B}\right] \frac{R_A}{R_A + R_B} E_o \tag{4.17}$$

from the above equations. The range of resistance values that are to be measured is from 0 to ∞. As the above equation discloses the current is zero when R_x is zero and increases as R_x increases. In order to make the most efficient use of scale on the meter movement, the largest resistance to be measured (∞) should produce the current I_{mfs}. Therefore, taking the limit of the meter movement current as the resistance R_x tends to ∞ yields

$$I_{mfs} = G_m \frac{R_A}{R_A + R_B} E_o \tag{4.18}$$

which must be satisfied if $R_x = \infty$ is to result in full scale deflection of the pointer on the meter movement.

Using the expressions for I and I_{mfs} the ratio of θ/θ_{fs} becomes

$$\frac{\theta}{\theta_{fs}} = \frac{I}{I_{mfs}} = \frac{R_x}{R_x + R_A \parallel R_B} \tag{4.19}$$

where θ is the angular deflection of the pointer on the d'Arsonval meter movement and θ_{fs} is the full scale deflection angle. The resistance value $R_x = R_{ms} = R_A \| R_B$ causes the angular deflection of the pointer to be one half of the full scale deflection angle and is known as the mid scale resistance value.

4.2.12 Electronic Digital Voltmeter

Electronic voltmeters with digital readouts are simply termed digital voltmeters (DVM). If part of an instrument that can measure resistance, the rms value of an ac voltage, and current as well as voltage, the instrument is termed a digital multimeter (DMM). Instruments with digital displays are preferred in many applications because of ease and speed with which the parameter can be read, the accuracy, the repeatability of the measurement, and the ability to interface such instruments directly to computers.

A block diagram of a digital voltmeter is shown in Fig. 4.13. The input attenuator performs the same function as the input attenuator in the analog electronic voltmeter considered in previous sections, i.e. it converts the voltage to be measured to a normalized range. The circuitry that follows the input attenuator has a very high input impedance and draws negligible current which means that the input impedance of the voltmeter is set by the resistors in the input attenuator.

The block indicated as A/D is known as an analog to digital converter. The electronic circuitry inside the A/D converter converts the analog input voltage, kV_x, into a digital quantity, C_L, which is a digital code word. The input voltage is known as an analog quantity since it can assume any of an infinite number values in an interval, $0 < kV_x < V_{fs}$, where V_{fs} is the full scale or range value for the A/D converter (the type of A/D converter being considered is known as unipolar because the input voltage is of one polarity; there are A/D converters for which the input voltage can be either positive or negative and these are known as bipolar). The output quantity is known as a digital quantity because there are only a finite number of code words, K, are produced as the input voltage varies from 0 to V_{fs}, i.e. each of the K code words represents a subset of the range of input voltages.

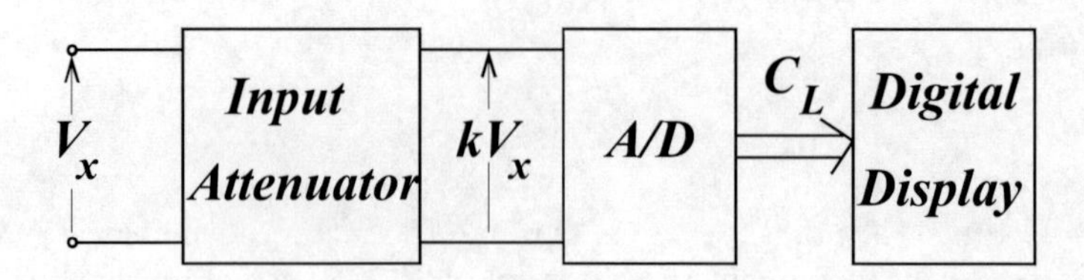

Figure 4.13: Digital voltmeter.

If the binary number system is used to represent the digital code words at the output of the analog to digital converter, the number of binary digits, N, that are required to represent a total of K code words is $N = \log_2 K$. Any analog to digital converter that uses code words that contain N binary digits is simply known as an N bit A/D converter. Thus a total of $K = 2^N$ code words are available to represent the range $0 - V_{fs}$.

Each of the output code words represents a subset of the range $0 - V_{fs}$. Normally, the voltage range $0 - V_{fs}$ is divided into K equal segments having a width of ΔV volts (except for the first and last segment in this range). This means if the voltage at the input to the analog to digital converter changes by less than ΔV volts the output code word may not change which means that the analog to digital converter cannot resolve or measure voltage changes of less than ΔV volts. The quantity ΔV given by

$$\Delta V = \frac{V_{fs}}{2^N} \tag{4.20}$$

is known as the resolution of the A/D converter and determines the smallest voltage change that a digital voltmeter can measure. Another important parameter associated with A/D converters is the dynamic range which is given by

$$DR_{dB} = 20 \log_{10} \frac{V_{fs}}{\Delta V} \tag{4.21}$$

which (expressed in decibels) determines the ratio of the largest to the smallest voltage change that can be measured with the A/D converter.

Analog to digital converters are the heart of all digital multimeters as well as data acquisition systems in which voltages are to be digitized for input to computers. Numerous techniques for performing this conversion are used such as integration, counting, successive approximation, flash or parallel, servo loop tracking, synchro to digital, and voltage to frequency conversion. The two techniques used most commonly in digital multimeters are dual slope integration and successive approximation.

4.3 Equipment

Fluke Model 73 Hand-Held Digital Multimeter (HHDMM). This is a battery powered digital multimeter similar in shape and size to a pocket calculator. A liquid crystal display is used which places a minimum power drain on the battery. It is shown in Fig. 4.14. This instrument is auto ranging and the range cannot be manually set except for a 300 mV range. The input terminals for the measurement of voltage and resistance are located in the lower right of the plastic case.

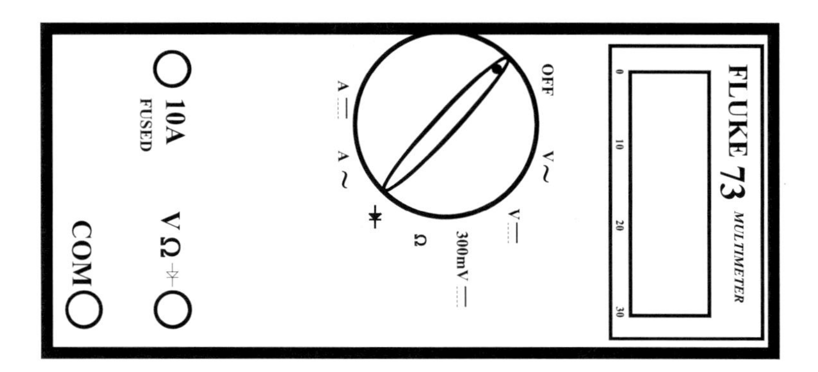

Figure 4.14: Fluke 73 hand held digital multimeter.

This instrument has both an analog and digital readout. The digital readout uses 3 and one half digits. (The one half digit means that the most significant digit cannot take on the full range of decimal digits. Instead the most significant digit will be blank, one, two, or three.) The analog readout is a sliding scale located below the digital readout.

When used as a ac voltmeter this instrument has an input impedance of 10 $M\Omega$ which is standard for digital voltmeters. This is equal to the input impedance of the **Simpson** meter on its 500 Volt range and larger than the input impedance of the Simpson meter on its other ranges. (The input impedance of the **Agilent 34401A** DMM is at least 10 $M\Omega$ on all scales and larger on some scales when used as a dc voltmeter.)

This type of meter will measure correct rms voltages only for sinusoidal waveforms with a dc level of zero volts. Just as is the case for the Simpson meter this instrument is not a true rms voltmeter. (The **Agilent 34401A** digital multimeter is a true rms voltmeter.)

4.4 Examples

4.4.1 DC Voltmeter Loading

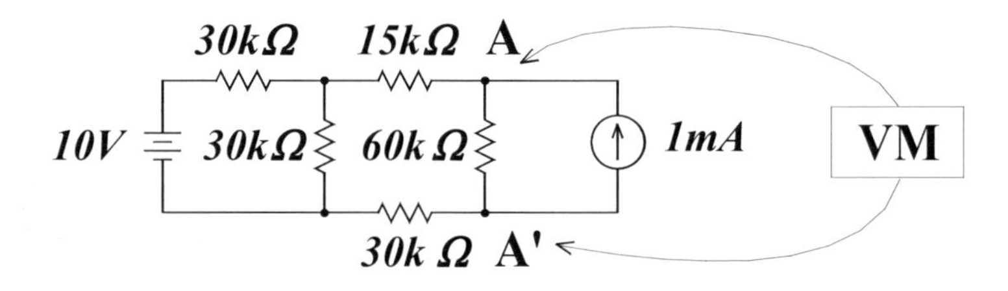

Figure 4.15: DC voltmeter loading example circuit.

Problem: Determine the percentage error due to dc voltmeter loading in the measurement of the dc voltage $V_{AA'}$ for the circuit shown in Fig. 4.15 if the dc voltmeter is a: (a) **Simpson Meter Model 260-7** (b) **Agilent 34401A** DMM and (c) **Fluke 73** HHDMM.

Solution: The first step is to find the Thévenin equivalent circuit with respect to the voltmeter. The voltmeter is removed from the circuit and the voltage that appears across the terminals where the voltmeter was connected is determined. This is the Thévenin voltage, E_{th}. It is obtained by simple superposition as

$$E_{th} = 10 \times \frac{30\|(15+60+30)}{30+30\|(15+60+30)} \times \frac{60}{15+30+60} + 1 \times 60\|(15+30\|30+30) = 32.5\,\mathrm{V} \tag{4.22}$$

The next step is to find the Thévenin equivalent resistance. All of the sources are turned off. All voltage sources are replaced with short circuits and all current sources with open circuits. The Thévenin equivalent resistance is then the resistance between the two nodes where the voltmeter was connected. Thus, the Thévenin equivalent resistance is

$$R_{th} = 60\|(15+30\|30+30) = 30\,\mathrm{k\Omega} \tag{4.23}$$

Now that the Thévenin equivalent circuit has been determined the percentage error due to dc voltmeter loading can be determined. For the **Simpson** meter the input impedance is given by

$$R_{in} = SV_{fs} \tag{4.24}$$

where S is the dc sensitivity ($20\,\mathrm{k\Omega/V}$) and V_{fs} is the range voltage. The appropriate range to select is the 50 V range since it is the range just larger than the open circuit voltage. If a smaller range were selected the pointer would peg and if a larger range were selected it would be more difficult to read the position of the pointer. The input impedance is then

$$R_{in} = 20 \times 50 = 1\ M\Omega = 1{,}000\,\mathrm{k\Omega} \tag{4.25}$$

which makes the percentage error

$$\%\ error = -100 \times \frac{R_{th}}{R_{th}+R_{in}} = -2.91\% \tag{4.26}$$

The input impedance of both the **Agilent 34401A** and **Fluke 73** is 10 $M\Omega$ which makes the percentage error for either

$$\%\ error = -100 \times \frac{R_{th}}{R_{th}+R_{in}} = -0.299\% \tag{4.27}$$

4.4.2 AC Voltmeter Loading

Problem: Determine the percentage error due to ac voltmeter loading in the measurement of the rms magnitude of the ac voltage $\overline{V}_{AA'}$ for the circuit shown in Fig. 4.16 if the ac voltmeter is a: (a) **Simpson Meter Model 260-7** (b) **Agilent 34401A** DMM and (c) **Fluke 73** HHDMM. The frequency of the sinusoidal source is $f = 3\,\mathrm{kHz}$.

Solution: The first step is to find the Thévenin equivalent circuit with respect to the voltmeter. The voltmeter is removed from the circuit and the open circuit voltage is then

$$\overline{E}_{th} = 20\angle 0^\circ \times \frac{30\|(j37.7+47+68)}{-j53.1+30\|(j37.7+47+68)} \times \frac{47}{j37.7+47+68} = 3.31\angle 50.3^\circ\ \mathrm{V} \tag{4.28}$$

which is the voltage that would be measured with an ideal voltmeter (one with an infinite input impedance). The next step is to turn off all the sources and find the Thévenin equivalent impedance

$$\overline{Z}_{th} = 47\|(j37.7+30\|[-j53.1]+68) = 31.6\angle 5.09^\circ\ \mathrm{k\Omega} \tag{4.29}$$

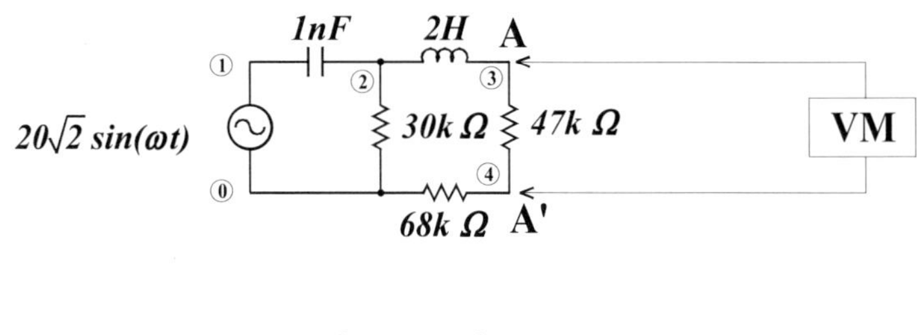

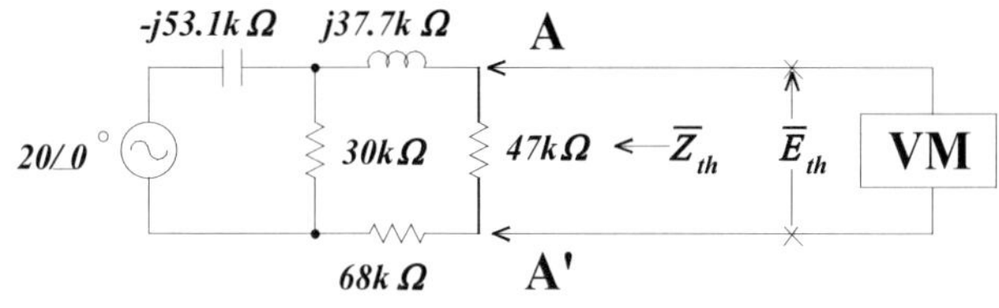

Figure 4.16: AC voltmeter loading example circuit.

The percentage error due to ac voltmeter loading in the measurement of the rms magnitude of the ac voltage $\overline{V}_{AA'}$ can now be determined if the input impedances of the ac voltmeters is known. For the **Fluke 73** HHDMM the input impedance is the same as that for the measurement of dc voltages, viz. $10\,\mathrm{M\Omega}$. For the **Agilent 34401A** the input impedance is only $1\,\mathrm{M\Omega}$. For the **Simpson** meter the input impedance is again given by $R_{in} = SV_{fs}$ but the ac sensitivity is only $5\,\mathrm{k\Omega}/Volt$. The appropriate range to select would be the 10 V range since a smaller range would peg the meter movement and a larger one would be more difficult to read. Thus the input impedance of the **Simpson** meter is $\overline{Z}_{in} = SV_{fs} = 5\,\mathrm{k\Omega}/\,\mathrm{V} \times 10\,\mathrm{V} = 50\,\mathrm{k\Omega}$.

The percentage error is given by Eq. 4.14

$$\%\,error = 100 \times \left[\frac{1}{\left| 1 + \dfrac{\overline{Z}_{th}}{\overline{Z}_{in}} \right|} - 1 \right] = \begin{cases} -38.7\,\% & \textbf{Simpson 260-7} \\ -3.05\,\% & \textbf{Agilent 34401A} \\ -0.31\,\% & \textbf{Fluke 73} \end{cases} \tag{4.30}$$

4.5 SPICE

SPICE can be used to analyze either dc or ac meter loading. For dc circuits the problem is particularly simple and the node voltages are printed in the file with extension `.OUT`.

For ac circuits SPICE can be used to determine and plot both voltages across nodes in a circuit as well as the percentage error due to ac voltmeter loading as functions of frequency. The input deck SPICE code to determine and plot the voltages as a function of frequency for the circuit shown in Fig. 4.16 is:

```
TITLE LINE
VI 1 0 AC 20
C 1 2 1N
R1 2 0 30K
L 2 3 2
R2 3 4 47K
R3 4 0 68K
.AC DEC 30 10 100K
.PROBE
.END
```

The start frequency is 10 Hz and the stop frequency is 100 kHz with 30 points being plotted per decade. To obtain the voltage magnitude of the voltage $\overline{V}_{AA'}$ the SPICE function to be plotted is `VM(3,4)`. The "M" after the "V" specifies the magnitude of the voltage between the node numbers. (Note that this is different from `VM(3)-VM(4)` due to the phase differences between the voltages at nodes 3 and 4).

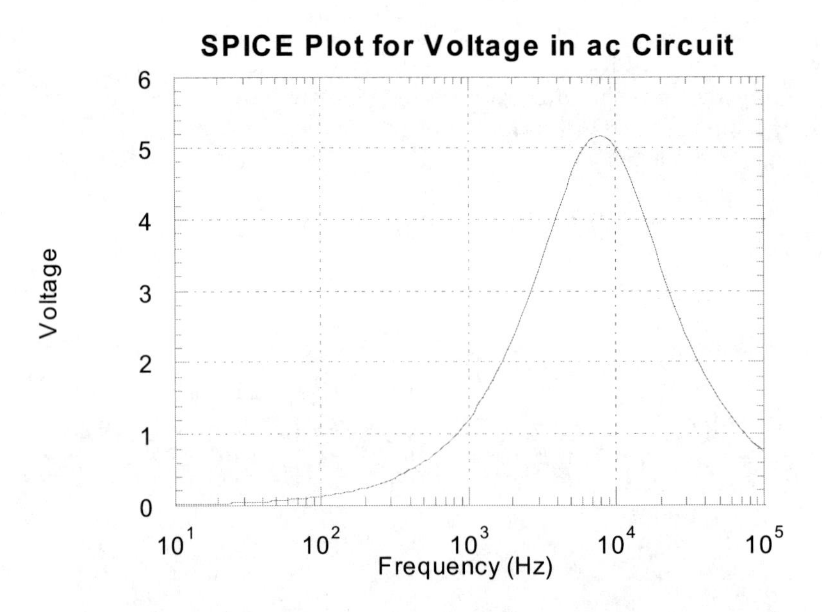

Figure 4.17: SPICE plot of magnitude of voltage across nodes AA' for circuit in Fig. 4.16,

The rms magnitude of the ac voltage measured across nodes AA' is shown in Fig. 4.17. The plot is made with no voltmeter across nodes AA' using the above SPICE code.

The percentage error due to ac voltmeter loading can be determined by adjoining the above circuit with an identical one with the voltmeter added. If the voltmeter were a **Simpson** meter set to its 10 V the SPICE code would be:

```
TITLE LINE
VI 1 0 AC 20
C 1 2 1N
R1 2 0 30K
L 2 3 2
R2 3 4 47K
R3 4 0 68K
VIP 1P 0 AC 20
CP 1P 2P 1N
R1P 2P 0 30K
LP 2P 3P 2
R2P 3P 4P 47K
R3P 4P 0 68K
RIN 3P 4P 50K
.AC DEC 30 10 100K
.PROBE
.END
```

The appropriate SPICE variable to plot would then be `100*(VM(3P,4P)-VM(3,4))/VM(3,4)`.

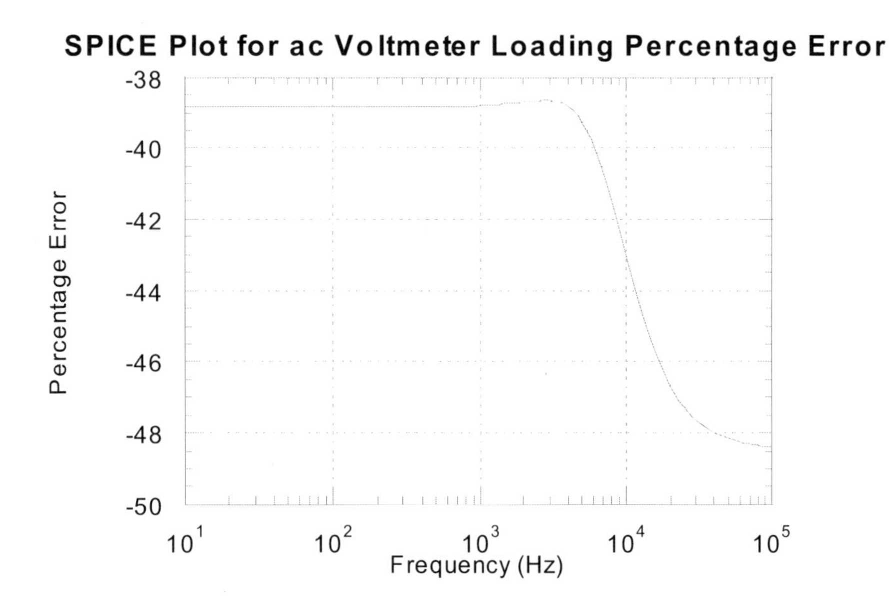

Figure 4.18: Percentage error due to ac voltmeter loading for circuit shown in Fig. 4.16.

The plot for the percentage error due to ac voltmeter loading using the above SPICE code is given in Fig. 4.18.

4.6 Procedure

4.6.1 Input Impedance Measurement

Simpson Meter as DC Voltmeter

Turn the function select on the **Simpson Model 260-7** to "+*DC*" and the range switch [rotary knob] to 500 V. Turn on the **Agilent 34401A** DMM wait until it boots and press "Ω 2*W*". Connect the *HI Input* of the **Agilent 34401A** DMM to the "+" input on the Simpson meter [located on lower left on the plastic case] and the *LO* input of the **Agilent 34401A** to the "*COMMON*" input on the **Simpson** meter. Measure and record the resistance indicated on the **Agilent 34401A** as the DC input impedance of the Simpson meter for the 500 V range. Change the range switch on the **Simpson** meter to 250 V, 50 V, 10 V, and 2.5 V and record the resistance indicated on the **Agilent 34401A** as the input impedances for these ranges.

V_{fs}	R_{in} (kΩ)
500 V	
250 V	
50 V	
10 V	
2.5 V	

Simpson Meter as DC Ammeter

Switch the range on the **Simpson** meter to 10 mA and record the resistance reading on the **Agilent 34401A** as the internal resistance of the Simpson meter when used as a DC ammeter on the 10 mA range. Switch the range on the Simpson meter to 1 mA and repeat. (If reasonably small resistances are not obtained, the fuse inside the **Simpson** meter is probably blown. Should this be the case, have a skilled individual such as the laboratory instructor check the instrument.) Disconnect the **Simpson** meter from the **Agilent 34401A**.

I_{fs}	R_{in} (Ω)
10 mA	
1 mA	

Fluke 73 as DC Voltmeter

Turn the **Fluke 73** HHDMM to DC volts "$\overset{==}{V}$" and connect it (red V scale and *Common*) to the input of the **Agilent 34401A** DMM. Record the resistance reading on the **Agilent 34401A** as the input impedance of the **Fluke 73** when used as a dc voltmeter.

R_{in} =______________________________

HP34401A as DC Ammeter

Set the **Agilent 34401A** to measure dc current (push the blue *Shift* button and then *DC I*). Connect the lead from the "*I 3A*" input on the **Agilent 34401A** to the high input on the **Fluke 73**. Set the **Fluke 73** to the "Ω" scale and record its resistance reading as the internal resistance of the **Agilent 34401A** DMM when used as a dc ammeter.

R_{in} =______________________________

HP34401A as DC Voltmeter

Set the **Agilent 34401A** to measure dc volts (push the button labeled "*DC V*". Connect the lead from the high input of the **Agilent 34401A** DMM to the high input of the **Fluke 73**. Set the **Fluke 73** to the "Ω" scale. If the resistance is too high to be read enter $\geqq 10\ M\Omega$ as the resistance.

R_{in} =____

ELVIS as DC Voltmeter

Launch **ELVIS** with the desktop icon on the pc or the START PROGRAMS sequence. Set the **Agilent 34401A** DMM to measure resistance and measure the input resistance of the **ELVIS** DMM at the banana jacks labeled voltage (V) and common (COM).

R_{in} =____

4.6.2 DC Voltmeter Loading

Turn on the **HP3630A** dc power supply. Set the power supply to monitor the +20 V supply and set it to 10 V. Set the **Agilent 34401A** to measure dc volts (press button labeled *DC V*) and connect it to the output binding posts on the power supply (the +20 V and *COM*). Set the power supply to as close to 10 V as can be obtained.

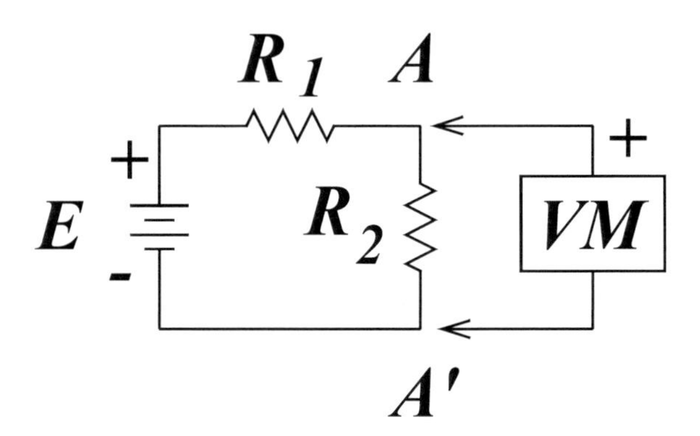

Figure 4.19: DC voltmeter loading circuit.

Assemble the circuit shown in Fig. 4.19 with $R_1 = 1\,\mathrm{k\Omega}$ [brown, black, red] and $R_2 = 2\,\mathrm{k\Omega}$ [red, black, red]. Set the voltage E to 10 V dc (it should be already set to this by the previous steps). Measure the voltage across terminals AA' with the **Agilent 34401A**, **Fluke 73**, **ELVIS**, and the **Simpson** meter. (Make the measurements using one voltmeter at a time. Two or more voltmeters should never be simultaneously placed in parallel since it would reduce their input impedance.) Change R_1 to $10\,\mathrm{k\Omega}$ [brown, black, orange] and R_2 to $20\,\mathrm{k\Omega}$ [red, black, orange] and repeat. Change R_1 to $100\,\mathrm{k\Omega}$ [brown, black, yellow] and R_2 to $200\,\mathrm{k\Omega}$ [red, black, yellow] and repeat. Change R_1 to $1\,\mathrm{M\Omega}$ [brown, black, green] and R_2 to $2\,\mathrm{M\Omega}$ [red, black, green] and repeat.

R_1	R_2	$V_{AA'}$ (*Volts*) Simpson	*Fluke* 73	*HP*34401*A*	ELVIS
$1\,\text{k}\Omega$	$2\,\text{k}\Omega$				
$10\,\text{k}\Omega$	$20\,\text{k}\Omega$				
$100\,\text{k}\Omega$	$200\,\text{k}\Omega$				
$1\,\text{M}\Omega$	$2\,\text{M}\Omega$				

4.6.3 DC Ammeter Loading.

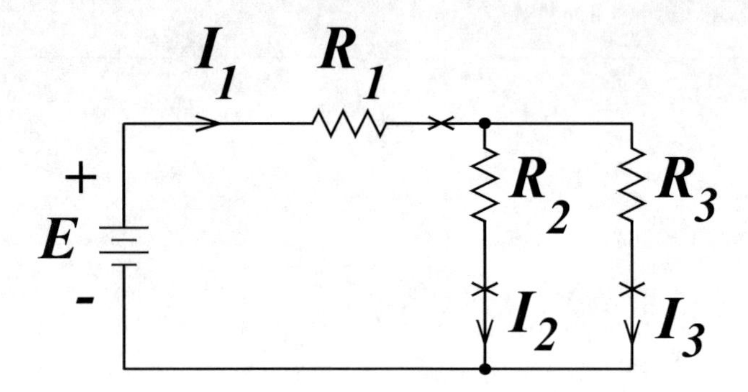

Figure 4.20: DC ammeter loading.

Assemble the circuit shown in Fig. 4.20 with $E = 10\,\text{V}$, $R_1 = 10\,\text{k}\Omega$ [brown, black, orange], $R_2 = 20\,\Omega$ [red, black, black], and $R_3 = 200\,\Omega$ [red, black, brown].

The currents I_1, I_2, and I_3 are to be measured using the **Agilent 34401A** and **Simpson 260-7** as dc ammeters. The "x" indicates the position in the circuit where the dc ammeter is to be inserted for each measurement.

Using the **Agilent 34401A** as a dc ammeter (press the blue *Shift* button and then the button *DC I* and use as the plus input the input labeled "3*A I*" and as the minus input the input labeled "*LO*"), measure first the current I_1 by inserting the ammeter where the appropriate x is located and the other two $x's$ are short circuits. Then measure the current I_2 and finally the current I_3.

Measure the currents I_1, I_2, and I_3 using the **Simpson** meter as a dc ammeter. Set the function select to $+DC$ and the range to $10\,\text{mA}$. If the measured current is less than $1\,\text{mA}$, switch the range switch on the **Simpson** meter to $1\,\text{mA}$. Always begin on the largest range scale. (If the **Simpson** meter indicates zero current, the internal fuse is probably blown. Should this be the case, ask the laboratory instructor for assistance.)

Currents (mA)	**Simpson Model 260-7**	**Agilent 34401A** DMM
I_1		
I_2		
I_3		

Turn off the dc power supply and disconnect all leads connected to it. It will not be used for the rest of this experiment.

4.6.4 AC Voltmeter Loading

Signal Generator Adjustment

Turn on the **Hewlett-Packard 3311A** function generator. Set the function to sine. Connect the output of this instrument to the "*HI Input*" and "*LO*" inputs of the **Agilent 34401A** DMM. Press the button "*DC V*". Vary the dc Offset of the function generator until the dc component of the output voltage is less than 5 mV in magnitude. Press the button labeled "*AC V*". Vary the Amplitude of the 600 Ω output of the function generator until the output is $7\,\mathrm{V}_{rms}$.

First Circuit

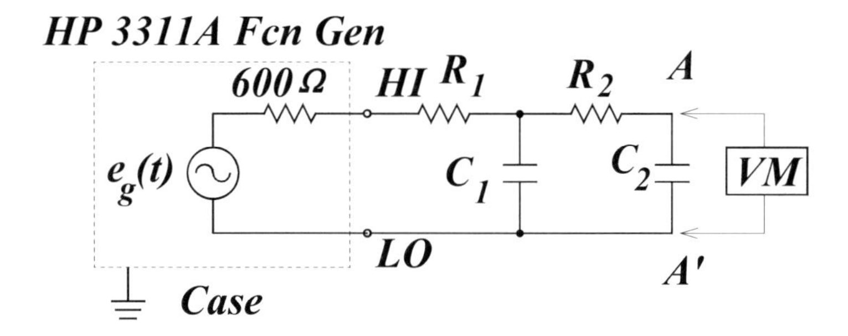

Figure 4.21: AC voltmeter loading.

Assemble the circuit shown in Fig. 4.21 using $R_1 = 10\,\mathrm{k}\Omega$ [brown, black, orange], $R_2 = 20\,\mathrm{k}\Omega$ [red, black, orange]; the values of the capacitors C_1 and C_2 will be specified by the laboratory instructor. The rms ac voltage across the terminals AA' will be measured with all three voltmeters as the frequency is varied from 100 Hz to 10 kHz. (Remember, don't place the voltmeters in parallel.) If the measured voltage is too small to be accurately measured with a particular voltmeter at a particular frequency, it should be recorded on the rough data sheet as UM (unmeasurable) and not plotted on the graph that will be subsequently requested. Data should be taken at frequencies: 100 Hz, 300 Hz, 700 Hz, 1 kHz, 3 kHz, 7 kHz, and 10 kHz. Use the **Agilent 34401A** to set the frequency of the function generator rather than the dial setting of the function generator.

C_1 = blue ____________________

C_2 = gray ____________________

f (Hz)	**Agilent 34401A**	**Fluke 73**	**Simpson Model 260-7**	**ELVIS**
100				
300				
700				
1 k				
3 k				
7 k				
10 k				

Second Circuit

Change the components in Fig.4.21 to: $R_1 = 100\,\mathrm{k\Omega}$ [brown, black, yellow], $R_2 = 200\,\mathrm{k\Omega}$ [red, black, yellow] and the capacitors to the values specified by the laboratory instructor and repeat the above measurements.

C_1 = C1= 220 pF

C_2 = 470 pF

f (Hz)	**Agilent 34401A**	**Fluke 73**	**Simpson Model 260-7**	**ELVIS**
100				
300				
700				
1 k				
3 k				
7 k				
10 k				

Checkout

Turn off all three of the voltmeters and the function generator and return to the laboratory instructor any items provided by the laboratory instructor.

4.7 Laboratory Report

4.7.1 DC Voltmeter Loading

A calculation of the percentage error due to voltmeter loading in the measurement of the voltage $V_{AA'}$ for the four circuits and four voltmeters used in step 4.6.2. Use as the input impedance of the voltmeters the values that were obtained in step 4.6.1. Compute the percentage error using both the expression for the measured and theoretical voltages and the circuit and meter resistances.

The expression for the percentage error using the measured voltage, V', and the theoretical voltage, V, is

$$\%\,error = 100 \times \frac{V' - V}{V} \tag{4.31}$$

where the theoretical voltage V is given by

$$V = E\frac{R_2}{R_1 + R_2} \tag{4.32}$$

and R_1 and R_2 are the nominal values of the resistors. The expression for the percentage error using the circuit and meter resistances is

$$\%\,error = -100\frac{R_{th}}{R_{th} + R_{in}} \tag{4.33}$$

where $R_{th} = R_1 \parallel R_2$.

$\mathbf{R_1}$	$\mathbf{R_2}$	**% error, voltage expression**	**% error, resistance expression**
1 kΩ	2 kΩ		
10 kΩ	20 kΩ		
100 kΩ	200 kΩ		
1 MΩ	2 MΩ		

Table 4.1: Simpson Model 260-7

$\mathbf{R_1}$	$\mathbf{R_2}$	**% error, voltage expression**	**% error, resistance expression**
1 kΩ	2 kΩ		
10 kΩ	20 kΩ		
100 kΩ	200 kΩ		
1 MΩ	2 MΩ		

Table 4.2: Fluke 73

$\mathbf{R_1}$	$\mathbf{R_2}$	**% error, voltage expression**	**% error, resistance expression**
1 kΩ	2 kΩ		
10 kΩ	20 kΩ		
100 kΩ	200 kΩ		
1 MΩ	2 MΩ		

Table 4.3: HP34401A

4.7.2 DC Ammeter Loading

A calculation of the percentage error due to ammeter loading in the measurement of the currents I_1, I_2, and I_3 in step 4.6.3 for the two dc ammeters that were used. Use as the internal resistance of the ammeters the values that were measured in step 4.6.1. Use the formula

$$\%\,error = -100 \times \frac{R_{in}}{R_{in} + R_{th}}$$

to compute the percentage error for dc ammeter loading.

4.7.3 AC Voltmeter Loading

Plots of the ac rms voltages experimentally measured in step 4.6.4 as a function of frequency. The plots should be made on log-log graph paper (the frequency data extends for two cycles the number of cycles required for the AC rms voltages depends on the data taken). A separate graph should be made for each set of circuit components for which data was taken. The plots may be made on the graph paper at the end of this experiment or using a spreadsheet.

If the laboratory experiment and/or report cannot be completed by the end of the three hour lab session, turn in what has been completed to the laboratory instructor.

$\mathbf{R}_1$	$\mathbf{R}_2$	**% error, voltage expression**	**% error, resistance expression**
1 kΩ	2 kΩ		
10 kΩ	20 kΩ		
100 kΩ	200 kΩ		
1 MΩ	2 MΩ		

Table 4.4: ELVIS

Current	R_{th} (expression)	R_{th} (value)	% error (**Simpson**)	% error (**Agilent 34401A**)
I_1	$R_1 + R_2 \| R_3$			
I_2	$R_2 + R_1 \| R_3$			
I_3	$R_3 + R_1 \| R_2$			

Table 4.5: DC Ammeter Loading

4.8 References

1. Coombs, C. F., *Electronic Instrument Handbook*, 2nd ed., McGraw-Hill, 1995.
2. Cooper, W. D., *Electronic Instrumentation and Measurements Techniques*, 2nd edition, Prentice-Hall, 1978.
3. Jones, L., and Chin, A. F., *Electronic Instruments and Measurements*, John Wiley, 1983.
4. Krenz, J. H., *Introduction to Electrical Circuits and Electronic Devices: A Laboratory Approach*, Prentice-Hall, 1987.
5. Kantrowitz, P., et. al., *Electronic Measurements*, Prentice-Hall, 1979.
6. Oliver, B. M., and Cage, J. M., *Electronic Measurements and Instrumentation*, McGraw-Hill, New York, 1971.
7. Miner, G. F., and Comer, D. J., *Physical Data Acquisition for Digital Processing*, Prentice-Hall, 1992.
8. Riaz, M., *Electrical Engineering Laboratory Manual*, McGraw-Hill, 1965.
9. Shiengold, D. H., *Analog-Digital Conversion Notes*, Analog Devices, 1977.
10. Wolf, S., *Guide to Electronic Measurements and Laboratory Practices*, Prentice-Hall, 1973.
11. Wolf, S., *Guide to Electronic Measurements and Laboratory Practices*, 2nd edition, Prentice-Hall, 1983.
12. Wolf, S., and Smith, R. F. M., *Student Reference Manual for Electronic Instrumentation Laboratories*, Prentice-Hall, 1990.

Chapter 5

The Oscilloscope

5.1 Introduction

The oscilloscope is one of the most powerful and useful instruments available to the engineer, technician, or scientist. It enables one to monitor the state of system or circuit as a function of time. This provides an almost complete characterization of the system which facilities the design and analysis processes. Lower levels instruments such as meters can provide only gross characteristics of signals. Its uses are limited only the imagination of the user.

Many physical parameters can be accurately and rapidly measured with an oscilloscope. Since the oscilloscope measures voltages directly, it can be used to measure voltages in an electrical or electronic circuit. Current can be measured by measuring the voltage across a resistor through which the current passes. Sensors (transducers) can and are used to convert physical parameters such as pressure, force, displacement, velocity, sound level, etc. into voltages; unless great precision is required, the oscilloscope can be used to provide meaningful real time measurement of these parameters.

When used in the time sweep mode, the oscilloscope can be used to display one or more voltage waveforms. From these plots of amplitude versus time, time intervals can be measured and, therefore, the frequency of periodic waves as well as the phase difference between periodic waves. (Phase shifts and frequency can also be measured from Lissajous patterns). From this data, the amplitude and phase transfer functions of systems, impedance, or power can be obtained. Impedance measurements and the response of circuits for excitations such as sinusoidal or square wave can be obtained.

If the voltages to be measured are not periodic functions of time, then the oscilloscope must be capable of storing the occurrence of these non-periodic events. This requires a special type of oscilloscope known as a storage scope. These types of oscilloscopes use either analog or digital storage techniques.

When an oscilloscope is used to measure a voltage in a circuit, a loading effect occurs. Unlike the loading effects that occur with DC and AC voltmeters, the circuit loading produced by an oscilloscope is frequency dependent. In order to evaluate loading effects, a comparison must be made between the input impedance of the oscilloscope and the impedance of the circuit to which it is connected.

The bandwidth of an oscilloscope places a limitation on the types of waveforms that can be accurately measured with an oscilloscope. The rise time of the oscilloscope is inversely related to the bandwidth. Unless the bandwidth of the oscilloscope is significantly larger than the bandwidth of the waveform being observed, the display will be distorted.

5.2 Oscilloscope Hierarchy

The hierarchy of oscilloscope architectures is shown in Fig. 5.1. There are two basic architectures: analog and digital. Hybrid oscilloscopes can function in either the analog or digital mode. An analog oscilloscope displays a waveform by amplifying or attenuating the input waveform or signal by connecting it directly to

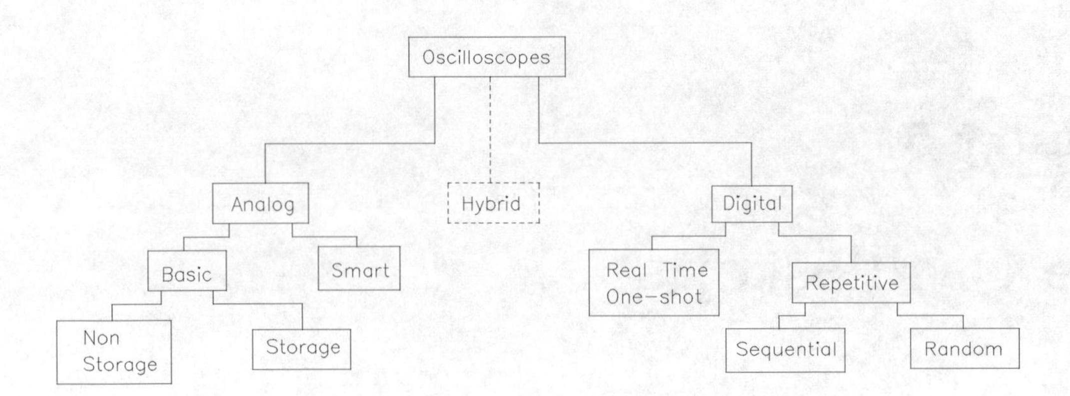

Figure 5.1: Oscilloscope hierarchy.

the display device. A digital oscilloscope digitizes the input waveform, stores it in a memory, and the stored digitized waveform is sent to the display device. Both architectures are in use and have their advantages and disadvantages.

During the last two decades of the Twentieth Century the development of cheap, powerful, small, lightweight, high-speed microprocessors, analog-to-digital converters, and memories have led to an increased usage of digital oscilloscopes and they will undoubtable dominant the market during the next millennium. However, there is a Praetorium Guard who prefer analog instruments which will give them a permanent niche in measurement science.

5.3 Analog Oscilloscopes

The simplest form of analog oscilloscope is the basic nonstorage oscilloscope. An elementary block diagram is shown in Fig. 5.2. This instrument can be used to display one voltage as a function of another voltage ($X - Y$ mode) but its primary use is to display periodic or repetitive waveforms or signals as functions of time.

5.3.1 Cathode Ray Tube (CRT)

The heart of any oscilloscope is the display device which is almost always a cathode ray tube (CRT). [Some modern handheld oscilloscopes use LCD displays instead of a CRT.] The number of inputs for a modern oscilloscope is at least two and a typical number is four. There are two major signal paths: from the input to the vertical deflection plates and from the input to the horizontal deflection plates of the CRT; voltages applied to these plates determine the vertical and horizontal position of the beam produced in the CRT.

A rudimentary block diagram of a CRT of the type used in laboratory oscilloscopes is shown in Fig. 5.3. It consists of an evacuated or sealed glass tube which contains various electrodes on one end and a flat surface covered by a fluorescent material known as phosphors on the other end. Phosphors are substances which emit light when struck by an electron beam. The electrodes are used to create and position the electron beam. One set of electrodes is known as the heater; a low ac voltage (typically 6.3 V) is applied to this

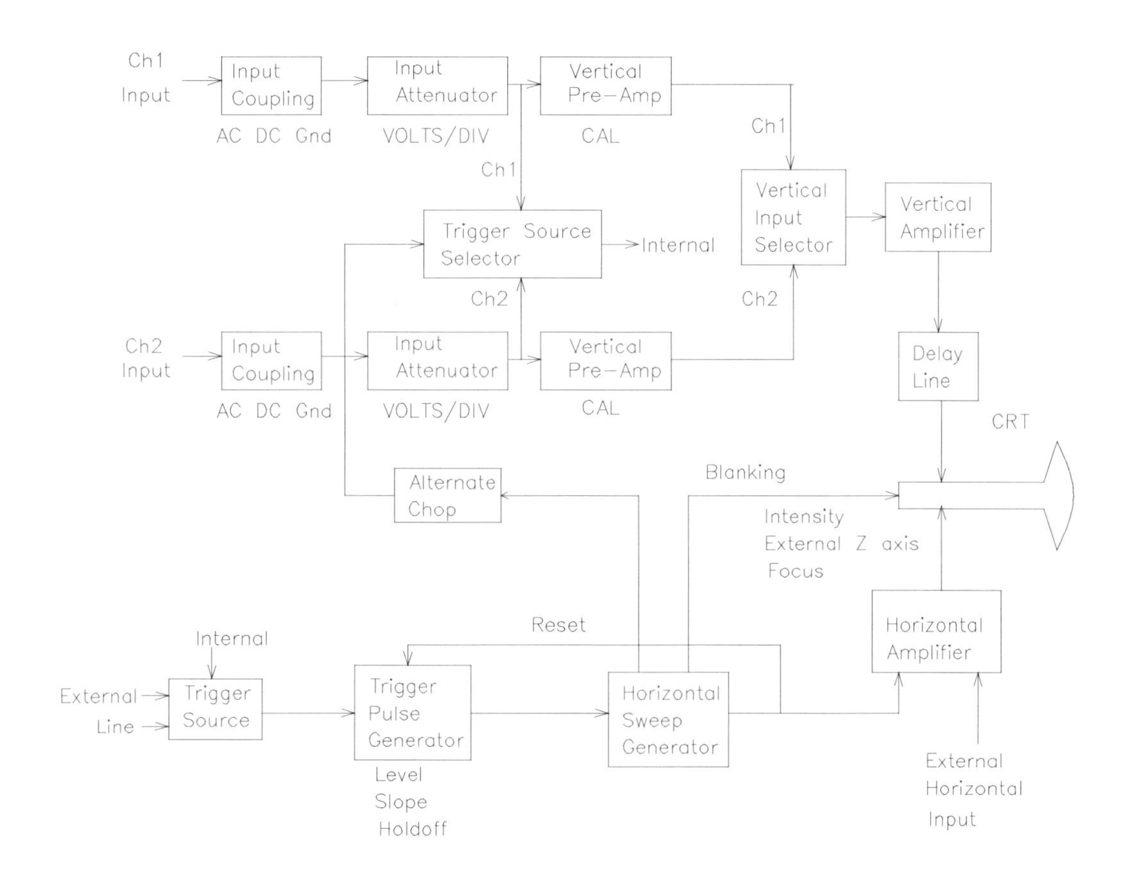

Figure 5.2: Nonstorage analog oscilloscope.

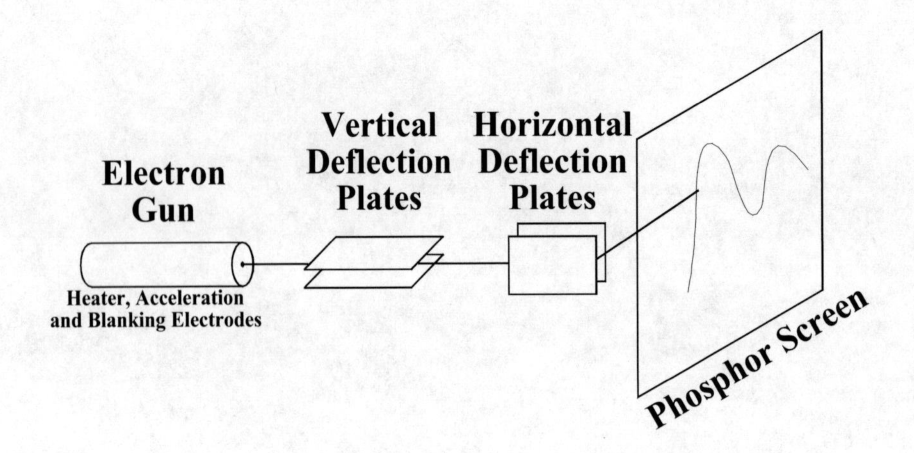

Figure 5.3: Cathode ray tube.

set of electrodes which heats up another electrode known as the cathode until it produces an electron cloud by thermionic emission. Electrodes with positive voltages known as accelerating anodes are used hurl the electrons in the direction of the screen and electrodes known as focus anodes are used to form the electron cloud into a thin electron beam. The resulting current is returned to ground via a conducting surface known as aquadag. Voltages applied to the vertical and horizontal deflection plates are then used control the point that the electron beam strikes the screen. When the electron beam hits the fluorescent screen the phosphors emit light. A table of the type of phosphors used and the color of light that they produce is given in Table 1; the parameter known as the persistence is how long the phosphors emit light after being struck by the electron beam.

The method of positioning the electron beam for CRTs used in analog oscilloscopes is known as vector scanning. This simply means that the position of the beam is determined by voltages applied to the vertical and horizontal deflection plates. This differs radically from the method used in TV, computer monitor, and digital oscilloscope CRTs which employ raster scanning and the beam is positioning used magnetic deflection coils.

The CRTs used in digital oscilloscopes use raster scanning. Lines are scanned from left to right beginning at the top of the CRT and when the beam arrives at a particular point on the screen the pixel information stored in the video RAM determines whether it should be fully or partially illuminated or not illuminated. The scanning may be non interlaced (the lines are drawn in one sweep from top to bottom of the CRT screen) or interlaced (half the lines are drawn in one sweep and then another sweep begins and the other half are drawn between them). The number of lines used determines, along with the screen size, dot spacing and refresh rates and whether the scanning is interlaced or non interlaced, the quality of the picture on the display.

A typical VGA computer monitor using a dot spacing of 0.31 inches and displays 640 $\times$ 480 lines and a refresh rate of 60 Hz. This means the physical spacing between pixels (picture element) is 0.31 inches, there are 640 pixels horizontally, 480 vertically, and the complete screen is drawn 60 times per second. Interlaced scanning is used which mean 240 lines are drawn on the first vertical sweep and the remaining 240 on the next sweep.

Phosphor	Color	Persistence	Application
P1	Yellow-Green	Medium	Scopes,Radar
P2	Blue-Green	Medium	Scopes
P4	White	Medium to Medium Short	Black and White TV
P7	Blue-White	Medium Short	Radar, Medical
P11	Blue-Violet	Medium Short	Photostatic Recording
P15	Blue-Green	Very Short	Flying Spot Scanners for TV
P16	Blue-Violet	Very Short	Flying Spot Scanners for TV
P18	White	Medium	Low Frame Rate TV
P19	Orange	Long	Radar
P22	3-Color Dot	Medium	Color TV
P26	Orange	Very Long	Radar
P28	Yellow-Green	Long	Radar,.Medical
P31	Green	Medium Short	Scopes
P33	Orange	Very Long	Radar
P39	Green	Medium to Medium Long	Computer Graphics

Table 5.1. Phosphor Types

5.3.2 Vertical Section

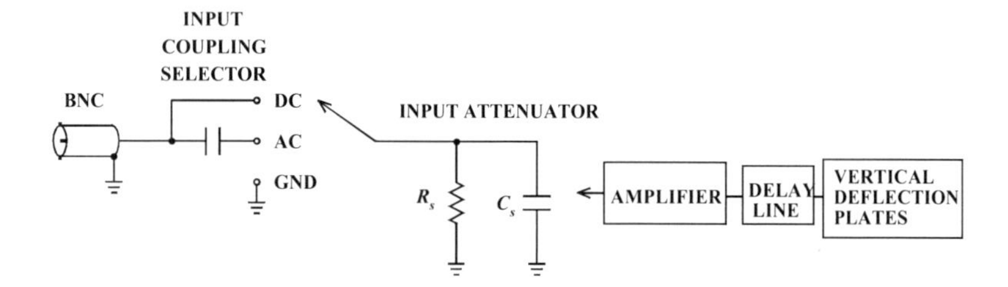

Figure 5.4: Vertical section.

The vertical section of an analog oscilloscope is the path taken by the input signal to the vertical deflection plates of the CRT as shown in Fig. 5.4. An input coupling selector precedes the input attenuator. The input coupling selector has three settings: ground, which grounds the input to the oscilloscope and is used to set a reference for zero volts; ac, which places a coupling capacitor in the signal path which removes the dc component of the input signal (useful for viewing the ac component of a signals which have a small ac component superimposed on a large dc component); and dc which couples the entire input signal to the vertical input section. The vertical pre-amplifier has a variable gain and is used in conjunction with the input attenuator to set the vertical sensitivity of the oscilloscope,i.e. the $VOLTS/DIV$ for that channel. A switch is used to select which input signal is to be displayed and another switch is used to determine the input to the trigger circuit in the horizontal section. A delay line is used to insure that the signal produced by the horizontal sweep generator and the input signal arrive at the CRT at the same time.

5.3.3 Horizontal Section or Time Base

If the desired display is one or more voltages as a function of time, then the waveform connected to the horizontal deflection plates must be a linear function of time. This linear waveform is produced by the

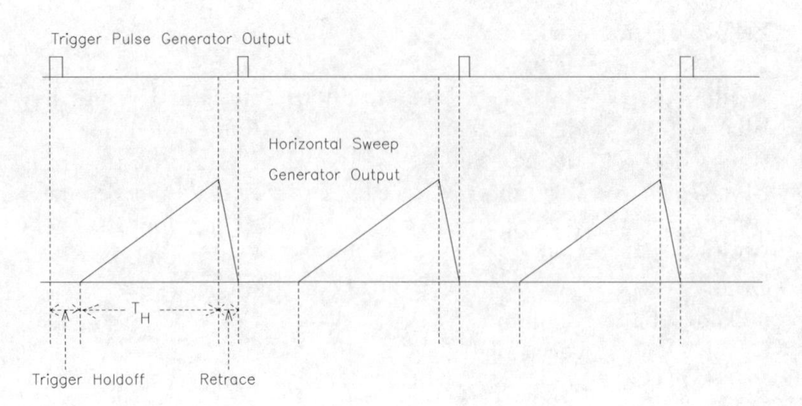

Figure 5.5: Horizontal sweep generator output.

horizontal sweep generator. The ramp output produced by the horizontal sweep generator sets (shown in Fig. 5.5) the $TIME/DIV$ setting for the horizontal sensitivity ($TIME/DIV = T_H/10$ where there are 10 horizontal major divisions on the screen of the CRT). When the ramp is rising the electron beam is being swept from the left hand side to the right hand side of the screen of the CRT. When the ramp is falling the electron beam is being quickly being retraced from the right hand side to the left hand side; during the retrace a blanking pulse is sent to the CRT which turns the electron beam off during the retrace. Electronic circuits that produce the ramp waveforms cannot produce a perfect ramp and the deviations from the linear slope introduce a distortion in the time axis plot of about 3 % (digital oscilloscopes can improve on this by more than an order of magnitude.

If the period of the input signal is equal to T_H, then one cycle or period of the input signal will be displayed on the screen (assuming that the trigger level is properly set). If T_H is greater than T, then more than one cycle will be displayed and if T_H is less than T then only part of a cycle of the input will be displayed.

In order for a stable display to occur the waveforms connected to the vertical and horizontal deflection plates must be synchronized. When this is the case the same display will be drawn on exactly the same place on the screen of the CRT for each horizontal sweep of the electron beam and it will appear as a picture of a portion of the input signal. If the two were not synchronized, the sweep would begin on a different portion of the period of the waveform connected to the vertical input for each output of the horizontal sweep generator. The purpose of the trigger section of the oscilloscope is to synchronize these two waveforms (*The time is out of joint: O cursed spite, That ever I was born to set it right!* Hamlet, Act. I, Scene V). The input to the trigger circuit can be one of the input signals (internal triggering), an external signal, or the ac line voltage which is used to provide ac power to the oscilloscope. There are two inputs to the trigger pulse generator: the trigger level and the slope which are set by the user. When the input to the trigger pulse generator crosses a certain level with a certain slope (either positive or negative, aka a rising or falling signal) it produces a trigger pulse which is supplied to the input of the horizontal sweep generator. If the horizontal sweep generator has completed a sweep, it will then produced a ramp waveform at a later time known as the trigger holdoff (one of the user controls on the oscilloscope). If a trigger pulse is received while the horizontal sweep generator is producing an output, it will ignore it and respond to the first trigger pulse received after the retrace of the current sweep has been completed.

A typical input to the trigger pulse generator is shown in Fig. 5.6; each time the input crosses the positive

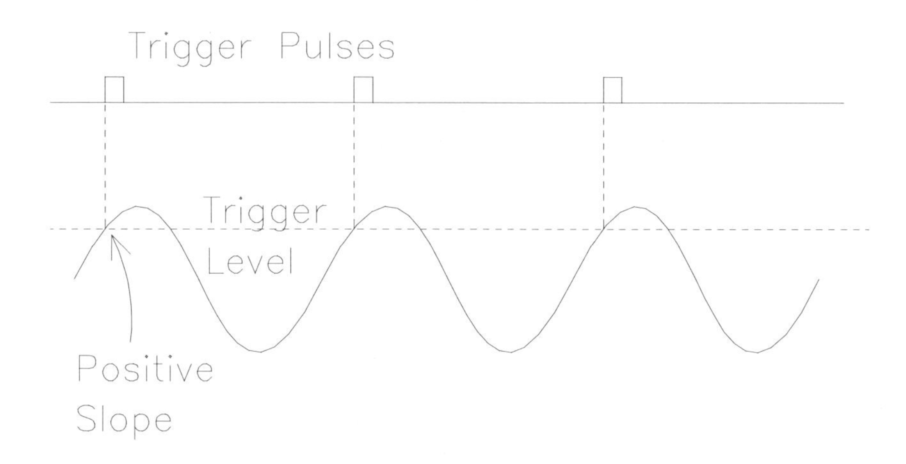

Figure 5.6: Typical input to trigger pulse generator.

level indicated with a positive slope a trigger pulse is produced. The resulting display is shown in Fig. 5.7 where it is assumed that $T = T_H$ which results in one cycle of the input sine wave being displayed. Also, it is assumed that the trigger holdoff is zero. Note that trigger level and slope set the point on the input waveform at which the display begins.

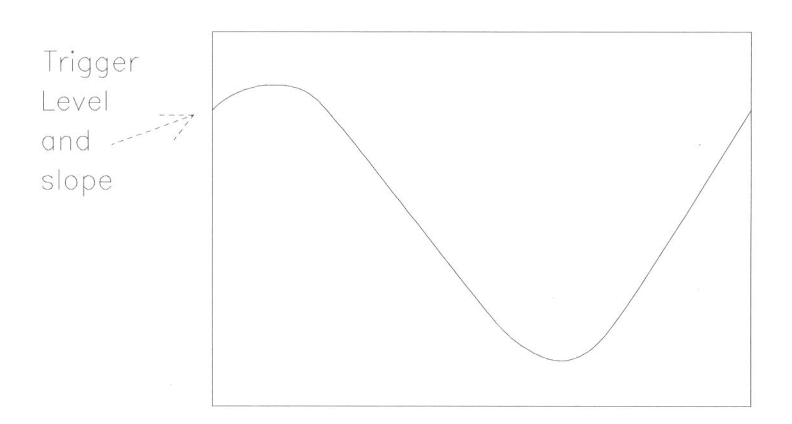

Figure 5.7: Display for trigger level and slope illustrated above.

If the trigger level is set above or below the maximum or minimum level of the signal connected to the input of the trigger pulse generator, then no trigger pulses are produced. If the basic oscilloscope is operating in what is known as its normal trigger mode ($NORMAL$), then there will be no output from the horizontal sweep generator and the display will be blank. There is another mode of operation known as the auto mode ($AUTO$) for which the horizontal sweep generator acts as a free running oscillator when no trigger pulses are being received; if this is the case, then the waveform displayed on the screen of the CRT will appear to be a maze of waveforms with the shape of the input signal with random starting times. The auto trigger mode display for a trigger level which has not been properly set is shown in Fig. 5.8; note that the display appears to be "running across the screen". If the oscilloscope is a "smart" instrument it has a microprocessor to assist the user in making measurements which also provides an auto level triggering mode (*AUTO LVL*). When the auto level mode is selected, the microprocessor will reset the trigger level to a value which will permit a stable display if the user attempts to set the trigger level to value that is too high or low to produce trigger pulses.

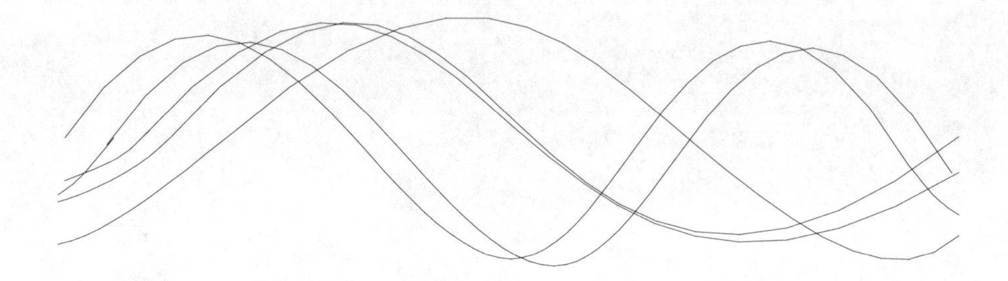

Figure 5.8: Auto mode unsynchronized display.

Which of the input waveforms is displayed on the screen of the CRT is controlled by the vertical input selector. The user can select which of the waveforms is to be displayed with a mechanical switch, e.g. $Ch1$, $Ch2$, $Ch3$, $Ch4$. If more than one channel is to be displayed simultaneously, then the input to the vertical amplifier is electronically multiplexed. There are two forms of electronic multiplexing known as alternate and chop. In the alternate mode, each input is connected to the vertical amplifier on alternate sweeps of the timebase (horizontal sweep generator). In the chop mode the input is switched back and forth between the channels to be displayed at a high rate compared with the frequency of the waveforms being produced by the horizontal sweep generator. The chop mode is normally preferred when phase measurements are to be made between the input signals. At certain sweep speeds the chopping is visible on the display and the alternate mode may be preferred if this is the case.

The input to trigger pulse generator can be AC or DC coupled. Modern oscilloscopes also feature low frequency reject coupling (a high pass filter), high frequency reject (a low pass filter), and noise reject (low amplitude random signals). If the input to the trigger coupling selector is one of the channels to be displayed, then this is known as internal triggering and is the most commonly used mode. If the input is taken from an external signal, then this is known as external triggering, and if the input is taken from the AC power line, this is known as line triggering. Line triggering is most useful when the signal to be displayed is a low amplitude replica of the AC line voltage or a harmonic thereof.

Smart analog oscilloscopes use analog to digital converters and microprocessors to assist the user is making measurements. Proper trigger levels and slopes as well as the $VOLTS/DIV$ and $TIME/DIV$ can be automatically set. Measurement cursors can be used to automatically measure voltages, frequency, and time intervals. However, these oscilloscopes are still analog in that the waveform that is displayed on the screen is never digitized. Most modern analog oscilloscopes are smart.

Analog nonstorage oscilloscopes are useful for viewing repetitive or periodic signals with frequencies above 50 Hz. For frequencies lower than this the flicker becomes noticeable and the display cannot be monitored and must be photographed with an oscilloscope camera. This type of oscilloscope cannot be easily used to observe nonperiodic or random events. To observe these types of signals a storage oscilloscope is usually used.

Analog storage oscilloscopes require a special type of CRT. These types of tubes are expensive and can retain a trace on the screen for only a few hours. Today most storage oscilloscopes are digital oscilloscopes.

All digital oscilloscopes capture the input signal and store it in a digital memory where is can not only be recalled but various measurements can be made on the data by a microprocessor. Digital oscilloscopes also have other advantages over their analog brethren. Digital oscilloscopes have only become possible in the last two decades of the Twentieth Century with the emergence of low cost and reliable solid state microprocessors, memories and analog to digital converters.

5.4 Digital Oscilloscope

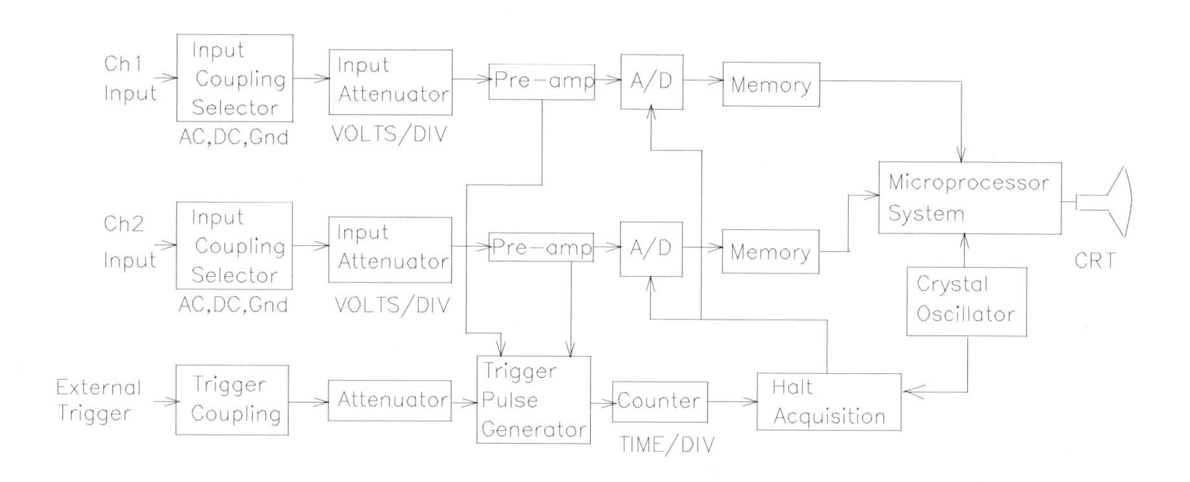

Figure 5.9: Block diagram of a digital oscilloscope.

A block diagram of a digital oscilloscope is shown in Fig. 5.9. The input couplers and attenuators perform the same task as those in the analog oscilloscopes. The timing is controlled by a highly accurate crystal oscillator. The A/D converters convert data from analog to digital form by sampling and quantizing and then the digitized data is stored memory. Separate A/D converters are used for each channel so that there is no need to multiplex the input to the CRT as is done with analog oscilloscopes. Indeed, such terms as chop and alternate are meaningless for digital oscilloscopes.

The time base of a digital oscilloscope consists of a trigger section, a digital counter, and associated circuitry. The purpose of the counter is to introduce a time delay. If the delay were set to zero by omitting the counter, the occurrence of a trigger pulse would cause the acquisition to be halted and the digitized values displayed. Thus all of the displayed values would precede the trigger. By introducing a delay data that occurs after the trigger can also be displayed. This is one of the principle advantages of a digital oscilloscope over an analog one. An analog oscilloscope can only display information that follows the trigger; a digital oscilloscope can display either or both.

Once a trigger event has been detected, the trigger pulse causes the counter to begin counting and counting ceases once the count corresponds to the desired time delay (the count times the period of the clock). If the delay is set to zero, all of the displayed data precedes the trigger and the trigger instant is on the right hand edge of the screen. If the delay is set equal to the time window to be displayed, the trigger instant is on the left hand edge of the display and all of the displayed data occurs after the trigger as is the case with an analog oscilloscope. The delay is normally set equal to half the time window to be displayed with puts the trigger point at the middle of the screen which means that the display in the left half of the screen precedes the trigger and the portion in the right half follows the trigger. Thus the ramp wave of the analog oscilloscope time base has been replaced with a digital counter counting up. This time base is more accurate than that of an analog oscilloscope since the time points are set by multiples of the system clock which is produced by a highly accurate crystal oscillator.

Once the data has been acquired, the microprocessor then writes the data to the CRT. The CRT of a digital oscilloscope operates in a different manner than that of an analog oscilloscope. Namely, it uses raster scanning like a TV set or computer terminal. Raster scanning means that the electron beam moves from left to right beginning at the top of the screen and moves sequentially down the screen in the manner of someone painting a house. Once the beam reaches the bottom of the screen the beam is positioned to the top of the screen and the process begins anew. If a point on the screen is supposed to be blank, a blanking pulse is used to turn the electron beam off (the intensity is set low so that the phosphors do not emit light) and, conversely, if a point is to be written the electron beam is turned on by disabling the blanking pulse. This is one of the real advantages of digital over analog oscilloscopes because this type of CRT is much cheaper than that required for the analog version.

In an analog oscilloscope each component in the vertical and horizontal section as well as the CRT has to operate at a speed set by the specified bandwidth. This means that each component has to have a bandwidth larger than the oscilloscope bandwidth. Oftentimes the most expensive component in an analog oscilloscope is the CRT because of the requisite high bandwidth requirement. With a digital oscilloscope, the CRT has to operate at only the speed at which the microprocessor is writing data to it which means that need not have a high bandwidth. Another difference between analog and digital CRTs is the method used to steer the electron beam. With analog oscilloscopes the beam is positioned by electric fields while CRTs intended for use in only digital oscilloscopes use magnetic deflection.

Digital oscilloscopes differ in the manner used to sample the input signals. Real time sampling is used to examine nonperiodic signals. This sampling scheme requires that the sampling be done on one pass of the input signal and requires that the sampling rate be high. This, of course, requires fast and expensive A/D converters. With this type of oscilloscope the displayed waveform can exhibited before (pre trigger) the trigger event since sampling occurs constantly; this would be impossible with an analog oscilloscope.

Digital oscilloscopes used to examine periodic waveforms can use a lower sampling rate and sample the waveform once per cycle (repetitive sampling). The sample point is moved either sequentially (sequential sampling) or at a random point on each cycle (random sampling). The sampled waveform can then be reconstructed by placing the samples back together in their proper order. These types of digital oscilloscopes can use much cheaper A/D converters because the sampling is done at a much lower rate. (This doesn't violate the revered Nyquist sampling theorem because the sampling is not done in real time.) There is no simple relation between the bandwidth and the rise time of digital oscilloscopes.

All digital oscilloscopes are storage oscilloscopes because the digitized signals are stored in memory. This means that the data can be recalled at any time, the microprocessor can make measurements on the data, and the data can be sent to computers, printers, plotters, etc. This is the major advantage of digital oscilloscopes compare with analog instruments.

Digital oscilloscopes have some disadvantages compared with their analog counterparts. Although the time base of a digital oscilloscope is more accurate, the vertical display is not because the input signal must be quantized by the A/D converter. For instance, if an 8 bit A/D converter is used, then only $2^8 = 256$ voltage levels can be displayed where the analog oscilloscope can display an infinite number. If the input signal is undersampled, then a form of distortion known as aliasing occurs which cannot occur in analog instruments. With analog oscilloscopes the user has more of a feeling of "what you see is, more or less, what you've got" whereas the digital display may occasionally resemble chaos. For these reasons, the ultimate oscilloscope is the hybrid oscilloscope which can function as either a digital or analog oscilloscope.

5.5 Hybrid Oscilloscope

A block diagram of a hybrid oscilloscope is shown in Fig. 5.10 (the analog section is omitted since it would be identical to that of Fig. 5.2). This type of oscilloscope must use the type of CRT used in analog scopes, i.e. high bandwidth, electric field deflection, CRTs. D/A converters are used to converted the digital signals back to analog form. This type of oscilloscope is most useful in that it exploits the advantages of both types

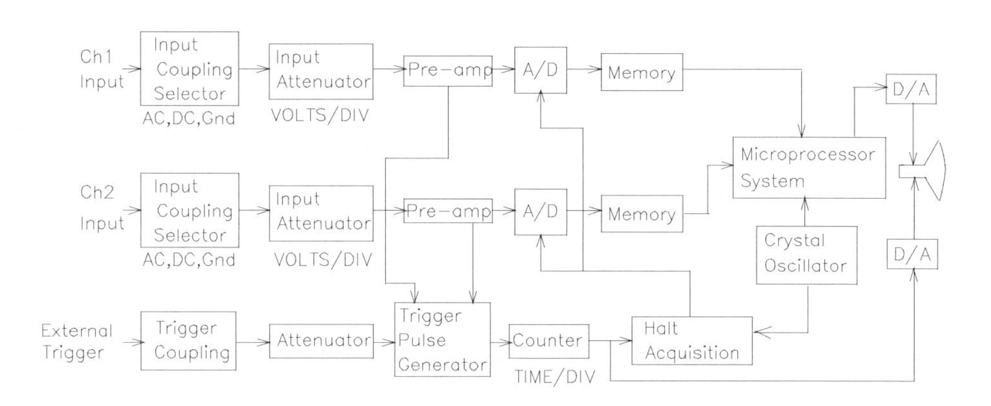

Figure 5.10: Block diagram of a hybrid oscilloscope.

of architectures.

5.5.1 Non CRT Displays

Many newer modern digital oscilloscopes use non CRT displays. These are either LCD (Liquid Crystal Displays) or plasma displays. They are thinner, lighter, and consume less electricity and produce less heat than CRT displays and do not require the high dc voltages that CRTs must have to produce an electron beam. They are essentially the same display devices used in personal computers displays and large screen tvs. Electronic circuitry similar to the video card of a personal computer is used to determine whether each pixel on the screen should given a certain intensity and color.

5.6 Voltage Measurement

5.6.1 Analog Oscilloscope

Since the vertical deflection of the electron beam in the cathode ray tube (CRT) of an oscilloscope is linearly proportional to the voltage applied to the vertical deflection plates, which in turn is linearly proportional to the voltage applied to the vertical input of the oscilloscope, the measurement of voltage simply involves measuring this vertical deflection. Unknown voltages can be measured if it is known how much deflection a known voltage would produce.

A plastic grid is permanently attached to the screen of a CRT with 8 major vertical divisions and 10 major horizontal divisions. Each of these major divisions is subdivided into 5 minor divisions to assist in the measurement of vertical and horizontal deflections. The oscilloscope has a vertical sensitivity selector which (when in the calibrated position) directly indicates the voltage that each major vertical division represents.

The basic steps involved in voltage measurement are:

- Setting the scope up for best trace definition.
- Determining the zero volts level by putting the input coupling selector on GND (ground) and adjusting the vertical position control for the electron beam to establish the zero volts reference on the screen.
- Setting the input coupling selector to either DC or AC, i.e. taking the input coupling out of the GND position.

- Calibrating the oscilloscope.
- Connecting the oscilloscope to the circuit in which the measurements are to be made and adjusting the $VOLTS/DIV$ until a suitably large deflection is obtained.
- Reading the deflection in major divisions on the screen of the CRT and converting that into a voltage measurement.
- If the oscilloscope has measurement cursors, a voltage difference may be measured by aligning the cursors with the points on the waveform that correspond to the voltage difference.

In measuring dc or low frequency time-varying signals it is important that the input coupling selector be set to DC. In the AC coupled position, the capacitor in series with the input blocks dc and alters the waveform of low frequency signals. Normally, AC coupling is used to view the ac or time-varying component of periodic waveforms whose frequency is 2 kHz or greater.

Proper oscilloscope grounding cannot be overemphasized. Most oscilloscopes are grounded, i.e. one of the input leads (usually the outer cylindrical shield of a BNC connector) is connected to the ground prong of the ac power line cord. Some oscilloscopes have differential inputs and, therefore, float above ground, i.e. neither input lead is connected to the ground prong of the ac power line cord. It is important to know how the instrument being used is configured with respect to ground and to employ caution in all ground connections.

5.6.2 Digital Oscilloscope

One may use the same techniques used to make the measurements with an analog oscilloscope.. A grid with 10 major horizontal divisions and 8 major vertical divisions is also provided with a digital oscilloscope and the $VOLTS/DIV$ and $TIME/DIV$ controls have the same significance. However, with a digital oscilloscope the grid is drawn on the screen with along with the displayed waveforms.

Alternatively, digital oscilloscopes have cursors which can be used to measure the voltage between any two horizontal lines, peak-to-peak voltages, rms voltages. Waveform math functions are available that permit extensive functions of the stored data to be calculated and displayed.

5.7 Current Measurement

Current measurements are usually made with an oscilloscope by measuring a voltage that is proportional to the unknown current. This is done by measuring the voltage across a resistor through which the current is passing. If the value of the resistor is known, then the current is simply the measured voltage divided by the resistance. This resistor is referred to as a current sampling resistor because it is used to obtain a sample of the current flowing through the circuit.

If it is necessary to place a current sampling resistor in a branch of a circuit to measure a current, then this produces circuit loading. Also if the branch is such that neither end of the resistor can be connected to ground potential, then the measurement cannot be made unless the oscilloscope floats with respect to ground or a differential measurement is made.

5.8 Time Measurements

5.8.1 Analog Oscilloscope

A time interval measurement may be made with an oscilloscope by simply measuring the number of major horizontal divisions corresponding to the time interval and multiplying by the $TIME/DIV$ setting of the time base of the oscilloscope. If the time base of the oscilloscope is properly calibrated, then the desired time interval is obtained. Time intervals of paramount importance with periodic signals are the period and the phase shift between periodic waveforms with the same frequency.

Period Measurement

A periodic waveform is one that repeats itself every T seconds where T is known as the period of the waveform. If a plot of such a waveform is being viewed as function of time, then the waveform could be shifted along the time axis by an integral multiple of periods and the same display would be obtained. Expressed mathematically, if $x(t) = x(t + nT)$ where n is any integer and T is the smallest number for which this is true, the $x(t)$ is periodic with period T and frequency $f = 1/T$.

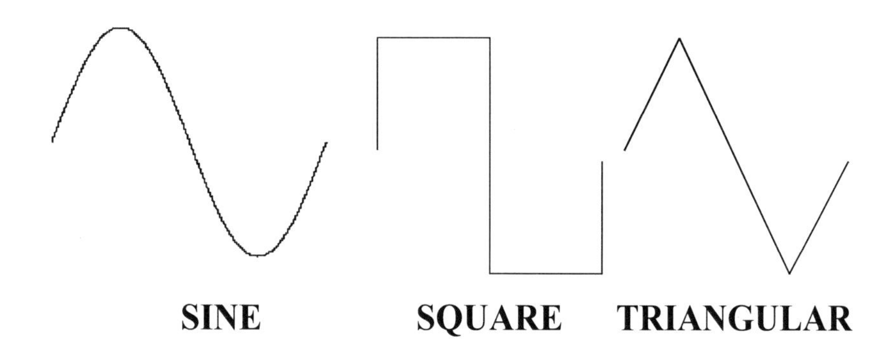

Figure 5.11: Sine, square, and triangular waveforms.

Shown in Fig. 5.11 are three periodic waveforms that are commonly used as test signals for electrical circuits. They are known as sine waves, square waves, and triangular waves for obvious reasons. The three most commonly used test signals are sine, square, and noise; the first two will be extensively examined but noise will be relegated to latter courses. The period of each of these waveforms is T and two cycles are displayed. The horizontal displacement corresponding to one period is measured and multiplied by the $TIME/DIV$ setting. Once the period has been determined, the frequency is then simply obtained from

$$f = 1/T \tag{5.1}$$

where the units for f are Hertz if the units used for T are seconds.

Some oscilloscopes have cursors to assist in the measurement of time intervals. These are vertical lines that can be placed at two spots on the time axis of the display. The time interval is directly displayed on the screen of the oscilloscope. If the cursors are placed so that they cover one period, then the period or the frequency can be displayed on the screen of the scope.

Phase Shift Measurement

Shown in Fig. 5.12 are two periodic waveforms with the same frequency. Mathematically they could be described as

$$v_1(t) = A \sin \omega t \tag{5.2}$$

and

$$v_2(t) = A \sin(\omega t + \phi) \tag{5.3}$$

where $\omega = 2\pi f$ and ϕ is the phase angle by which $v_2(t)$ leads $v_1(t)$. The phase angle or phase shift ϕ may be measured by measuring the time interval t_o and then using the expression

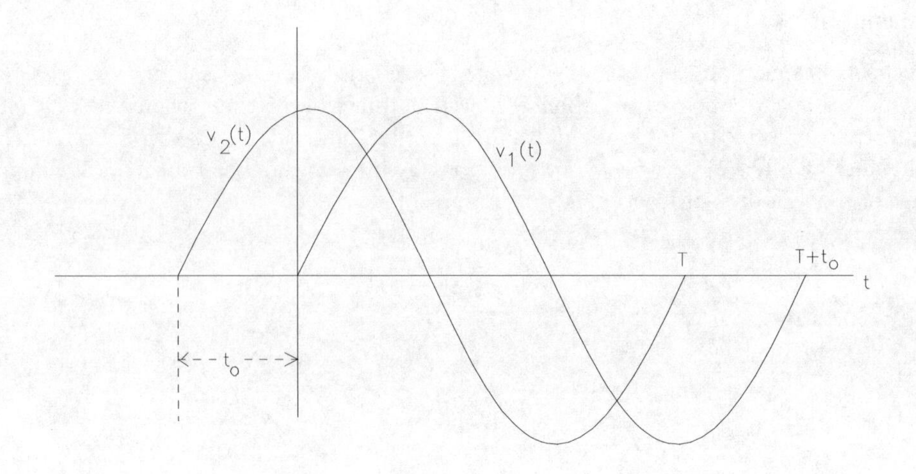

Figure 5.12: Phase measurement.

$$\phi = \frac{t_o}{T} \times 2\pi \text{ [radians]} \tag{5.4}$$

or

$$\phi = \frac{t_o}{T} \times 360^o \text{ [degrees]} \tag{5.5}$$

for the phase shift with the desired units.

The basic steps involved in phase shift measurements are:

- Ground both the inputs for $Ch1$ and $Ch2$ on the oscilloscope (obviously a dual trace oscilloscope is required).
- Use the vertical position controls for $Ch1$ & $Ch2$ to position the references for 0 volts for these two channels to the same horizontal line on the screen grid.
- Unground both channels.
- Remove any dc offset from either waveform.
- Measure the time interval t_o and use the above formula to determine the phase shift ϕ.

Lissajous Phase Measurement

The phase difference between two waveforms with exactly the same frequency may be measured with a Lissajous pattern. A Lissajous pattern is a stable (nonrotating) geometric figure that appears on the screen of an oscilloscope when the waveforms connected to the vertical and horizontal deflection plates have frequencies that can be expressed as the ratio of integers. To make this type of measurement the time sweep mode of the oscilloscope is disabled and the oscilloscope is used in its $X - Y$ mode in which the inputs to the horizontal and vertical deflection plates are both external voltages. If the frequency of both the X and Y inputs to the oscilloscope are the same the phase shift between the two waveforms can be measured.

If $v_1(t)$ given in Eq. 5.2 is connected to the vertical deflection plate and $v_2(t)$ given in Eq. 5.3 is connected to the horizontal deflection plate, the Lissajous pattern shown in Fig. 5.13 results. This Lissajous pattern is an ellipse. If the pattern is centered on the screen, then top of the pattern can be projected to the vertical

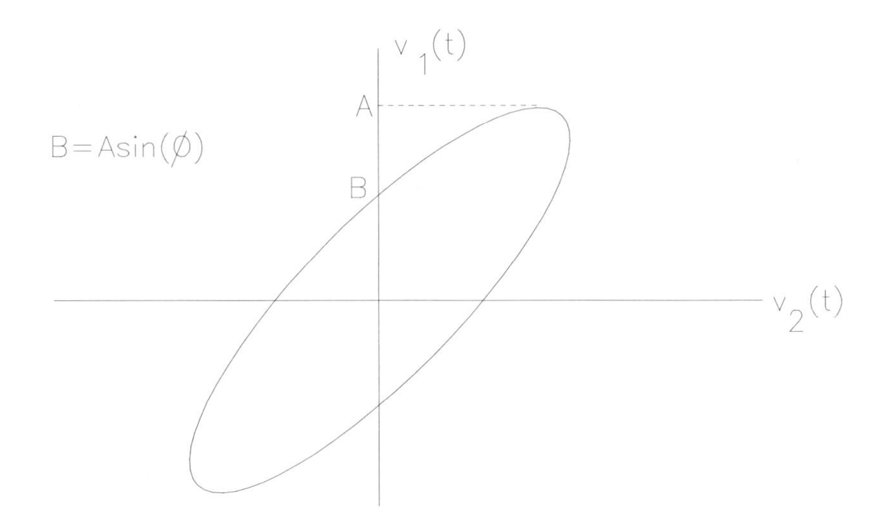

Figure 5.13: Lissajous measurement of phase shift.

axis which intersects it at a height proportional to A. The ellipse intersects the vertical axis at a height $B = A\ \sin(\phi)$. Therefore the ratio of B to A is given by $\sin(\phi)$ from which ϕ can be obtained. Unfortunately the Lissajous pattern cannot be used to determine which of the two waveforms is leading or lagging.

Lissajous Frequency Measurement

If the oscilloscope is set to its $X - Y$ mode and the waveform connected to the horizontal axis is given by

$$v_x(t) = A\ \cos[M\omega_o t] \tag{5.6}$$

and the waveform connected to the vertical axis is given by

$$v_y(t) = A\ \cos[N\omega_o t + \pi/(2M)] \tag{5.7}$$

this results in a Lissajous pattern as shown in Fig. 5.14. Each of the Lissajous patterns can be inscribed in a square having sides of length A. The number of times that the Lissajous pattern touches the vertical side is known as the number of vertical loops in the pattern and the number of times that the pattern touches the horizontal side of the square is known as the number of horizontal loops. Therefore, if the ratio of the frequency of the waveform connected to the horizontal axis to the frequency of the waveform connected to the vertical axis is given by

$$\frac{M}{N} \tag{5.8}$$

this results in a Lissajous pattern with M vertical loops and N horizontal loops. This technique can be used to measure the frequency of a signal if a standard signal is available for which the frequency is precisely known. If the ratio of the frequencies of the two signals does not exactly satisfy Eq. 5.8, then the pattern on the screen of the oscilloscope rotates in either the vertical or horizontal plane.

Circular Lissajous Patterns

If the waveform connected to the horizontal axis is given by

$$v_x(t) = A\ \cos[\omega t] \tag{5.9}$$

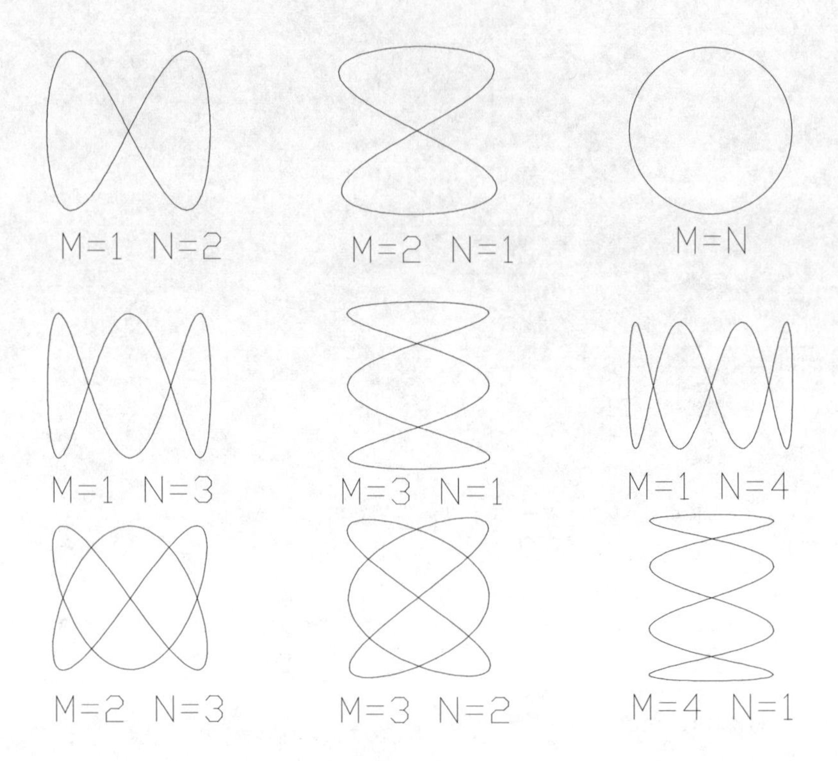

Figure 5.14: Lissajous measurement of frequency.

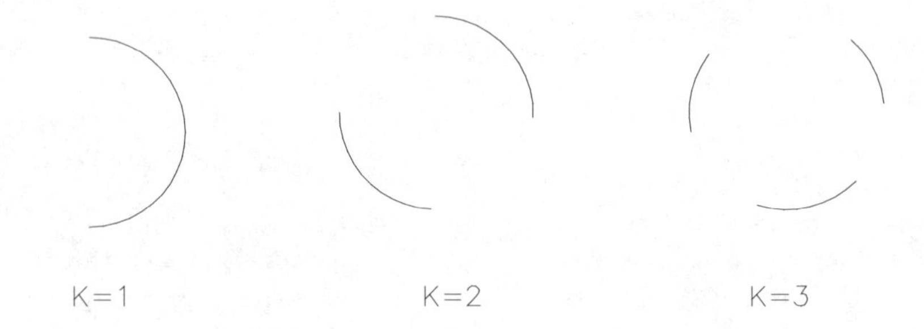

Figure 5.15: Circular Lissajous patterns.

and the waveform connected to the vertical axis is given by

$$v_y(t) = A\ \cos[\omega t + \pi/2] = -A\ \sin[\omega t] \tag{5.10}$$

then a circular Lissajous pattern results. If a waveform is then connected to the external intensity input of the oscilloscope,

$$v_z(t) = B\ \cos[K\omega t] \tag{5.11}$$

a Lissajous pattern is obtained with K bright and dark bands. The waveforms given by Eqs. 5.9 & 5.10 cause the electron beam in the oscilloscope to rotate in a circular pattern f times per second $[f = \omega/(2\pi)]$. By connecting a waveform given by Eq. 5.11 to the external intensity input (aka the Z axis input), the intensity of the beam is enhanced or diminished at exactly the same points on each cycle of $v_x(t)$ and $v_y(t)$. Since the frequency of $v_z(t)$ is N times that of the waveforms that cause the circular pattern, this results in a Lissajous pattern with N bright (intensity enhanced) and N dark (intensity diminished) bands. This technique can also be used to measure the frequency of an unknown waveform. If the amplitude of the waveform connected to the external intensity input is large enough, the beam will be blanked when the intensity is diminished. Circular Lissajous patterns are illustrated in Fig. 5.15.

5.8.2 Digital Oscilloscope

All of the time domain parameters may be measured with a digital oscilloscope in exactly the same fashion as with an analog oscilloscope. But with the digital instrument measurement cursors are available that can automatically measure most of these parameters.

5.9 Impedance Measurement

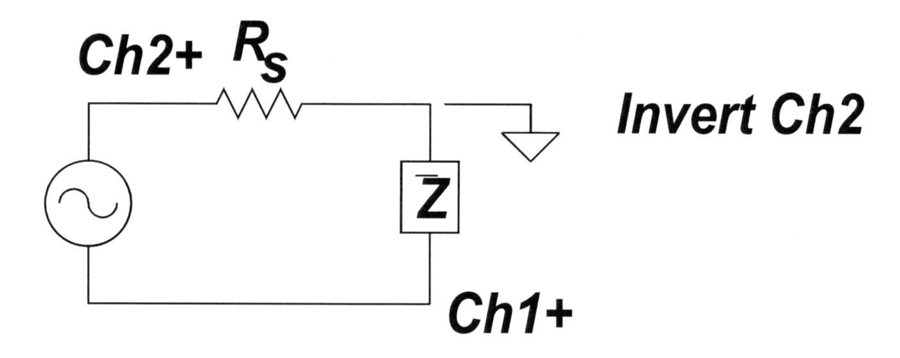

Figure 5.16: Impedance measurement.

Since the oscilloscope can be used to measure both the amplitude and phase relationship between waveforms of the same frequency, impedance measurements can be indirectly made. A sinusoidal excitation is used for the circuit and a dual trace oscilloscope is used to simultaneously display the voltage across the impedance element in question and the voltage across a current sampling resistor in series with it. This technique is illustrated in Fig. 5.16 where *Ch*1+ and *Ch*2+ refer to the placement of the oscilloscope lead high terminals and the ground symbol indicated the placement of the shield or ground terminal of the oscilloscope leads. (It should be borne in mind that the grounds are connected together inside the oscilloscope. Therefore, it is only necessary to connect one of the grounds for the oscilloscope leads to the circuit.) The *Ch*2 *Inv.* means that the *Ch*2 *Invert* button should be pushed. The measured impedance is then

$$\bar{Z}(j\omega) = R_s \frac{V_{p1}}{V_{p2}} \angle\, \phi \tag{5.12}$$

where V_{p1} is the peak voltage of the sinusoid displayed on $Ch1$, V_{p2} is the peak value of the sinusoid displayed on $Ch2$, and ϕ is the angle by which the inverted $Ch2$ waveform leads the $Ch1$ waveform.

This technique requires that, since the oscilloscope is grounded, the function generator float, i.e. that neither terminal of the function generator be connected to the ac power line ground. If the both the function generator and the oscilloscope were grounded and the above configuration used, then either the current sampling resistor, R_s, or the impedance element $\bar{Z}$ would be shorted to ground and this technique would not work.

5.10 General Considerations

Oscilloscope measurement quality can be specified in terms of the general concepts of accuracy, sensitivity, resolution and dynamic response. Loading effects introduced by the scope are also important. In addition, since the oscilloscope is used most frequently in the time base mode of operation, the stability of the trigger circuitry is an important determinant of oscilloscope quality.

Environmental effects can severely limit the performance of the oscilloscope in measuring the low level and/or high frequency signals. Principal among these are the effects of stray electric and magnetic fields, and the capacitance of the input leads. Such effects are minimized by using shielded coaxial input leads of minimum length.

5.10.1 Loading Effects–Voltage Probes

The input impedance of an oscilloscope is finite and usually frequency dependent. When an oscilloscope is connected to a circuit to measure a voltage, loading occurs, i.e. the voltage measured differs from the voltage that was present in the circuit prior to the connection of the measuring instrument. This input impedance manifests itself as a resistor in parallel with a capacitor–these are the equivalent impedance of the input attenuator of the oscilloscope as seen from the input of the oscilloscope. If the voltages measured with an oscilloscope are to be meaningful, then the percentage error due to oscilloscope loading must be small.

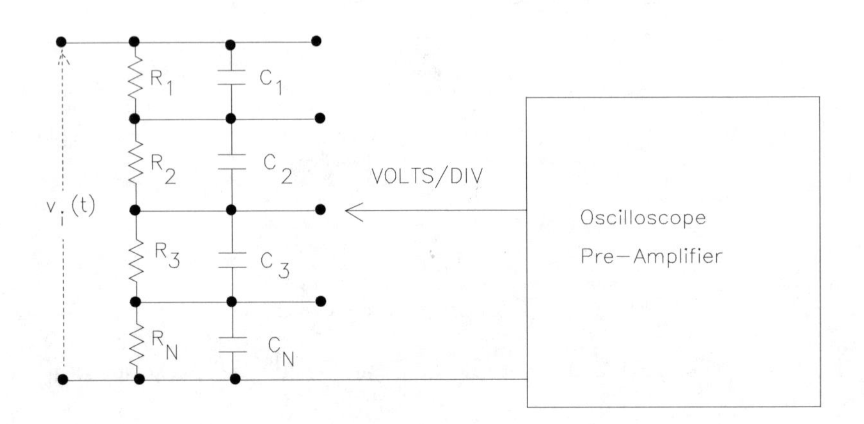

Figure 5.17: Input attenuator.

The input attenuator of an oscilloscope consists of a chain of resistors and capacitors as shown in Fig. 5.17. The purpose of the input attenuator is to match the high input impedance of the voltage probe to the low input impedance of the electronic amplifier that follows the voltage probe and to scale the signal level of the input. These are present in all oscilloscopes and set the input impedance of the oscilloscope. The input impedance $\bar{Z}_{in}$ can be reduced to a single resistor R_s in parallel with a single capacitor C_s and is the impedance of interest of loading calculations. The transfer function $\bar{T}$, however, is just a constant, k, which is independent of frequency and is the attenuation of the input attenuator that sets the $VOLTS/DIV$. If the transfer function were not constant, then the input attenuator would change the shape of waveforms connected to the oscilloscope's input and would make the instrument essentially useless. The relationships between the resistors and capacitor are carefully maintained to insure that the attenuation is the same for all frequencies.

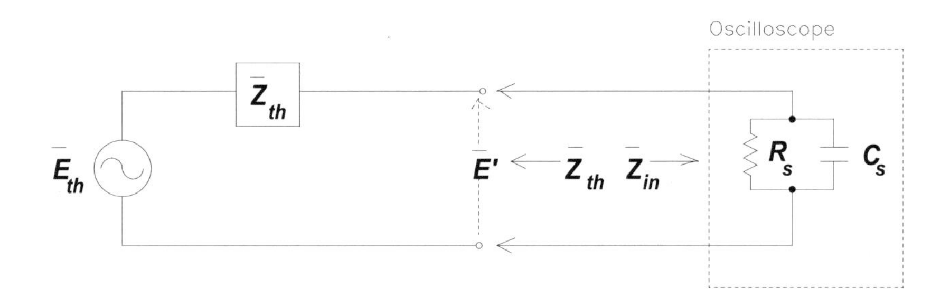

Figure 5.18: Oscilloscope loading.

The combination of the oscilloscope and the Thévenin equivalent circuit with respect to the terminals to which the oscilloscope is connected is shown in Fig. 5.18. Only the input attenuator is shown for the oscilloscope because it determines the degree of oscilloscope loading. The measured voltage will differ both in magnitude and phase from the voltage that exist prior to the connection of the oscilloscope. The percentage error due to oscilloscope loading in the measurement of the magnitude of $\overline{E}_i$ is defined as

$$\% \, error = 100 \times \frac{|\,\overline{E}_i'\,| - |\,\overline{E}_i\,|}{|\,\overline{E}_i\,|} \tag{5.13}$$

where $\overline{E}_i$ is the voltage that exists in the circuit before the oscilloscope is connected and $\overline{E}_i'$ is the voltage that exists after the scope is connected. Examination of the circuit shown in Fig. 5.18 discloses that

$$\bar{E}_i = \bar{E}_{th} \quad \bar{E}_i' = \frac{\overline{Z}_{in}}{\overline{Z}_{in} + \overline{Z}_{th}} \bar{E}_{th} \tag{5.14}$$

as the expressions for the unloaded and loaded voltage. Using these expressions, the expression for the percentage error due to oscilloscope in the measurement of the magnitude of the voltage $\bar{E}_i$ is given by

$$\% \, error = 100 \times \left[\frac{1}{\left| 1 + \dfrac{\overline{Z}_{th}}{\overline{Z}_{in}} \right|} - 1 \right] \tag{5.15}$$

which illustrates that to minimize oscilloscope loading that $|\bar{Z}_{in}| >> |\bar{Z}_{th}|$.

If oscilloscope loading is excessive, it may be decreased by using a 10 X or times 10 oscilloscope probe. This essentially adds another step to the input attenuator by placing and RC parallel combination in series with the oscilloscope. This increases the input impedance of the oscilloscope by a factor of 10. The capacitor in the oscilloscope probe is variable which permits it to be adjusted until the oscilloscope leads, oscilloscope probe, and oscilloscope amplifier combination have a flat frequency response.

A $10X$ oscilloscope probe can be used to increase the input impedance of the oscilloscope. This simply adds another parallel combination of a resistor ($9R_s$)and a capacitor ($C_s/9$) in series with the input attenuator. This increases the impedance seen from the circuit by a factor of 10. The capacitor is usually a variable capacitor so that the effect of the cable connecting the probe to the oscilloscope can be compensated so that the frequency transfer function is unity. The probe is first connected to the oscilloscope's calibrator loop (an output producing a low frequency square wave) and the capacitance of the probe is adjusted until the waveform seen on the screen is a square wave with edges that are sharp.

5.10.2 Oscilloscope Bandwidth

Since the transfer function of the input attenuator is a constant, the frequency selective characteristics of the oscilloscope are set by the behavior of the amplifiers in the vertical deflection subsystem of the oscilloscope. These amplifiers behave as low pass filters with a gain that is constant up to a frequency known as the bandwidth of the oscilloscope. The bandwidth of an oscilloscope is one of the more crucial performance criteria.

The rise time of an oscilloscope is defined as the time it takes the electron beam to go from 10% to 90% of its final vertical position on the screen when the input to the scope is a perfect step function. An inverse relationship exists between the rise time of an oscilloscope and its bandwidth; namely, the rise time t_r in seconds is given by

$$t_r = 0.35/(BW) \tag{5.16}$$

where BW is the bandwidth in Hertz. Eq. 5.16 assumes that the oscilloscope can be modeled as a first order system and is only approximate. If the rise time of the waveform connected to the input of the oscilloscope is smaller than the rise time of the oscilloscope, then the waveform displayed will be distorted. Thus, to insure that the waveforms are displayed without appreciable distortion, the bandwidth of the oscilloscope must be large enough so that the rise time of the oscilloscope is smaller than the rise time of the input waveform. The bandwidths of commercially available oscilloscopes range from 100 kHz to 1 GHz.

5.10.3 Intrinsic Performance Parameters

Oscilloscope sensitivity is determined largely by the gain of the output amplifiers, which drive the deflection plates, and by the electrical noise characteristics of the oscilloscope circuitry. The higher the output amplifier gain, the smaller is the required input to achieve a given beam deflection. Unfortunately, high gain also results in high noise levels and a compromise must be reached between gain and noise levels. Extremely high sensitivity oscilloscopes incorporate extensive design features to minimize electrical noise.

Oscilloscope resolution is influenced by amplifier gains, calibration, size of the screen, and by trace definition. A small change in input signal results in beam deflection as determined by the input channel gains. The beam deflection can be calibrated on the screen if it is large enough to be resolved on the screen graticule. For a given oscilloscope, the resolution can be maximized by setting up the trace for best definition and employing control settings which result in utilization of the whole screen.

Accuracy of the oscilloscope is again influenced by design features. Scale calibration, linearity of control potentiometers, amplifier balance and internal electrical noise all affect accuracy. Most oscilloscopes provide calibration signals and balance controls so that accuracy can be optimized by proper calibration before measurements are taken.

5.11 Equipment

5.11.1 Leads

The type of leads that will be used to establish connections from the function generator to circuits constructed on solderless breadboards and/or oscilloscopes are known as coaxial leads or shielded leads. This means that

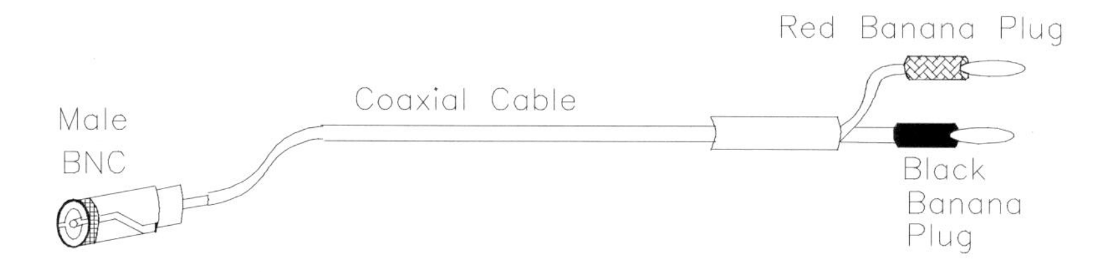

Figure 5.19: BNC to banana plug coaxial cable lead.

the two conductors of the lead consist of a thin metal wire surrounded by a metal cylinder known as the shield. Coaxial leads are preferred over parallel wires because they are totally shielded with respect to interfering electrostatic fields and offer a minimum cross section to stray magnetic fields. A flexible dielectric insulator is used to maintain the wire at the geometric center of the metal cylinder and the shield is covered on the outside with a thin insulator. It is imperative that as the lead is bent that this geometry is maintained so that the characteristic impedance of the cable is constant. [Commonly used impedances for coaxial cables are 50 Ω which is used for most electronic test leads and 75 Ω which is used in cable TV].

The type of lead that will be referred to as a BNC to banana plug lead is shown in Fig. 5.19. It consists of a section of RG 58 coaxial cable (characteristic impedance 50 Ω) terminated on one end with a BNC connector and on the other end with a banana plug connector. The length of the wires connecting the banana plugs is made as short as possible to minimize possible stray electromagnetic field pickup.

There are two type of banana jack and BNC connectors, as is the case for most connectors, known as "male" and "female" connectors. The "male" must be connected to or mated to "female" connectors. The type of input connectors found on oscilloscopes and grounded function generators are known as "female" BNC connectors and the BNC connector on the BNC to banana plug lead is known as a "male" connector. Any attempt to mate two connectors of the same "sex" will prove futile. After using leads with these connectors, the reason for this terminology should be obvious.

5.11.2 Tektronix AFG 3022B Arbitrary/Function Generator

The function generator that will be used in this experiment is the **Tektronix AFG 3022B Arbitrary/Function Generator** shown in Fig. 5.20. Two separate, independent function outputs are available as well as a triggering signal. The user selects the type of waveform desired from the front panel buttons. At the BNC *OUTPUT* connectors it produces standard sine, square, triangular, and ramp waves. It also produces Gaussian noise that is white (flat power spectral density) to 10 MHz. It can also produce user defined arbitrary waveforms (these must be entered through a computer port) and has five stored arbitrary waveforms available: sinc, exponential rise, exponential fall, negative ramp, and cardiac. The user can enter the frequency and amplitude of the waveform using either the keyboard or the rotatory adjustment knob. Another BNC output is available at the front panel which is the *SYNC* which is a TTL compatible signal with the same frequency as the function output. The instrument is grounded which means the shield of the BNC connector is internally connected to the earth ground wire in the ac power cord for the instrument. The frequency for sine and square outputs is from 1 μHz to 25 MHz. The maximum frequency for the other outputs is limited to 250 kHz.

The **Tektronix AFG3022B** function generator can also be amplitude and frequency modulated. Shown in Fig. 5.21 are typical AM and FM modulated waveforms. The modulation frequency and amplitudes are controlled from the front panel setting. Tone bursts, FSK (frequency shift key), and swept frequency modulation are also available.

A block diagram of a typical function generator is shown in Fig. 5.22. This instrument produces output

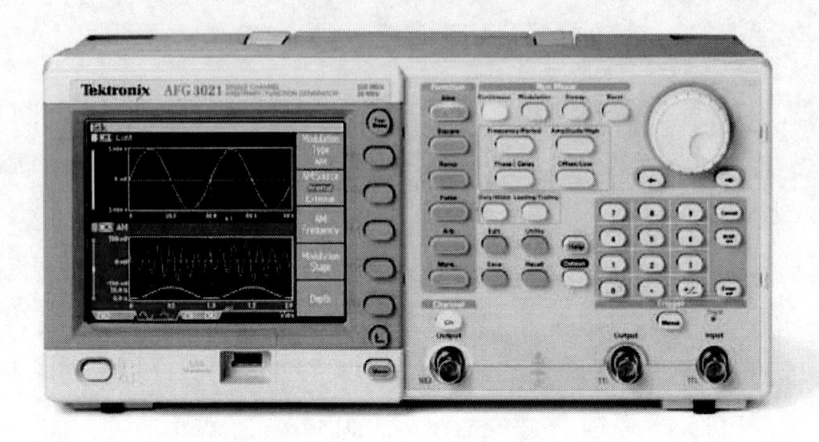

Figure 5.20: Tektronix AFG 3022B Arbitrary/Function Generator.

analog waveforms using DSP principles. The method employed is known as direct digital synthesis (DDS). Once the waveshape has been selected, one cycle of the digitized waveform is stored in the waveform RAM (Wfm RAM). A total of $2^{14} = 16,328$ time points (rounded off to 16,000) and $2^{12} = 4,096$ amplitude points are used. Thus the waveform RAM has 2^{14}addresses each of which contain a 12 bit word.

A subset of the total of 16,000 points stored in the waveform RAM is connected on each cycle to the input of a 12 bit digital-to-analog converter. At the lowest frequency all 16,000 points are used and at the highest frequency as few as 8 points can be used. At higher frequency some of the stored points are simply jumped over in memory.

Once the user selects the frequency from the keyboard, the CPU (not shown) sets the contents of a 48 bit register known as the Phase Increment Register (PIR). On each clock cycle the contents of the PIR are added to the current contents of the Phase Register (PR) counter. The 14 most significant bits of the output of the PR are used as the address bits for the waveform RAM. The amount of time required for the contents of the PR to go through one complete cycle is determined by the size of the constant in the PIR and the clock frequency. The clock frequency is 40 MHz. This means that frequency increments of 40 MHz/$2^{48} \approx 100$ mHz can be produced.

The stored digitized values of the waveform amplitude are sequentially connected to the input of a 12 bit DAC (digital-to-analog) converter which converts the digital code into an analog voltage. This is then followed by an anti-aliasing filter. For sine and square outputs, the filter is a 17 MHz elliptical low-pass. And, for the other outputs, a 10 MHz Bessel low-pass filter is used. If these filters weren't there the output

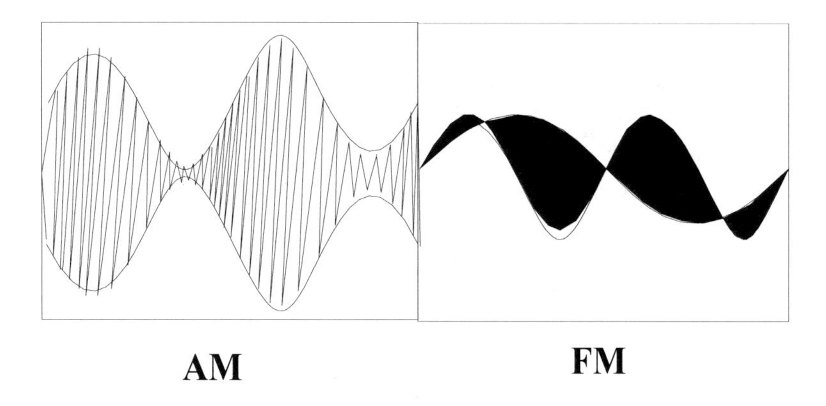

Figure 5.21: Typical AM and FM signals.

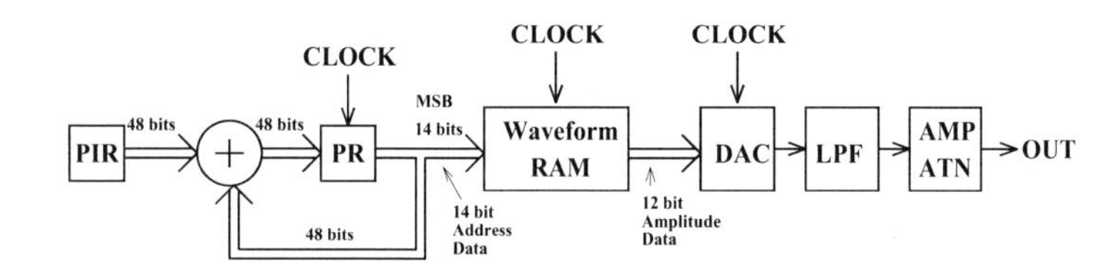

Figure 5.22: Block diagram of Digital Function/Arbitrary Waveform generator.

would be corrupted with high frequency glitches.

The output amplitude is selected from the keyboard and then the CPU sets a series of relays in the output attenuator to produce the desired output.

5.11.3 Tektronix TDS3012B Oscilloscope

The oscilloscope that will be used in this experiment is the **Tektronix TDS 3012B** shown in Fig. 5.23. This is a two channel digital oscilloscope which means that up to two different waveforms may be simultaneously plotted as functions of time. The input channels are denominated $CH1$ and $CH2$. There is also a channel for an external trigger known as $EXT\ TRIG$. The three input connectors are female BNC connectors. It is a grounded instrument since the shield on each BNC connector is connected internally to the ground wire in the ac power cord for the instrument.

Specifications

The general specifications for this instrument are:

- Bandwidth with Random Repetitive Sampling. 100 MHz for $CH1$ & $CH2$.
- Vertical Sensitivity. 1 mV/div to 10 V/div.
- Rise Time. ≤ 2.33 ns for $Ch1$ & $Ch2$.
- Input Impedance. 1 $M\Omega$ resistor in parallel with a 13 pF capacitor.
- Input Coupling. DC, AC, and GND for $CH1$ & $CH2$ and DC for the $EXT\ TRIG$.

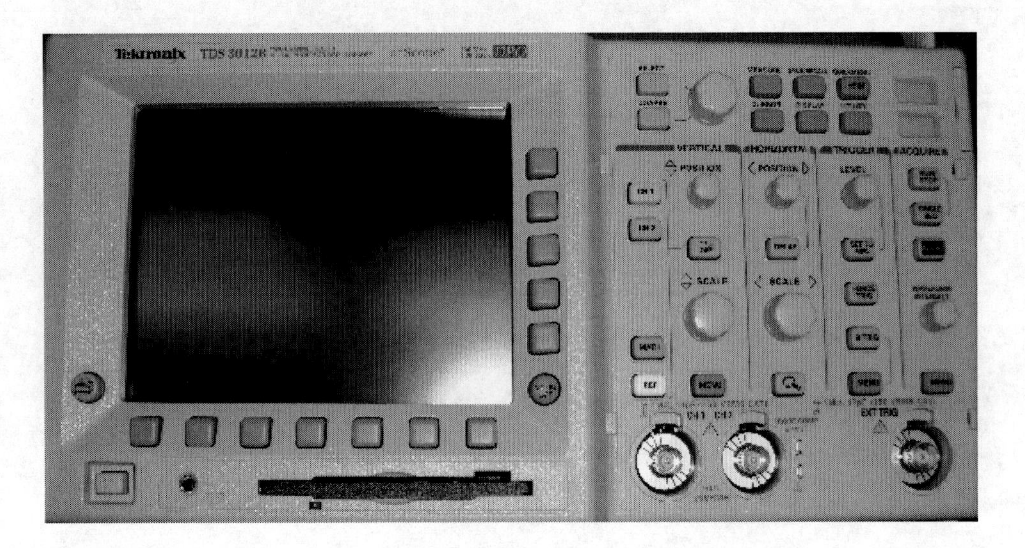

Figure 5.23: **Tektronix TDS 3012B Two Channel 100 MHz** oscilloscope.

- Maximum Input 150 V rms.
- Timebase Range (Main & Delayed). 10 s/div to 4 ns/div.
- Trigger Sources. $CH1$, $CH2$, EXT, and $LINE$.
- Vertical Resolution. 9 bits.
- Maximum Sample Rate. 1.25 GSa/s.
- Color LCD .

A single MPC680DC Power PC microprocessor is used to operate the entire instrument. A color LCD (liquid crystal display) is used instead of a CRT. LCDs are smaller, consume less electricity, operate at lower voltages, and produce less heat than CRTs. The instrument may be battery operated or from an ac outlet. In addition to the BNC connectors that are used to connect waveforms to be measured various computer connectors are also present that permit the instrument to be controlled by a personal computer, the internet, or to print to a printer.

5.12 Procedure

5.12.1 Function Generator Adjustment

Turn on the **Tektronix AFG 3022B** function generator and wait until it finishes its self-test on power-on; this should only take a few seconds. Press the On buttons above both CH1 and CH2 BNC connectors so that the LCD display does not indicate that the output is disabled. The rocker switch between CH1 and CH2 enables the display to switch from CH1 to CH2. Select CH1 and press the Output Menu soft button. Press Load Impedance and select High Z. Repeat for CH2. The function generator is now configured so that it will display the Thévenin source voltage when the amplitude display is selected.

This procedure will have to be repeated each time the power is turned off and back on if one wishes that the open circuit or Thévenin source voltage amplitude be displayed. Alternatively, one could configure the function generator to boot up in the power-down state. This is done by pressing the Utility, System and switching Power On from the Default to Last. But this would assume that the previous user had configured the instrument in the state the current user desires.

Set the function generator to produce a sine wave with a peak-to-peak amplitude of 2 V and a frequency of 1 kHz. for both channels. Do this by using the channel select button and then select frequency and amplitude. It may be directly entered from the keyboard or the rotatory knob on the upper right may be used.

To enter the amplitude from the keyboard first press the Amplitude button. Next enter the value from the keyboard and then select the appropriate unit. The display should now indicate a 2 V peak-to-peak sine wave.

Hard button short cuts are available for the more commonly used parameters such as Frequency and Amplitude. These are located to the right of the Function hard buttons. Parameter values can be changed using either the keyboard or a rotary knob located in the upper right of the function generator.

Set the frequency to 1 kHz (the default or boot up value is 1 MHz). Press the *Frequency* button and change it in the same manner that the amplitude was changed.

Note the right and left arrows under the rotary knob. These determine which of the digits is being changed when the knob is rotated.

5.12.2 Initial Oscilloscope Adjustment

Turn the **Tektronix TDS 3012B** oscilloscope and wait until it boots. (The on/off switch is located in the lower left hand corner.) Use a BNC to BNC connector to connect the CH1 and CH2 outputs of the **Tektronix AFG 3022B** function generator to the CH1 and CH2 inputs on the **TDS 3012B** oscilloscope. Turn CH2 and the Math Functions OFF if they are on; this is done by pressing the undesired channel and the OFF button on the oscilloscope. Press *AUTOSET* (on the right under the acquire menu). A stable display of three and one half cycles of the sine wave which occupies 4 major vertical divisions should be displayed.

5.12.3 Display

Press the *DISPLAY* button. (Located top center.) Note that the labels above the soft keys to the right of the LCD display change each time a different hard key is pressed. Press the Graticule Full and note the choices that are available on the right. Press Grid, Cross Hair, and Frame and note the effect. Press *Autoset*.

Note the top above the graticule on the display. At the lower left it indicates with a high lighted $CH1$ that *Channel* 1 is being displayed. To the right of this is 500 mV which indicates that each of the 8 major vertical divisions represent 500 mV. At the center of the top of the display it indicates that this is the location of the $t = 0$ point (which is set by the trigger controls).To the right of this it indicates that each of the 10 major horizontal divisions is 200 μs. At the far right at the top of the display is 1 $\uparrow$ which indicates that the trigger source is $CH1$ and the trigger slope is selected as positive. At the extreme top right RUN is printed which indicates that the oscilloscope is continually updating the display. Press $STOP$ and note the effect on the display.

___

___

___

Press *RUN* again. Manually change the $Volts/div$ for $Ch1$ and the $Time/div$ rotatory knobs and note the effect on the display.

5.12.4 Vertical Position

Vary the vertical position for $CH1$ and note the effect on the display. The control for the vertical position is the small rotatory knob above the BNC connector input for $CH1$. Note that a ground symbol appears at the far right of the display which informs the user where 0 V is located.

5.12.5 Horizontal Position

Vary the horizontal position for the display and note the effect. The control for the horizontal position is the small rotatory knob above the $Time/DIV$ (Scale) knob.

5.12.6 Trigger

Press *AUTOSET*. Press the MENU button in the trigger section of the front panel keyboard. Note that the soft keys displayed under the display have changed. Press the trigger slope button followed by the icon for a negative slope in the soft key on the right and note the effect on the display. Vary the trigger level rotary knob and note the effect on the display.

Switch the trigger source from $CH1$ to Ext examine the effect on the display. Since there is no signal connected to the external input the scope can't synch the display and it scrolls across the screen. Under the ACQUIRE menu toggle the RUN/STOP button and note the effect on the display. Press the FORCE TRIG button and note the effect on the display.

____

____

____

____

____

____

Press *AUTOSET*.

5.12.7 Measurements

Voltage

Measure the peak-to-peak value of the waveform by measuring the number of major vertical divisions that it occupies and multiple by the $Volts/div$ for $CH1$. Measure the period in same manner. Turn on the voltage measurements menu by pressing the MEASURE button in the top middle on the front panel and scroll through the menus on the right of the display until Pk-Pk is found. Also press the softkeys that measure the mean and rms value. After the measurements have been made press Remove Measurements followed by All Measurements.

Division Method ______________________________

V_{p-p} ______________________________

V_{rms} ______________________________

V_{mean} ______________________________

Time

Measure the number of major divisions represented by one cycle of the signal and compute the period. Press Select Measurement for $CH1$. Use the measure menu to determine the period, frequency, and positive duty cycle. Measure and record these parameters. After the measurements have been made press Remove Measurements followed by All Measurements.

Division Method for Period ______________________

$Freq$ ______________________

$Period$ ______________________

$Duty\ Cy$ ______________________

5.12.8 Functions

Switch the function on the function generator to square, ramp, and pulse and note the effect. When set to ramp, press the Ramp Parameter menu and then Symmetry button on the function generator and use the large rotary knob to set the symmetry to 0% and 100%. Select noise output by pressing More, More Waveform Menu, and then Noise. Manually change the Volts/DIV on CH1 of the oscilloscope to 100 mV and the time per division to 400 ns. With the function set to noise examine the effect of switching the storage mode from *RUN* to *STOP*.When set to *RUN* press DISPLAY followed by Waveform Display and rotate the knob on the scope next to the MEASURE button clockwise (to the right) until the Persist Time is ∞ Seconds. Note the display. Set the function back to sine and press AUTOSET.

____

____

____

Press More on the **Tektronix AFG 3022B** function generator followed by Arb Waveform Menu, and select sin(x)/x, and then observe the display.

____

____

____

5.12.9 DC Level Measurement

Set the **Tektronix AFG 3022B** function generator to produce a 2 V peak-to-peak square wave with a frequency of 1 kHz. Use the button labeled Pulse rather than Square. Press the Pulse, Pulse Parameter, Duty soft buttons on the function generator and then change the duty cycle to 75% with the rotatory knob on the function generator or the keyboard. Press the $CH1$ button followed by MENU on the oscilloscope. Press the soft key that switches the coupling from DC to AC. Note that the displayed waveform moves downward. The amount that it moves downward is the dc level of the waveform. Measure and record this dc level

DC Level = ____

Set the duty cycle back to 50 % and set the function to sine.

5.12.10 Dual Inputs

Connect the TTL Output on the Function Generator to $CH2$ on the scope. Leave the CH1 Output of the Function Generator connected to $CH1$ of the oscilloscope. Press AUTOSET. Press the blue $CH2$ button. Press AUTOSET again. Note that there are now two waveforms on the display. The waveform on $CH2$ is the Sync output of the Function Generator. It is used primarily for synchronization purposes. The Sync output is a TTL signal which means that it varies from 0 V to 5 V. Turn off CH2 on the oscilloscope.

5.12.11 AM

Set the **Tektronix AFG 3022B** to produce a sine wave with frequency of 100 kHz and a peak-to-peak value of 2 V. Switch the Run Mode on the function generator to Modulation, select Modulation Type AM, Modulation Source Internal, AM Frequency to 100 Hz, Modulation Shape to sine, and Modulation Depth to 100 %. Press *AUTOSET* on the oscilloscope. Manually set the $Time/Div$ on the oscilloscope to 2 ms/div with the rotatory knob on the scope on the HORIZONTAL menu (SCALE). Press MENU on the TRIGGER section followed by Source to $CH2$, Coupling HF Reject, and Trigger Level to TTL. Press DISPLAY, and use the rotary knob at the top of the scope to turn the persistence to ∞.

Press the *CURSORS* button on the oscilloscope. Select H bars. Use the rotary knob at the top of the scope to position the first cursor to the top of the envelope. Press select to the left of the rotary knob which will then switch to the second cursor and position the second cursor to the trough of the envelope. Note that the position of the cursors are indicated on the display as well as the difference between them. Examine the effect of toggling the SELECT button on the scope on the cursor readouts.

Printouts

Next the display will be printed; this requires using the computer. The image will be transferred from the oscilloscope to the computer and then a standard Windows print procedure will be used. A card known as a GPIB card plugged into a PCI slot on the pc is used to enable the computer to communicate with the oscilloscope and the software that is used is the Tektronix software Wavestar.

Log onto the computer using the standard Windows login procedure. Enter the experimenter's User No. and Password in the correct Domain. If the user accounts can remember profiles, some of the rigmarole may have to be done only once.

- Open the Lab Apps folder on the Desktop and select Scope Capture
- Double Click the Scope Capture Icon. It should say Scope Ready.
- Click Capture Screen Image.
- Click Save Screen Image.
- Use the saved screen image for the laboratory report.

Press AUTOSET. Turn the cursors off and the modulation off.

5.12.12 FM

With the function generator set to produce a 100 kHz sine wave with a peak-to-peak value of 2 V, press Modulation, set the Type to FM. Set the FM source to Internal, Deviation to 10 kHz and the FM Frequency to 1 kHz. Press the DELAY button on the HORIZONTAL section of the oscilloscope which will switch $t = 0$ near the left of the display. Press DISPLAY and change the Persist time to ∞.

Print the display. Use the procedure previously described.

Turn the modulation off and press AUTOSET.

5.12.13 Storage

Disconnect the lead connected to $CH2$ from both the function generator and scope. Disconnect the lead from the output of the function generator but leave it connected to $CH1$ on the oscilloscope. Manually set the $Volts/DIV$ to 10 mV and the $Time/DIV$ to 400 μs. Press the *Display* button and set the Persist time to ∞. Set the Trigger Level to 0 V. Press Single SEQ under the Acquire Menu.

Gently tap the free end of the BNC connector on the laboratory table top. A transient waveform may have been captured and the display will switch from RUN to STOP.. The trigger may be rearmed by pressing SINGLE SEQ again.

Print the display.

5.12.14 Tone Burst

Set the **Tektronix AFG 3022B** function generator to produce a 2 V peak-to-peak sine wave with a frequency of 100 kHz. Press the Burst button and change the number of cycles to 2 and the burst period (Trigger Interval) to 1 ms.

Press *AUTOSET* on the **TDS 3012B** oscilloscope. Manually set the TIME/DIV to 40 μs/DIV. Press the zoom button (the one with the picture of a magnifying glass on it). Rotate the rotary knob on the HORIZONTAL scale section of the scope until the brackets in the upper display have a width slightly larger than the duration of a pulse. Change the Horizontal position control until the brackets enclose a pulse. Turn the cursors on and use the time cursors to measure the width in time of the tone burst (the amount of time that it is nonzero per cycle).

Tone Burst Time Duration____

Turn the cursors off. Turn the tone burst off by pressing the Continuous button on the function generator.

5.12.15 Lissajous Patterns

Connect the CH1 output of the **Tektronix AFG 3022B** function generator to the $CH1$ input on the **TDS 3012B** oscilloscope with a BNC to BNC lead or connector and the CH2 output of the function generator to CH2 of the oscilloscope. Set both channels of the function generator to produce a 100 Hz sine wave with a peak-to-peak value of 2 V. Sequentially press the VIEW button at the bottom of the function generator and note the effect on the display on the function generator.

Set both channels of the oscilloscope to dc coupled. Turn both channels on the oscilloscope and press AUTOSET. Use the vertical position controls on the oscilloscope to overlay both sine wave so that the zero volts lines coincide. On the function generator, press phase for CH1 Change the Phase to 90 degrees and press Align Phase on the function generator. Change the phase to zero degrees and repeat. Change the phase back to 90 degrees and press align phase.

On the oscilloscope, press $AUTOSET$. Press $DISPLAY$, followed by XY DISPLAY and then Triggered XY. Use the vertical position controls for $Ch1$ & $Ch2$ to center the display which should be a rotating circle. Change the phase to zero degrees and then press align phase and note the display. Set the frequency of the the CH1 output of function generator to 200, 300, and 400 Hz and repeat for phase shifts of both zero and ninety degrees. Change the frequency of the CH1 output to 300 Hz and the CH2 output to 200 Hz and repeat. Print a few of the displays for the laboratory report.

Select CH1 on the function generator, Modulation, FM, FM Frequency 0.1 Hz, and Deviation 1 Hz. Describe the display in the laboratory report.

5.13 Laboratory Report

1. Turn in all the print outs that were made.
2. Compare the dc level measured in step 5.12.9 with the theoretical value given by

$$E_{dc} = \frac{1}{T}\int_0^T e(t)dt \tag{5.17}$$

3. Answer all questions posed in the procedure as well as any posed by the laboratory instructor.

5.14 References

1. Bigelow, S. J., "All About Monitors", *Electronics Now*, July 1997.
2. Coombs, C. F., *Electronic Instrument Handbook*, Prentice-Hall, 1995.
3. Cooper, W. D., *Electronic Instrumentation and Measurements Techniques*, 2nd edition, Prentice-Hall, 1978.
4. R. P. Dolan, et. al., HP 54600 Series Oscilloscopes, *Hewlett-Packard Journal*, vol. 43, no.1, pp. 1-59, February 1992.

5. T. A. Dye and E. Teose, "Digital Storage Scopes Advance", *IEEE Spectrum*, vol. 29, no. 2, Feb. 1992, pp. 38-41.
6. Hewlett-Packard, *HP54600-Series Oscilloscopes*, Hewlett-Packard, 1992.
7. Hewlett-Packard, *HP 54657A, HP54658A, and HP 54659B Measurement/Storage Modules*, Hewlett-Packard, 1996.
8. Hewlett-Packard, *HP 54600-Series Oscilloscopes Programmer's Guide*, Hewlett-Packard, 1996.
9. Hewlett-Packard, *8 Hints for Making Better Scope Measurements, Hewlett-Packard*, 1995.
10. Hewlett-Packard, *Feeling Comfortable with Digitizing Oscilloscopes*, Hewlett-Packard, 1987.
11. Hewlett-Packard, *HP 33120A Function Generator/ Arbitrary Waveform Generator User's Guide*, Hewlett-Packard, 1996.
12. Hewlett-Packard, *HP 33120A Function Generator/ Arbitrary Waveform Generator Service Guide*, Hewlett-Packard, 1996.
13. Hewlett-Packard, *HP 33120A/Opt. 001 Phase Lock Assembly User's and Service Guide*, Hewlett-Packard, 1994.
14. Jones, L., and Chin, A. F., *Electronic Instruments and Measurements*, John Wiley, 1983.
15. Krenz, J. H., *Introduction to Electrical Circuits and Electronic Devices: A Laboratory Approach*, Prentice-Hall, 1987.
16. Kantrowitz, P., et. al., *Electronic Measurements*, Prentice-Hall, 1979.
17. Murphy, J., "Ten Pointers to Picking a Scope", *IEEE Spectrum*, November 1996.
18. Oliver, B. M., and Cage, J. M., *Electronic Measurements and Instrumentation*, McGraw-Hill, New York, 1971.
19. Philips, "Basic Principles of DSOs", *Philips*, 1988.
20. Prentice, S., "A Look at Digital Stoarage Oscilloscopes and Bandwidths from 20-200 MHz", *Radio Electronics*, November 1991.
21. Riaz, M., *Electrical Engineering Laboratory Manual*, McGraw-Hill, 1965.
22. Roth, C. H., *Use of the Dual Trace Oscilloscope*, Prentice-Hall, 1982.
23. K. Rush and T. Lucero-Hall, "Two Ways to Catch a Wave", *IEEE Spectrum*, vol. 30, no. 3, Feb. 1993, pp. 38-41.
24. M. M. Lee, *Winning with People: The First 40 Years of Tektronix*, Tektronix Inc., 1986.
25. Phillips, *Basic Principles of DSOs using the Philips PM 3350*, Phillips, June 1988.
26. S. Prentiss, "Digital Storage Oscilloscopes", *Radio Electronics*, vol. 62, no. 11, pp. 31-42.
27. Wolf, S., *Guide to Electronic Measurements and Laboratory Practices*, Prentice-Hall, 1973.
28. Wolf, S., *Guide to Electronic Measurements and Laboratory Practices*, 2nd edition, Prentice-Hall, 1983.
29. Wolf, S., and Smith, R. F. M., *Student Reference Manual for Electronic Instrumentation Laboratories*, Prentice-Hall, 1990.

Chapter 6

Computer Control of Instruments

6.1 Objective

The objective of this experiment is to examine the control of laboratory instruments with a personal computer. Graphical programming languages and software applications are used to establish control.

6.2 Theory

Digital instruments can be controlled by computers. The computer is simply an extension of the instrument. This is a considerable enhancement of the capabilities of the instrument since even small computers have more memory and a more flexible, faster, and robust microprocessor. Data can be directed ported into applications such as spreadsheets and word processors. By connecting the personal computer to the Internet remote control of instruments is possible.

6.2.1 Hardware

Computer

The computer used in this experiment uses the Windows operating system. It is a networked personal computer. It cannot be used in the stand-alone mode and must be used on a network.

Interface

The interface between the computer and the laboratory instruments is either a GPIB card or a USB connection. Ethernet connections are also available but will not be used.. The GPIB card is installed in the PCI bus of the computer. The particular GPIB card used in the laboratory PC is manufactured by either National Instruments or Agilent Technologies.

Control of instruments can also be established with other different types of interfaces. For instance the RS 232 port of the computer could be used. However, the RS 232 port is a serial interface whereas the GPIB is a parallel interface and is, therefore, much faster. Another popular interface is VXI which is also a parallel interface but requires that the instruments be placed in the same chassis.

Communications between the instruments and the computer are established through the GPIB card on the computer over daisy chained connectors known as GPIB connectors. Each instrument has a GPIB connector and the system consisting of the computer and instruments are joined with GPIB connectors. If the USB connection is used the PC will assign a different name to each USB connection.

A unique GPIB bus address is assigned to each instrument on the system. This enables the computer to direct the appropriate commands to the appropriate instrument.

Oscilloscope

The oscilloscope is the **Tektronix TDS 3012B** two channel 100 MHz digital oscilloscope.. In this system the GPIB address is 1; it should not be changed on either the instrument or the software.

6.2.2 Function /Arbitrary Waveform Generator

The function / arbitrary waveform generator is the **Tektronix AFG 3022B ARBITRARY/FUNCTION.**. It has several connectors on the back, one of which is a GPIB. The bus address of this instrument is this system is 11; it should not be changed either on the instrument or the software used to control the instrument. A USB connection will also be used for flexibility.

Digital Multimeter

The digital multimeter used in the experiment is the **Agilent 34401A DMM**. It has a GPIB connector on the back; the bus address is 22 and should not be changed on the instrument or the software.

6.2.3 Software

Three software packages are used to communicate with or control instruments in this experiment: ArbExpress, VEE, and LabVIEW. The first is manufactured by Tektronix, the second by Agilent Technologies (aka Hewlett Packard) and the last by National Instruments.

ArbExpress

ArbExpress is used to control a Tektronix function generators. The specific use that will be made of it in to program arbitrary functions in conjunction with Mathcad.

VEE

This is a graphical programming language that can be used to control laboratory instruments. The acronym VEE stands for Visual Engineering Environment. The programmer essentially draws a block diagram or flow chart for the program and the compiler produces the program from it. Texts commands are kept to a minimum and data and program flow are easier to follow than in text based languages.

LabVIEW

LabVIEW is a graphical programming language manufactured by National Instruments. Graphical programming is a concept originally conceived and developed by National Instruments. Text program steps are replaced by functional blocks. There are at least two windows used with it; one is know as the Front Panel and the other the Block Diagram. This is also known as the G programming language.

6.3 Procedure

6.3.1 Waveform Editor

- Log onto the computer.
- Turn on the **Tektronix AFG 3022B** Arbitrary/Function Generator. Note that it boots up with a GPIB address of 11. Do not change this either on the instrument or any of the software to be used or communication between the computer and the instrument will be impossible. Set the output for High Z, enable the CH1 output and set it to produce a sine wave with a frequency of 1 kHz and a peak-to-peak value of 2 V. In order for the amplitues indicated by the function generator to be correct the High Z mode must be used.

- Double left click on the icon ArbExpress which is for the software that programs the **Tektronix** function generator.

- Turn on the **Tektronix TDS 3012B** oscilloscope. Turn CH1 on and CH2 off. This instrument will use a GPIB address of 1 which should not be changed either on the instrument or the software. Connect the CH1 function output of the function generator to CH1 on the oscilloscope and the TTL out of the function generator to EXT TRIG input of the oscilloscope with BNC to BNC connectors. Press AUTOSET. Set the Trigger Source for the scope to Ext/10 by pressing the MENU button on the trigger section of the front panel and then select source for CH2 and then set the Trigger Level to 50 %.

- Click Waveform, Standard Waveform on ArbExpress, and then select Multi Tone. Click Preview and then OK.Click Send to Arb on the upper right toolbar. If it responds "No Arb Detected" follow the sequence of commands to add the Arb. In the Arb List select the ARB 3022B at GPIB address 11 or the USB port and then click Connect followed by Close. Click Send to Arb again.

- Use the Mathcad code used in Experiment 2 to program the ELVIS Arb to create a text file for the Tektronix Arb. Use 16,000 points this time. Use the function specified by the laboratory instructor. Make sure Word Wrap is enabled in the text file. On the ArbExpress program click open and import the text file. For the file format use Volts, <cr>, <lf>. Click Send to Arb. Print the display with Screen Capture and insert it into the laboratory report.

6.3.2 VEE

- Double left click the icon for VEE. A page for programming should open.

Function Implementation

- Implement the function specified by the laboratory instructor. Three basic types of blocks will be needed: data, functions, and displays. These categories are found on the toolbar. Evaluate the function for the set of input parameters specified by the laboratory instructor. Obtain notepad from Display, Notepad. (Left click on display on the toolbar on the top and then select notepad.) Position Notepad near the function. Type the name of the experimenter in the space on Notepad. Select File and then Printscreen. Clear the screen by selecting Edit, Select All, and Scissors (or Delete). (If it is necessary to add data input or output pins this is done by placing the pointer within the block and right clicking the mouse. The select add terminal.)

Virtual Instruments

- Implement the Lissajous Pattern specified by the laboratory instructor. This will require the virtual function generator (found under device) and a X Y display (found under displays). Place a notepad with the name of the experimenter in it near the display and print the screen. Clear the screen by selecting Edit, Select All, and Scissors.

Plots

- Implement the plot specified by the laboratory instructor. This will require formulas, a ramp for the independent variable, and an X Y or possibly polar display. Place a notepad containing the name of the experimenter near the display and print the screen. Clear the screen by selecting Edit, Select All, and Scissors.

Control of Instruments

- Left click I/O from the top toolbar and then Instrument Manager. Note that there is list or the type of instrument interfaces one of which is a GPIB bus which will be listed as GPIB0 with three instruments: **Agilent 34401A** Digital Multimeter @ address 22, **Tektronix TDS** Oscilloscope @ address 01, and a **Tektronix AFG 3022B Arb/FG** at address 11. If it finds bogus instrument on the COM ports and LPT ports, delete them. The addresses coincide with the addresses on the instruments. Neither the address in the software nor the instrument should be changed or communications will not be possible between the computer and the instrument. The instrument address may be shown as XXX22 but the last 2 number will identify the instrument. A USB and ethernet port will also be shown. The conntection to the function generator may be made with either the GPIB port or the USB port.

- Left click the list component for the **Tektronix** function generator so that it is highlighted. (Don't double click it since this will bring up the Device Configuration panel.) With the function generator highlighted drag the icon to the center of the screen and left click. Double click in the VEE box where it states "Double Click to Add a Transaction." Enter this code:<"sour1:freq ", f, " hz">; do not enter the brackets (but do enter the quotation marks) and there must be a space after the q in freq and before the f and before and after the quote mark in front of the h in hz. Check the box to create an input terminal named f if one is not automatically created. Select Data, Constant, Real 64 and position the box to the left of the box for the function generator. Connect the output of the Real 64 box to the input of the FG box. Enter a value such as 3 kHz (just $3k$ the software doesn't use units) for the frequency in the Real 64 Box. Press Run on the top toolbar and note the effect.

____

Left click edit, select all, and delete to clear the screen.

6.3.3 Frequency Response

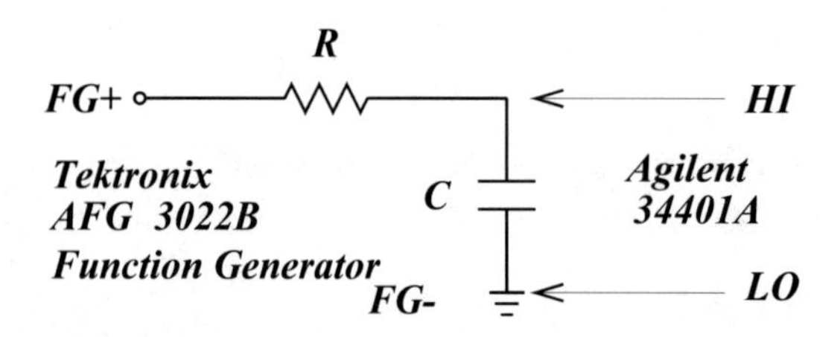

Figure 6.1: Circuit for Frequency Response

Assemble the circuit shown in Fig. 6.1. The resistor $R = 12\,\mathrm{k}\Omega$ (brn-red-org) and the capacitor $C = 0.01\,\mu\mathrm{F}$ (indicated as either 0.01 or 103 which is the capacitor code for $0.01\,\mu$F.) Connect the function generator to the input of the circuit and the digital multimeter to the output. Connect the red banana plug for the function generator lead (FG+) to the left side of the resistor and the black banana plug to the ground (FG-). Connect the HI lead from the DMM to the top of the capacitor and the LO lead to the system ground.

Gain

Locate the VEE file called "**VEE_Gain**" on the class web page using the IE and open it. This program will obtain the amplitude frequency response. The amplitude of the output of the function generator will be kept constant while the frequency is swept and output voltage from the circuit will be measured by the digital voltmeter. To use this the input to the circuit should be CH1 of the function generator and the output should be the input to the DMM.

Move the pointer to an open area on the screen (no object or block). Left click and hold down the mouse key and note that the entire program can be drug around the screen. Look at the left portion of the program and note that it contains the inputs to the program which set the amplitude of the function generator to 1 V and the start frequency to 100 Hz, stop frequency to 100 kHz, and the number of points per decade to 20. Drag the program to look at the right portion which contains the display where the frequency response plot will be made.

Start the program by left clicking on the triangle symbol on the tool bar. When the program stops position the pointer over the display object and right click the mouse. Select Properties from the pop up menu that will appear. Append to the end of title [Magnitude Frequency Response] in parentheses the name of the experimenter. Click for one marker to be displayed. Select OK. Use the pointer to move the marker to the point on the response where the output is 0.707 V which is $-3dB$ on a decibel scale (this can't be done exactly because it's a digital instrument). Position the pointer inside the display object and right click the mouse. Select print.

The above procedure produces a display of the magnitude frequency response. Although adequate for many applications, it is also desirable to be able to take the measured values and place them in a file so that it can be imported into a spreadsheet. Click MyFile and select a location to store the data on the pc. Minimize VEE and use the Windows Notepad to read the contents of the file. (The default location for MyFile is My Documents, VEE.) Copy the data array to the Windows clipboard. Start Microsoft Excel and paste the data array in a column. Plot the results. Annotate the plot with the experimenter's name. Print the plot. A similar procedure could be used to also obtain the frequency data points which could be placed in one column and the gain data points in another. This would permit it to be plotted in a more sophisticated software product such as KaleidaGraph. Close the VEE program.

Gain-Phase

Locate the VEE program "**VEE_Gain_Phase**" on the class web page with the IE and open it. This will obtain both the amplitude and the phase frequency response with the oscilloscope. Regrettably, the oscilloscope is much slower and more inaccurate than the digital multimeter. Disconnect the DMM from the circuit.

Connect a BNC T connector to the CH1 output of the function generator. Use a BNC to BNC connector (or banana plug to BNC) to connect the CH1 output of the function generator to CH1 input on the oscilloscope. Use a BNC to banana plug lead to connect the input of the circuit (portion on the left) to CH1 on the oscilloscope and Ch2 to the output of the circuit (the capacitor voltage). Make sure that the black banana plug for the scope leads is connected to the system ground. It is only necessary to connect one ground lead on the oscilloscope. (Why?)

On the Tektronix 3012B oscilloscope select Measure, Select Measurement for CH1, Cycle RMS. Press the Select Measurement button for CH1 and then the blue CH2 button followed by Cycle RMS and then Phase From CH2 to CH1. All three of these measurements must appear on the screen of the scope or this program will not work.

Start the program by left clicking on the triangle symbol on the toolbar. As before, change the properties of the two displays to append the name of the experiment to the title bar of the display and then print it. Accurate phase measurements can only be made with this program when the amplitude of the output is sufficiently large.

6.3.4 LabVIEW

6.3.5 Amplitude Response

The frequency response of the above circuit will now be obtained using LabVIEW. Connect the CH1 output of the function generator to the input of the circuit. Connect the output of the circuit to the input of the digital multimeter. Locate the LabVIEW program "**LV_Gain**" on the class web page and open it with the IE. Set the desired input voltage level, start and stop frequencies, and points per decade. Click the start arrow near the upper left hand corner. After the program runs the amplitude response will appear.

To print the results to a file click on the appropriate switch and re run the program.

6.3.6 ELVIS

The frequency response of the above circuit will now be obtained with ELVIS. Start ELVIS. From the NI ELVISmx Instrument Launcher double click Bode. Connect the ELVIS function generator output to the input of the circuit and to AI6+ on ELVIS. Connect the output of the circuit (the junction between the resistor and capacitor) to AI7+ on ELVIS. Connect AI6- and AI7- to the system ground. On the ELVIS panel switch the Stimulus Channel to AI6 and the Response Channel to AI7. Change the Steps per Decade to 50. Use a Start Frequency of 100 Hz and a Stop Frequency of 10 kHz.Click the Run arrow

After ELVIS has plotted the frequency response, click once and only once on Log. Open Notepad. Paste the data in and save as a file somewhere on the pc. Open Excel and import the contents of the file. This will start the Text Import Wizard. Click Next. Select all the delimiters including other and for the other box enter a colon. Click Next and then Finish. Plot the amplitude and phase responses as functions of frequency using KaleidaGraph. Change the frequency axis on the plots from linear to log. Insert name using the text feature. Print and include in laboratory report.

6.3.7 Swepth Frequency

Connect the CH1 output of the function generator to the input of the filter and the output of the filter to CH1 on the oscilloscope. Connect the Synch or TTL output of the function generator to the EXT TRIG on the oscilloscope. Set the Run Mode on the function generator to Sweep. Under the Sweep menu set the Start Frequency to 100 Hz, the Stop Frequency to 10 kHz, the Sweep Time to 100 ms, and the Sweep Type to Logarithm. On the oscilloscope manually set the Volt/Div to 500 mV, the Time/Div to 10 ms, the Trigger Menu to Source Ext/10, and the Trigger Level to TTL. Explain the display. Print it and include it in the laboratory report.

6.4 References

1. Robert Helsel, *Visual Programming with HP VEE*, 3rd ed., Prentice Hall, 1998.
2. Essick, John, *Advanced LabVIEW Labs*, Prentice-Hall, 1998.

Chapter 7

First-Order Circuits

7.1 Objective

Circuits for which the relationship between the excitation and the response is described by a first-order, linear, ordinary differential equation with constant coefficients are known as first-order circuits. Given the excitation and the circuit, the response may be obtained by solving this differential equation. Such circuits have responses that differ radically from zero-order or resistive circuits. With zero-order circuits the response changes almost instantaneously with changes in the excitation (at the speed of light) and has exactly the same functional form whereas with higher-order circuits there is a time delay between the response and excitation which quite often results in differing functional behaviors.

A first-order circuit contains one type energy storage element: either an inductor or a capacitor. Energy is stored in the magnetic field of the inductor or in the electric field of the capacitor. With nonimpulsive excitations energy stored in an inductor or a capacitor cannot be changed instantaneously. For such excitations the inductor current and the capacitor voltage must be continuous functions of time.

This experiment will examine the response of RC and RL first-order circuits. The response will be obtained for step function, square wave, triangular wave, ramp wave, and sinusoidal excitation. The principle of duality will be employed to obtain the response of RL circuits from the response of RC circuits.

Models for the physical circuits elements (resistor, capacitor, inductor, transformer) will be developed in terms of ideal components. Methods for measuring resistance, inductance, and capacitance will be examined.

7.2 Theory

7.2.1 RC Circuits

Shown in Fig. 7.1 is the series RC circuit. The voltage source $e(t)$ is the excitation and either the series current or the capacitor voltage can be considered the response. If the capacitor voltage is known the series current can be obtained from the constitutive relationship

$$i(t) = C\frac{dv_c(t)}{dt} \tag{7.1}$$

since the current that flows into the capacitor is also the series current. The differential equation for the capacitor voltage can be obtained by applying Kirchhoff's voltage law around the loop which yields

$$\tau\frac{dv_c(t)}{dt} + v_c(t) = e(t) \tag{7.2}$$

where $\tau = RC$ is the time constant of the circuit and has units of time or seconds. If $e(t)$ is given, Eq. 7.2 may be solved for $v_c(t)$ and then $i(t)$ can be obtained from Eq. 7.1. Since energy is stored in the electric

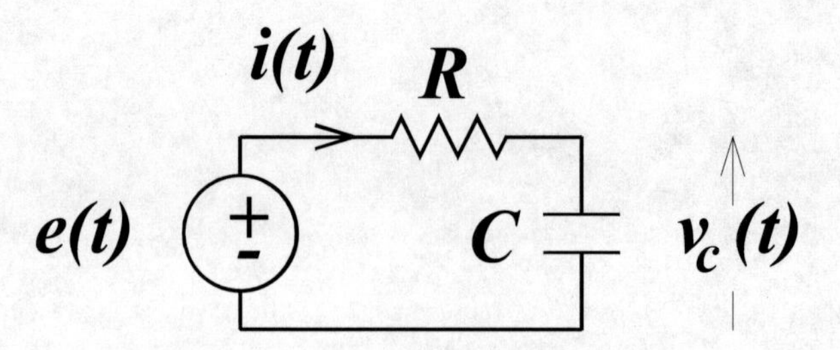

Figure 7.1: Series RC circuit.

field between the plates of the capacitor ($w_e(t) = Cv_c^2(t)/2$), the capacitor voltage must be a continuous function of time. (If the capacitor voltage were discontinuous then Eq. 7.1 would require that the current contain impulsives at the points of discontinuity which cannot be produced by a nonimpulsive source.)

7.2.2 Step Function Response

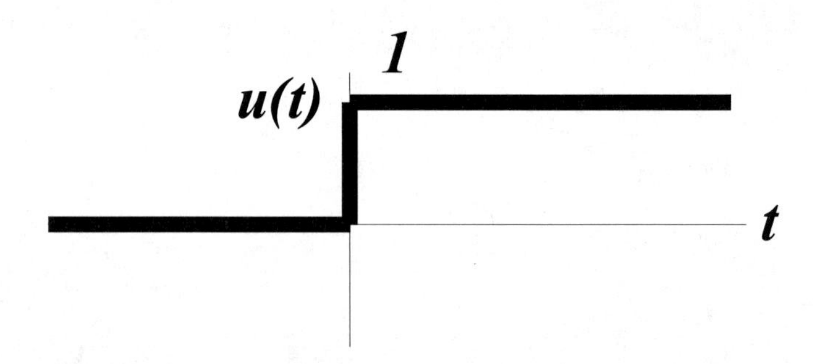

Figure 7.2: Unit step function.

Shown in Fig. 7.2 is the unit step function $u(t)$. This function is simply zero for negative t and unity (one) for positive t. (At the origin, $t = 0$, the step function can be taken as 0.5.) Step function excitations occur when a dc voltage source is placed in series with a switch which is closed at $t = 0$.

If the excitation for the circuit shown in Fig. 7.1 is $e(t) = Eu(t)$, then the response is known as the step function response and is given by the solution to the differential equation

$$\tau \frac{dv_c(t)}{dt} + v_c(t) = e(t) = Eu(t) \tag{7.3}$$

The solution to this elementary differential equation is

$$v_c(t) = \left[B\, e^{-t/\tau} + E\right] u(t) \tag{7.4}$$

where E is the particular solution and B is the constant of integration from the homogeneous solution to this differential equation. One boundary condition or initial condition is needed to solve for the constant B; the initial value of the capacitor voltage must be zero because the capacitor voltage is a continuous function of time and the circuit is unenergized prior to $t = 0$. Utilizing this initial condition on the capacitor voltage, the expression for the capacitor voltage becomes

$$v_c(t) = E\left[1 - e^{-t/\tau}\right] u(t) \tag{7.5}$$

and the series current is given by

$$i(t) = \frac{E}{R} e^{-t/\tau} u(t) \tag{7.6}$$

which is discontinuous at $t = 0$. The capacitor voltage is zero at $t = 0$ and charges exponentially toward its final value of E Volts. The current jumps discontinuously from 0 to E/R at $t = 0$ and decays exponentially toward its final value of zero. From Eqs. 7.5 and 7.6 it can be seen that the time constant is merely the instant of time for which the argument of the exponentials is -1.

Step functions are single event functions which are difficult to examine with laboratory instruments such, as non-storage analog oscilloscopes, that are geared to observing periodic events. Even digital oscilloscopes prefer periodic signals since they may be sampled at a lower rate. For this reason, the step function response is often obtained by observing the low-frequency square wave response. Indeed, the three most important signals for testing electrical and electronic systems are dc, sine, and square.

7.2.3 Square Wave Response

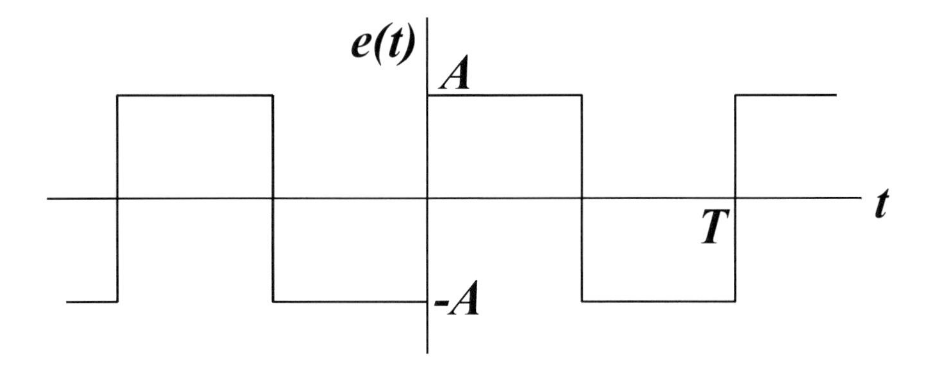

Figure 7.3: Symmetric square wave with dc level of zero volts.

Shown in Fig. 7.3 is the symmetric square wave with a frequency $f = 1/T$ Hertz where T is the period, peak value of A volts, peak-to-peak value of $2A$ volts, symmetry or duty cycle 50%, and dc level of zero volts. The duty cycle is the percentage of time that the waveform is at the upper voltage level of $+A$. The peak value of the waveform is the maximum value of the waveform and the peak-to-peak value of the waveform is the difference between the maximum and minimum value of the waveform. The dc level or average value of the waveform is given by

$$E_{dc} = \frac{1}{T}\int_0^T e(t)dt \tag{7.7}$$

which for the symmetric square wave is zero.

The time origin will be taken as that time at which the square wave switches from its lower to its upper level. The square wave switches from one level to the other instantaneously. This is a mathematical idealization since no physical source can instantaneously switch levels. However, if the time required to switch from one level to the other is sufficiently small compared to a period, this assumption leads to negligible error and greatly simplifies the solution.

If $e(t)$ is the excitation for the circuit shown in Fig. 7.1, then the capacitor voltage, $v_c(t)$, can be considered the response. The task of determining $v_c(t)$ might seem formidable. The capacitor voltage must be determined for all time t and the excitation switches from one level to the other ever $T/2$ seconds. The task is simplified with the observation that since $e(t)$ is a periodic waveform with period T and the circuit is a linear time-invariant system the response must also be a periodic waveform with period T. The shape of $e(t)$ and $v_c(t)$ may be radically different but both are periodic with period T. Therefore, $v_c(t)$ must be determined in only one interval of time having a length of T seconds; once this is known $v_c(t)$ is known for all time t because $v_c(t)$ is periodic with period T.

The first step in determining $v_c(t)$ is to select the interval of time having a length T in which $v_c(t)$ will be determined. The time interval that will be selected is $[-T/2, T/2]$. Eq. 7.2 then becomes

$$\tau\frac{dv_c(t)}{dt} + v_c(t) = \begin{cases} -A & -T/2 < t < 0 \\ A & 0 < t < T/2 \end{cases} \tag{7.8}$$

which requires that the differential equation be solved in each of the two subintervals of length $T/2$ seconds because it is not possible to write one analytical expression for $e(t)$ which is applicable for an entire period. The solution is simply

$$v_c(t) = \begin{cases} B_1 e^{-t/\tau} - A & -T/2 \le t \le 0 \quad (a) \\ B_2 e^{-t/\tau} + A & 0 \le t \le T/2 \quad (b) \end{cases} \tag{7.9}$$

where B_1 and B_2 are the constants of integration from the homogeneous solution. To solve for these two unknown constants two boundary value conditions are required. Since the capacitor voltage must be a continuous function of time, both Eqs. 7.9a and 7.9b must yield the same value when $t = 0$ which means that

$$B_1 - A = B_2 + A \tag{7.10}$$

which provides one equation for the two unknown constants B_1 and B_2. A second equation may be obtained by making use of the periodicity of $v_c(t)$; namely, Eq. 7.9a must yield the same value for the capacitor voltage when $t = -T/2$ as Eq. 7.9b does when $t = T/2$. This results in

$$B_1 e^{T/(2\tau)} - A = B_2 e^{-T/(2\tau)} + A \tag{7.11}$$

as the second equation for the two unknown constants B_1 and B_2. Simultaneous solution of Eqs. 7.10 and 7.11 yields

$$v_c(t) = \begin{cases} [A + v_c(0)]\, e^{-t/\tau} - A & -T/2 \le t \le 0 \quad (a) \\ [-A + v_c(0)]\, e^{-t/\tau} + A & 0 \le t \le T/2 \quad (b) \end{cases} \tag{7.12}$$

where the initial value of the capacitor voltage, $v_c(0)$, is given by

$$v_c(0) = -A \tanh\left[\frac{T}{4\tau}\right] \tag{7.13}$$

where the hyperbolic tangent is given by

$$\tanh x = \frac{e^x - e^{-x}}{e^x + e^{-x}} \tag{7.14}$$

where x is any real number. The capacitor voltage in any other time interval is obtained by simply translating Eq. 7.12 up or down the time axis to the desired time.

It should be noted that because the excitation has half-wave odd symmetry, the solution for $v_c(t)$ also has half-wave odd symmetry, i.e. $e(t) = -e(t - T/2) \rightarrow v_c(t) = -v_c(t - T/2)$. This means that the waveforms can be shifted by half a period and inverted and obtain the original function.

The series current $i(t)$ is given by substituting Eq.7.12 into Eq. 7.1 which yields

$$i(t) = \begin{cases} -\dfrac{A + v_c(0)}{R} e^{-t/\tau} & -T/2 < t < 0 \quad (a) \\ -\dfrac{-A + v_c(0)}{R} e^{-t/\tau} & 0 < t < T/2 \quad (b) \end{cases} \tag{7.15}$$

It should be noted that although the capacitor voltage is a continuous function of time, the capacitor current is under no such constraint. Indeed, the capacitor current is discontinuous at those instants of time at which $e(t)$ is discontinuous.

7.2.4 Triangular Wave

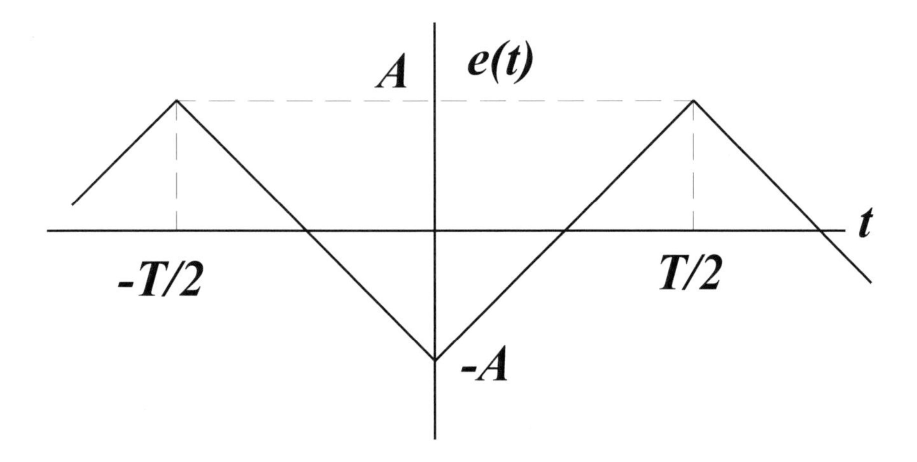

Figure 7.4: Symmetric triangular wave.

Shown in Fig. 7.4 is the symmetric triangular wave with frequency f Hertz ($f = 1/T$), a peak-to-peak value of $2A$, and a dc level of zero volts. The time origin is chosen as a instant of time for which the waveform is at its negative peak.

If $e(t)$ is the excitation for the circuit shown in Fig. 7.1, then Eq. 7.2 becomes for this waveform

$$\tau \frac{dv_c(t)}{dt} + v_c(t) = \begin{cases} -\dfrac{4A}{T} t - A & -T/2 \leq t \leq 0 \quad (a) \\ \dfrac{4A}{T} t + A & 0 \leq t \leq T/2 \quad (b) \end{cases} \tag{7.16}$$

which requires that the differential equation be solved for both the negative and positive sub intervals having a length of $T/2$ because it is not possible to write one analytical expression for $e(t)$ which is applicable for an entire period. The solution to Eq. 7.16 is

$$v_c(t) = \begin{cases} B_1 e^{-t/\tau} + p_1 t + p_2 & -T/2 \leq t \leq 0 \quad (a) \\ B_2 e^{-t/\tau} + p_3 t + p_4 & 0 \leq t \leq T/2 \quad (b) \end{cases} \tag{7.17}$$

where B_1 and B_2 are the constants of integration from the homogeneous solution. The unknown coefficients from the particular solution, p_i $(i = 1, 2, 3, 4)$ are evaluated by substituting Eq. 7.17a into 7.16a and 7.17b into 7.16b, respectively, and equating like powers of t. The constants B_1 and B_2 are evaluated in the same manner as for the square wave, i.e. by making use of the fact that $v_c(t)$ is periodic and continuous.

The solution for the capacitor voltage for triangular wave excitation then becomes

$$v_c(t) = \begin{cases} B_1 e^{-t/\tau} - \frac{4A}{T}t - A + \frac{4\tau A}{T} & -T/2 \leq t \leq 0 \quad (a) \\ B_2 e^{-t/\tau} + \frac{4A}{T}t - A - \frac{4\tau A}{T} & 0 \leq t \leq T/2 \quad (b) \end{cases} \tag{7.18}$$

where

$$\begin{aligned} B_1 &= v_c(0) + A - \frac{4\tau A}{T} \quad (a) \\ B_2 &= v_c(0) + A + \frac{4\tau A}{T} \quad (b) \end{aligned} \tag{7.19}$$

and the initial value of the capacitor voltage is given by

$$v_c(0) = -A\left(1 - \frac{4\tau}{T}\tanh\left[\frac{T}{4\tau}\right]\right) \tag{7.20}$$

where the hyperbolic tangent $\tanh(x)$ is defined in Eq. 7.13.

7.2.5 Ramp Wave

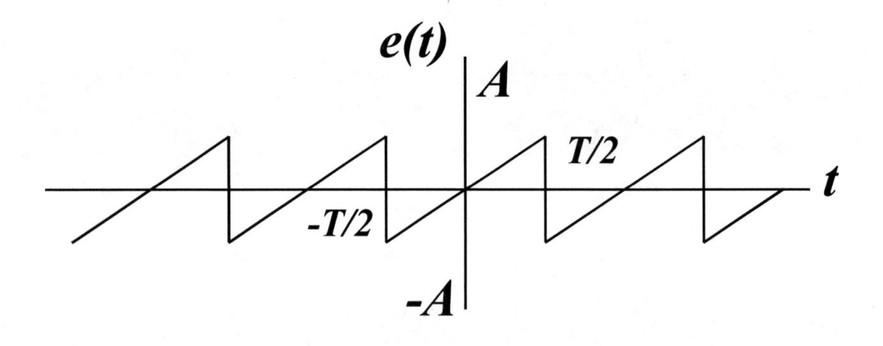

Figure 7.5: Ramp wave.

Shown in Fig. 7.5 is the ramp wave. The frequency is f Hertz ($f = 1/T$), the peak-to-peak value is $2A$ volts, and the dc level is 0 volts. The origin is taken as that instant of time for which the ramp is zero volts. The ramp switches from $+A$ volts to $-A$ volts at the origin; although any physical voltage source will require a finite time to change from $+A$ to $-A$, this assumption is valid if the time required to switch levels is negligible compared to a period, T.

If $e(t)$ is the excitation for the circuit shown in Fig. 7.1, then Eq. 7.2 becomes

$$\tau\frac{dv_c(t)}{dt} + v_c(t) = e(t) = \frac{2A}{T}t \qquad -\tfrac{T}{2} < t < \tfrac{T}{2} \tag{7.21}$$

which is simpler than the differential equation for the square and triangular wave because it is possible to write one analytic expression for $e(t)$ which is applicable for an entire period. The solution to this differential equation is

$$v_c(t) = Be^{-t/\tau} + p_1 t + p_2 \qquad -\tfrac{T}{2} \leq t \leq \tfrac{T}{2} \tag{7.22}$$

where B is the constant of integration from the homogeneous solution and p_1 and p_2 are the constants from the particular solution. The constants from the particular solution are evaluated by substituting Eq. 7.22 into Eq. 7.21 and equating like power of t and the constant from the homogeneous solution is then evaluated by making use of the fact that since $v_c(t)$ is periodic then $v_c(-T/2) = v_c(T/2)$.

The solution for the ramp wave then becomes

$$v_c(t) = \frac{Ae^{-t/\tau}}{\sinh(\frac{T}{2\tau})} + \frac{2A}{T}(t-\tau) \qquad -\tfrac{T}{2} \leq t \leq \tfrac{T}{2} \tag{7.23}$$

7.2.6 Sinusoidal Excitation

If the excitation is a sinusoidal function of time such as $e(t) = A\cos(\omega t)$, then the response may be obtained by solving the differential equation for $v_c(t)$ using classical techniques. However, for any problem with a sinusoidal excitation, it is far easier to use a complex phasor representation for the current and voltages. The real current and voltages may then be easily obtained from the corresponding phasors.

The complex phasor representation for $e(t) = A\cos\ \omega t$ is $\bar{E} = A/\sqrt{2}\angle 0^o$. This is known as a *rms* phasor since the magnitude of the complex quantity $\bar{E}$ is the peak value of the sinusoidal divided by the square root of 2 which is the rms value of the sinusoidal.

The complex phasor capacitor voltage in Fig. 7.1 is then

$$\bar{V}_c = \bar{E}\frac{1}{1+j\omega\tau} \tag{7.24}$$

and the complex phasor series current

$$\bar{I} = \frac{\bar{E}}{R}\frac{j\omega\tau}{1+j\omega\tau} \tag{7.25}$$

from which the real capacitor voltage and series current can be obtained. The driving point impedance presented to the excitation voltage source is

$$\bar{Z} = \frac{\overline{E}}{\overline{I}} = R + \frac{1}{j\omega C} = R\left(\frac{1+j\omega\tau}{j\omega\tau}\right) \tag{7.26}$$

7.2.7 Parallel GL Circuit

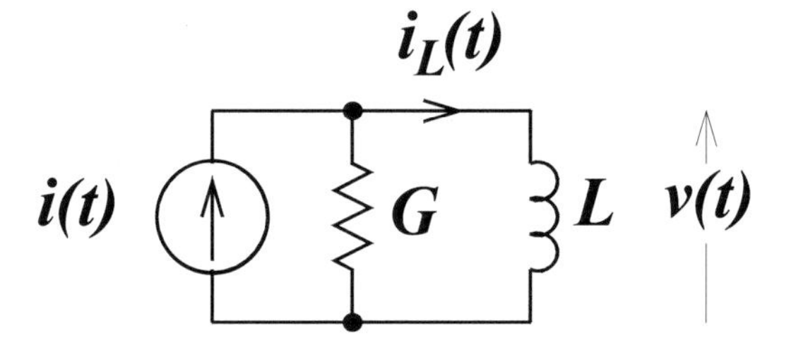

Figure 7.6: Parallel GL circuit.

Shown in Fig. 7.6 is the first order GL circuit. The conductance $G = 1/R$ is in parallel with the inductor and the current source $i(t)$. The current source is the excitation and either the inductor current or parallel voltage can be considered the response. Once the inductor current is known, the voltage can be obtained from the constitutive relationship

$$v(t) = L\frac{di(t)}{dt} \tag{7.27}$$

since the three circuit elements are in parallel. It will be assumed that the current source does not contain impulses which results in the inductor current having to be continuous.

Applying Kirchhoff's current law to the circuit of Fig. 7.6 yields

$$\tau\frac{di_L(t)}{dt} + i_L(t) = i(t) \tag{7.28}$$

where $\tau = GL$. This equation is the dual of Eq. 7.2 for the series RC circuit. Therefore, the solution for the parallel GL circuit can be obtained for the solution for the series RC circuit by making the following exchanges

$$\begin{array}{ccccccccccccc} \textit{Series} & i & v & E & I & G & R & C & L & \overline{Z} & \overline{Y} & \overline{V} & \overline{I} \\ \Updownarrow & \Updownarrow & \Updownarrow & \Updownarrow & \Updownarrow & \Updownarrow & \Updownarrow & \Updownarrow & \Updownarrow & \Updownarrow & \Updownarrow & \Updownarrow & \Updownarrow \\ \textit{Parallel} & v & i & I & E & R & G & L & C & \overline{Y} & \overline{Z} & \overline{I} & \overline{V} \end{array} \tag{7.29}$$

which is known as the principle of duality.

7.3 Circuit Components

7.3.1 Resistors

An ideal resistor is a circuit component for which the voltage across the component is equal to a constant, known as the resistance, times the current flowing through the component. No physical circuit component will have a current-voltage characteristic that can match that of an ideal resistor for all voltage and current levels. All circuit components that are used as resistors in electrical circuits have a parasitic inductance and capacitance associated with them. Any current carrying conductor produces a magnetic flux and therefore has an inductance associated with it and between any two conductors at different voltages there exists a capacitance.

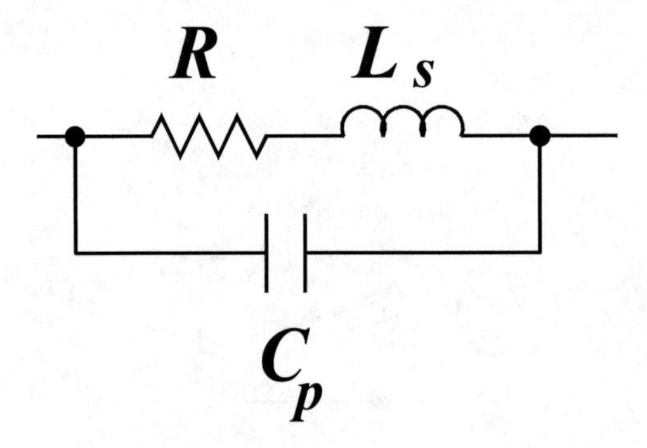

Figure 7.7: Lumped circuit model for physical resistor.

Shown in Fig. 7.7 is the lumped circuit model for a physical resistor. The resistor R is the desired resistor. The series inductance, L_s, is self inductance of the material from which the resistor is constructed. The parallel capacitance is due to the resistor's leads and internal construction.

The parasitic inductance and capacitance of the physical resistor may be neglected at low frequencies and the physical resistor behaves as an ideal resistor. At frequencies less than one hundredth of the resonant frequency of L_s and C_p the circuit shown in Fig. 7.7 can be reduced to the resistor R. At frequencies above this resonant frequency, the physical resistor behaves as a capacitor.

The resistance is a function of temperature. Normally, the resistance is specified at some temperature such as $20°C$ and the temperature coefficient of resistance is given so that the resistance can be determined at other temperature. As current flows through a resistor and power is dissipated the resistor will heat up and the resistance value will change.

All resistors have a maximum power dissipation capacity. If this is exceeded the resistor will be destroyed or damaged. Commercially available resistors have power ratings such as 1/4, 1/2, 1, or 2 watts.

The types of resistors that are commercially available are: carbon composition, wirewound, metal film, and carbon film. These types of resistors have different tolerances, power handling capacity, and parasitic inductance and capacitance.

Carbon composition resistors are the most common type in use in electronic circuits because they are cheap. They are made by mixing hot-pressed carbon granules with filler. Available resistance ranges are from $1\ \Omega$ to $22\ M\Omega$. Tolerances are relatively poor and range from 5% to 20%. The maximum power handling capacity is $2\ W$. The parasitic inductance ranges from $5-30\ nH$ and the parasitic capacitance ranges from $0.1-1.5\ pF$. The temperature coefficient of resistance value is $0.1\%/°C$.

Wirewound resistors are constructed by winding wire about an insulating cylindrical core. They have excellent tolerances (0.0005%), high power handling capacity ($200\ W$), and a low temperature coefficient of resistance ($0.0005\%/°C$). The resistance values range from $1\ \Omega$ to $100\ k\Omega$. The parasitic inductance is high ($47\ nH$ to $25\ \mu H$) as well as the parasitic capacitance ($2-14\ pF$).

Metal film resistor are made by depositing a very thin metal file on insulating materials to provide high resistance paths. Resistance values vary from $0.1\ \Omega$ to $10\ G\Omega$ with low tolerance values ($< 0.005\%$). The maximum power dissipation is $1\ W$ and the temperature coefficient of resistance is $0.0001\%/°C$. The parasitic inductance ranges from 15 to 700 nH and the parasitic capacitance ranges from 0.1 to 0.8 pF. These are low noise resistors which produce less thermal noise compared to other resistors. They are much more expensive than carbon composition resistors.

Carbon film resistors are fabricated in the same manner as metal film resistors. Resistance values range from $10\ \Omega$ to $100\ M\Omega$. The lowest tolerance is 0.5% and 5% is typical. The maximum power handling capacity is $2\ W$ and the temperature coefficient of resistance ranges from -0.015 to $0.05\%/°C$; the negative temperature coefficient of resistance is sometimes exploited in electronic design. The parasitic inductance and capacitance have the same ranges as those for metal film resistors.

Variable resistors are resistors for which the resistance between two leads can be changed by varying the position of a sliding or rotating contact. A variable resistor is shown in Fig. 7.8. The resistance between nodes or leads 1 and 2 is a constant, R_p. The resistance between nodes 2 and 3 is given by

$$R_{23} = (1-\theta)R_p \tag{7.30}$$

where the parameter θ is set by the position of the sliding or rotating contact which lies in the range $0 \leq \theta \leq 1$ The resistance between leads 2 and 3 can vary from 0 to R_p. This variable resistance may be used as a voltage divider

$$\frac{v_o}{v_i} = \frac{R_{23}}{R_p} = 1-\theta \tag{7.31}$$

where v_o is the output voltage and v_i is the input voltage. If only two of the leads are used in a circuit, the variable resistor is referred to as a rheostat and if all three are used the term potentiometer or "pot" is used. (The origin of the term potentiometer dates from the use of such a variable resistor to accurately

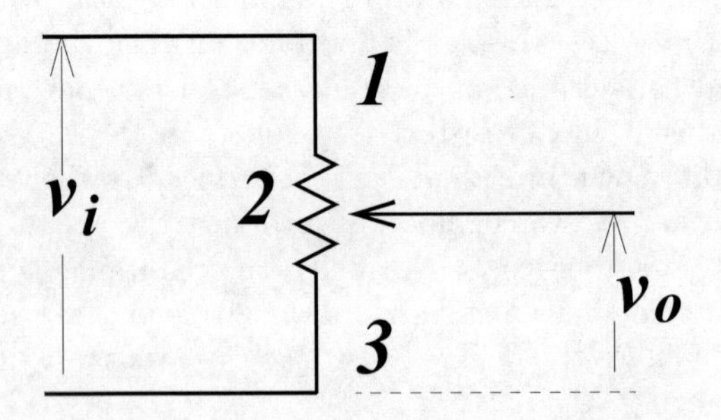

Figure 7.8: Variable resistor, rheostat, or pot.

measure an unknown voltage or potential by balancing the unknown voltage with a standard voltage.) Pots are commonly used in electronics for such applications as volume controls.

7.3.2 Capacitors

An ideal capacitor is a circuit component for which the current flowing into it is equal to a constant, known as the capacitance, times the rate of change of the voltage drop across it. No physical circuit component has exactly this behavior although many approximate it. All physical capacitors have an associated parasitic resistance and inductance.

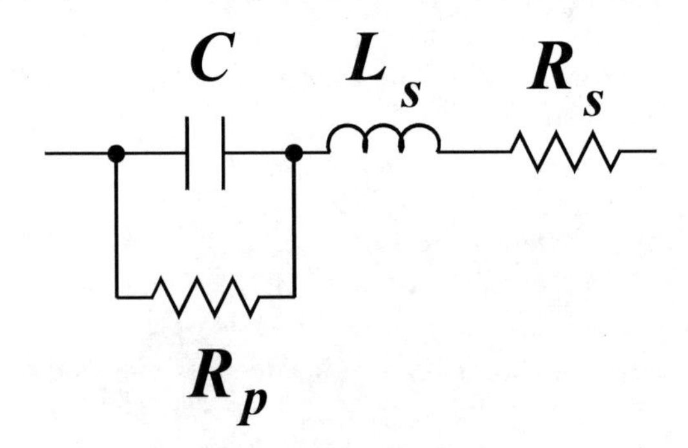

Figure 7.9: Lumped-impedance model for physical capacitor.

Shown in Fig. 7.9 is the lumped-impedance model for a physical capacitor. The parasitic series inductance, L_s, and resistance, R_s are due to the leads. The leakage resistance, R_p, is a measure of the insulating qualities of the dielectric material between the capacitor plates. At low frequencies the physical capacitor acts like an ideal capacitor in parallel with a resistor and at high frequencies the physical capacitor acts like an inductor. The boundary is the self resonant frequency set by C and L_s.

At low frequencies the capacitor is well modeled as an ideal capacitor in parallel with the resistor R_p which represents the leakage current flowing through the dielectric materials between the capacitor plates. If the dielectric were perfect, the leakage current would be zero because the conductivity of the dielectric would be zero and R_p would then be infinite; however, all physical dielectric materials have a nonzero conductivity which makes them to some extent lossy. A factor of merit for a capacitor is the dissipation or loss factor D which is defined as the ratio of the displacement current to the conduction or leakage current and is given by

$$D = \frac{1}{\omega C R_P} \tag{7.32}$$

which is a dimensionless quantity. For a perfect capacitor, the dissipation factor would be zero.

All capacitors have a maximum voltage rating that must not be exceeded. Voltages greater than this value will break down the dielectric materials between the plates and the resulting arc may well irreparably damage this material. All capacitors have a voltage rating and nominal capacitance value stamped on them.

The capacitance of an ideal parallel plate capacitor (neglecting field fringing and dissipation in the plates) is given by

$$C = \epsilon \frac{A}{d} \tag{7.33}$$

where ϵ is the permittivity of the dielectric between the plates of the capacitor, A is the area of the each plate, and d is the separation between the capacitor plates. Qualitatively, this equation can be used for other geometric shapes in that to increase the capacitance either the dielectric constant of the material between the plates must be increased, the area increased, or the separation between the plates must be decreased. Capacitors differ in their geometric shapes and the material used as the dielectric. Some of the more common types of capacitors are mica, ceramic, paper, plastic-film, and electrolytic.

Mica capacitors are fabricated by placing a strip of mica between two metal foils that act as conductors. Mica is a high dielectric constant material that can be formed into sheets that are as thin as 10^{-4} inches. These can then be rolled like a rug and inserted into a cylindrical package. Typical values for the parasitic elements of a mica capacitor are: $L_s = 0.52 - 25\ nH$, $R_s = 0.1 - 47\ \Omega$, and $R_p \geq 700\ M\Omega$. The capacitance range with these capacitors is 5 pF to 0.01 μF and the breakdown voltage ranges from 500 V to 10 kV. They are intended for general use and have tolerances no better than 5%. If the metal foil is silver, then these are known as silver mica capacitors.

Ceramic capacitors are made by placing a ceramic disc between a coating of metal. Two different types of ceramic materials are available: low-loss, low-permittivity and high-permittivity. The low-loss types have very small leakage currents while the high-permittivity types have a very high capacitance per unit volume. Tolerance is poor with 10 to 20% being typical. Typical parasitic values are 1 to 30 nH for L_s, 5 $m\Omega$ to 27 Ω for R_s, and 5 $G\Omega$ or greater for R_p. Breakdown voltages as high as 1 kV can be obtained. These capacitors are cheap and reliable and are used when precision is not a high priority.

Paper capacitors are cheap and widely used. They are made by rolling a sandwich of metal and paper sheets impregnated with oil, wax, or plastic into a cylindrical tube. They have high leakage currents and poor tolerances with parasitic parameters in the range: L_s = 6 to 160 nF, R_s = 1 to 16 Ω, and $R_p \simeq 100\ M\Omega$. Typical capacitance values are from 500 pF to 50 μF.

Plastic-film capacitors are fabricated in the same manner as paper capacitors except that plastic films such as mylar, teflon, or polyethylene are used as the dielectrics. They have lower leakage currents but are more expensive than paper capacitors. Typical parasitic element values are: L_s = 8 to 50 nH, R_s = 0.16 to 3.2 Ω, and $R_p \simeq 10\ G\Omega$.

Electrolytic capacitors are used when a large capacitance is desired in a small volume. These capacitors are polarized and must be inserted into the circuit with the proper polarity or they will be destroyed. Capacitance values from 1 μF to thousands of μFs are typical. Leakage currents are large and tolerances are poor. Typical parasitic element values are: L_s = 2 to 100 nH, R_s = 0.03 to 100 Ω, and $R_p < 1\ M\Omega$. Voltage ranges are from 6 V to 450 V. These capacitors are found in the filter section of DC power supplies and as coupling and bypass capacitors in electronic circuits.

Electrolytic capacitors are constructed using either aluminum or tantalum as the plates. The aluminum electrolytic capacitor is shown in Fig. 7.10. One of the two electrodes or plates is coated with a layer of aluminum oxide. Between the two plates is an electrolytic solution soaked into paper. This solution is

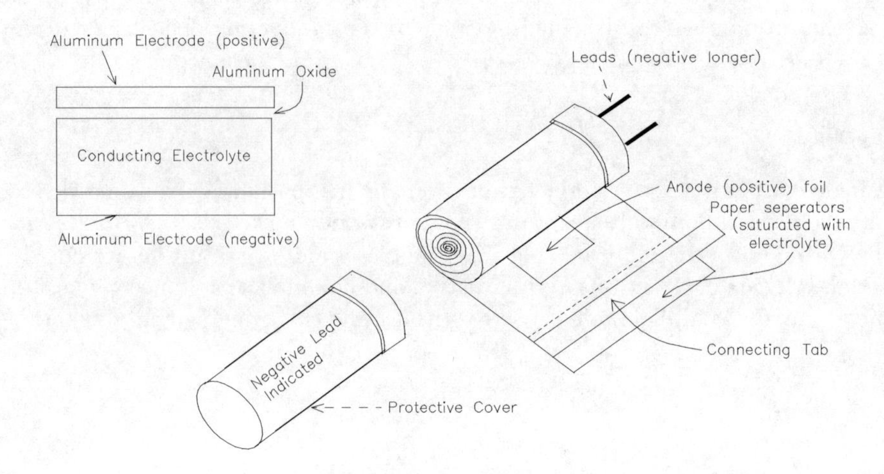

Figure 7.10: Aluminum electrolytic capacitor.

conducting and serves as an extension of the plate that is not coated with aluminum oxide. This results in a capacitor in which the plate separation is the thickness of the oxide layer which has a very large permittivity which results in a very large capacitance. If this type of capacitor is not properly biased, i.e. the + and − terminals are reversed, the oxide layer will be destroyed and the capacitor destroyed.

Variable capacitors are made by placing parallel plates so that the area between them can be varied by rotating a shaft. The dielectric that is used is air. Such passive variable capacitors were once an integral element in every communications receiver but have since been supplanted by electronic variable capacitors.

7.3.3 Inductors

An ideal inductor is a circuit element for which the voltage drop across it is given by a constant, known as the inductance, times the time rate of change of the current flowing through it. No physical circuit component exactly posses this characteristic. All physical inductors possess some parasitic resistance and capacitance and, depending on the core material, nonlinear properties. [The core of an inductor is the substance on which the wire is looped around to make the inductor.] Of the three passive two terminal circuit elements, inductors are the least well behaved and the most difficult to model.

Shown in Fig. 7.11 is the lumped circuit model for a linear physical inductor. At sufficiently high frequencies a linear physical inductor behaves like a capacitor. At low frequencies the linear physical inductor behaves like an ideal resistor in series with an ideal inductor. The boundary frequency is set by the resonant frequency between the inductor and the parallel capacitance. Air core inductors are well modeled as linear inductors.

Many inductors are constructed with metal, ferrite, or ceramic cores to increase the inductance per unit volume. These types of core materials are conducting and, therefore, have currents, known as eddy currents, induced in them by the time varying magnetic field produced by the inductor which results in power being dissipated in them. [An eddy current in a stream is a swirling current away from the main channel.] Although these types of inductors are nonlinear, for low currents levels they may be approximated as linear inductors in parallel with a resistor R_p which models core losses which is in turn in series with a resistor R_s which represents the resistance of the wire from which the inductor is made. Although many different types of materials are used as core materials, for simplicity the term iron core inductor will be used to describe this class of inductors.

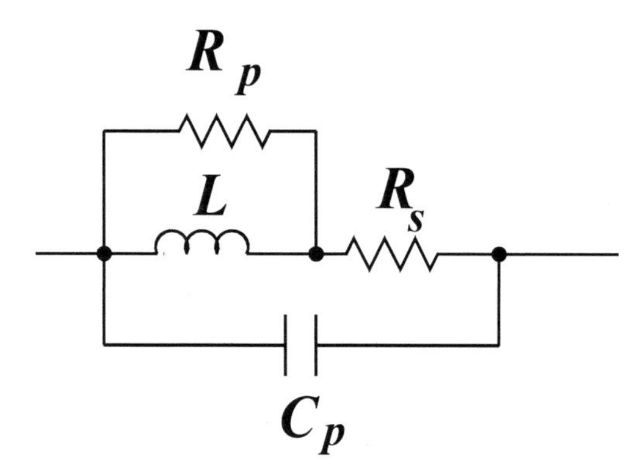

Figure 7.11: Lumped-circuit model for a linear physical inductor.

Inductors constructed with iron cores are nonlinear and the core saturates at high currents. Such inductors exhibit hysteresis which is not possible with linear circuit elements. Linear circuit models such as shown in Fig. 7.11 do not exactly apply although this circuit may be modified to approximate the nonlinear inductor behavior. The circuit symbol for an iron core inductor is shown in Fig. 7.12 which is just two parallel lines next to the symbol for a linear inductor.

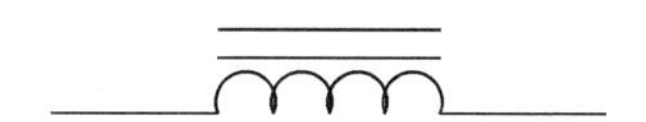

Figure 7.12: Circuit symbol for iron core inductor.

Shown in Fig. 7.13 is a representation of an iron core inductor for which N turns of wire have been wrapped around a toroidal iron core. The magnetic flux which is produced by the N turns of wire is essentially confined to the toroid. The resistor R_s represents the resistance of the N turns of wire and the voltage $e(t)$ represents the impressed voltage on this physical iron core inductor. The voltage $v_L(t)$ is given by Faraday's law

$$v_L(t) = N\frac{d\Phi(t)}{dt} \tag{7.34}$$

where $\Phi(t)$ is the magnetic flux passing through the toroid. In an iron core inductor, the flux is a nonlinear function of the current flowing through the inductor.

A typical plot of the magnetic flux as a function of the current for a physical inductor is shown in Fig. 7.14. (For a linear inductor this plot would be a straight line passing through the origin with a slope L/N.) This plot is known as a hysteresis curve for which the flux lags behind the current. (The term "hysteresis" comes from the Greek word "hysterein" which mean to lag.) The flux is not only a function of the magnitude of the current but whether the current is increasing or decreasing as well. The arrows indicate the portions

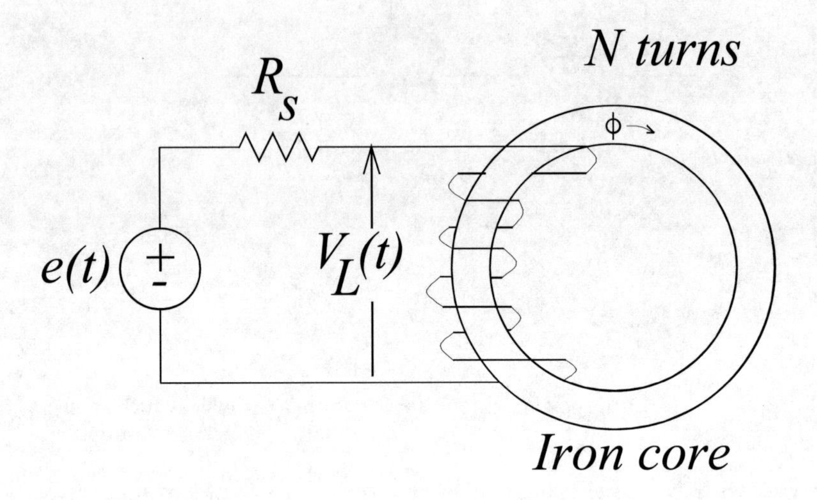

Figure 7.13: Iron core inductor.

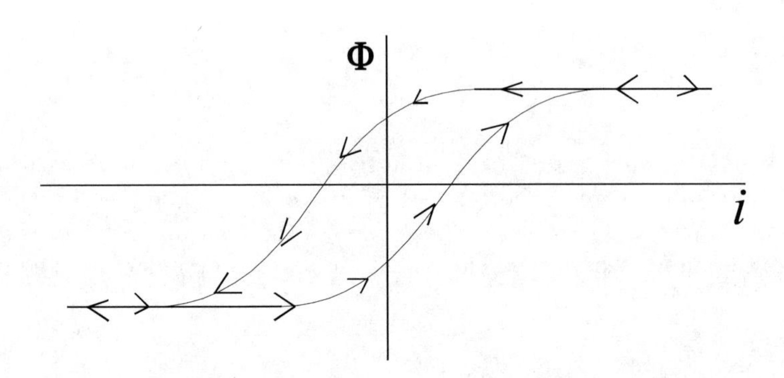

Figure 7.14: Magnetic flux magnetizing current for iron core inductor.

of the plot that apply for increasing and decreasing current. (The upper portion is for decreasing current and the lower for increasing current.) For sufficiently large currents, the core saturates when $\Phi = +\Phi_s$, i.e. increasing the current does not increase the flux once the core saturates. This hysteresis effect is a result of the magnetic domains in the core material being rotated in the presence of the time varying current flowing through the inductor.

The voltage $v_L(t)$ can be expressed as

$$v_L(t) = N\frac{d\Phi}{di}\frac{di(t)}{dt} \tag{7.35}$$

which means that the voltage $v_L(t)$ is given by the product of N, the number of turns of wire on the inductor core; $\frac{d\Phi}{di}$, the slope of the hysteresis curve at current $i(t)$; and $\frac{di(t)}{dt}$, the time rate of change of the current flowing through the inductor at time t.

Eq. 7.35 discloses that once the inductor core saturates, i.e. $\Phi = +\Phi_s$ and $\frac{d\Phi}{di} = 0$, the voltage $v_L(t) = 0$. Also if $v_L(t)$ is sinusoidal, then the current won't be sinusoidal and conversely, if the current is sinusoidal, then the voltage will not be sinusoidal. Therefore, if the voltage drop across R_s is negligible, when the iron core inductor is excited with a sinusoidal voltage source, the current will be nonsinusoidal, and when the excitation is a sinusoidal current source, the inductor voltage will be nonsinusoidal.

7.3.4 Transformers

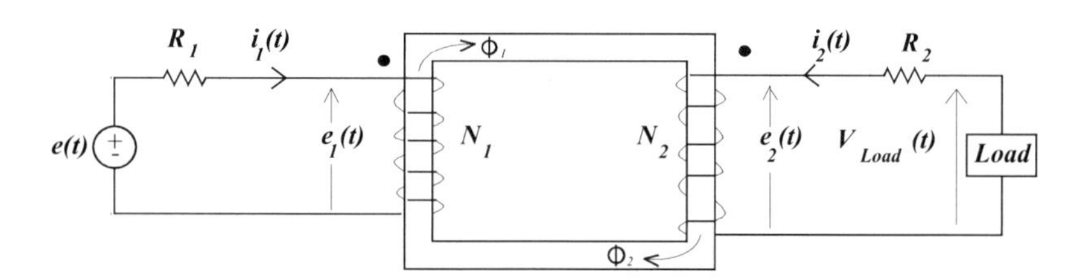

Figure 7.15: Two winding transformer.

When two or more coils or windings are placed in physical proximity, a transformer results. Shown in Fig.7.15 is an iron core transformer onto which two coils known as the primary and secondary have been wound. The dots indicate the side of the coils that the current is assumed to flow into. The winding connected to the circuit with $e(t)$ is known as the primary and the winding connected to the load is known as the secondary. Application of Kirchhoff's and Faraday's laws to this circuit yields

$$\begin{aligned} e(t) &= R_1 i_1(t) + e_1(t) = R_1 i_1(t) + N_1\frac{d\Phi_1(t)}{dt} \quad (a) \\ e_2(t) &= N_2\frac{d\Phi_2(t)}{dt} = -R_2 i_2(t) + v_{Load}(t) \quad (b) \end{aligned} \tag{7.36}$$

where Φ_1 is the magnetic flux linking coil 1, the primary and Φ_2 is the magnetic flux linking coil 2, the secondary. In general these two fluxes will not be equal because of leakage; they are related by $\Phi_2 = k\Phi_1$ where k is a constant known as the coefficient of coupling which lies in the range $0 \le k \le 1$. The fluxes $\Phi_1(t)$ and $\Phi_2(t)$ are functions of both currents $i_1(t)$ and $i_2(t)$.

If the coefficient of coupling, k, is 1, then the same flux links both coils. In this case

$$e_2(t) = \frac{N_2}{N_1} e_1(t) \tag{7.37}$$

which means that the primary and secondary voltages are in phase and differ in magnitude. If $N_2 > N_1$, then $|e_2(t)| > |e_1(t)|$ and this is known as a step-up transformer. Conversely, if $N_2 < N_1$, then $|e_2(t)| < |e_1(t)|$ and this is known as a step-down transformer. If $N_2 = N_1$, then $e_1(t) = e_2(t)$ and this is known as an isolation transformer, i.e. its only function is to provide electrical isolation between the circuit with coil 1 and the circuit with coil 2. The two coils are electrically isolated because there is direct path for current to flow between the two coils, i.e. they are magnetically coupled.

If the excitation, $e(t)$, is a sinusoidal and the load is a capacitor, C and is the frequency of excitation $f >> 1/(2\pi R_2 C)$, then the capacitor voltage, $v_{Load}(t)$, is given by

$$v_{Load}(t) = \frac{1}{RC} \int_{-\infty}^{t} e_2(u)du \quad = \frac{k}{RC} \frac{N_2}{N_1} \int_{-\infty}^{t} e_1(u)du \quad = \frac{k}{RC} N_2 \Phi_1(t) \tag{7.38}$$

which means that the capacitor voltage is proportional to the magnetic flux linking coil 1. Since the voltage across resistor R_1 is proportional to the current flowing into the primary coil, this provides a method of obtaining the $\Phi - i$ curve for coil 1 with an instrument such as an oscilloscope. The capacitor voltage is displayed on the vertical axis of the oscilloscope and the voltage across resistor R_1 is displayed on the horizontal axis.

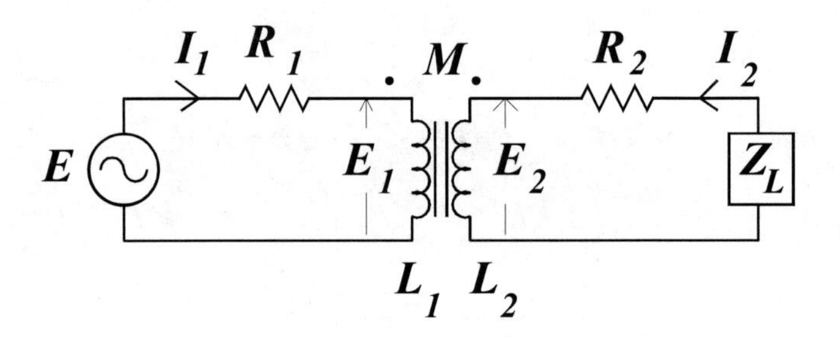

Figure 7.16: Linear model for two winding transformer.

If the currents are sufficiently small so that the core does not saturate, then the two winding transformer may be modeled as shown in Fig. 7.16. Coil 1 is modeled as an linear inductor L_1 and coil 2 as a linear inductor L_2. The two coils are connected via a mutual inductance, M. The mutual inductance, M, is defined by the equation

$$e_2(t) = M \frac{di_1(t)}{dt} \bigg|_{i_2(t)=0} \tag{7.39}$$

which is the voltage induced in coil 2 by the time-varying current in coil 1.

For sinusoidal excitation Eqs. 7.36 become

$$\begin{aligned} \bar{E} &= (R_1 + j\omega L_1)\bar{I}_1 + j\omega M \bar{I}_2 \quad (a) \\ j\omega M \bar{I}_1 &= -(R_2 + j\omega L_2 + \bar{Z}_L)\bar{I}_2 \quad (b) \end{aligned} \tag{7.40}$$

which assumes that the physical transformer is reasonably linear. The coefficient of coupling, k, is given by

$$k = \frac{M}{\sqrt{L_1 L_2}} \tag{7.41}$$

which is the ratio of the mutual inductance of the two coils to the geometric mean of their self inductances.

7.4 Measurement of R, C, and L

The accurate measurement of the component values of resistors, capacitors, and inductors is one of the classical electrical and computer engineering laboratory tasks. Numerous techniques have been developed such as voltmeter-ammeter, impedance bridges, and various types of digital meters.

7.4.1 Voltmeter-Ammeter

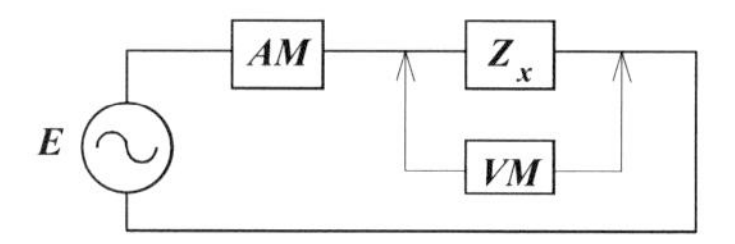

Figure 7.17: Voltmeter-ammeter impedance measurement.

The voltmeter-ammeter technique for impedance measurement is shown in Fig. 7.17. The experimental measured value of the impedance is just

$$|\bar{Z}| = \frac{V}{I} \tag{7.42}$$

where V is the reading of the voltmeter and I is the reading of the ammeter. If a resistor is being measured, a dc source is used and the meters are dc meters and V and I are dc quantities.

For the measurement of the capacitance of a capacitor, assuming that the parasitic components can be neglected, the experimentally measured impedance is given by

$$|\bar{Z}| = \frac{1}{\omega C} = \frac{V}{I} \tag{7.43}$$

where V is the rms value of the voltage as measured by the voltmeter and I is the rms value of the current as measured by the ammeter. Since the frequency of the source is known, the experimentally measured value of the capacitance can be solved for using Eq. 7.43.

Most inductors have nonnegligible series resistance due to the windings of the wire used to implement the physical inductor. The first step is to measure the series resistance by using a dc source in Fig. 7.17. Once the value of R_s has been ascertained, the source is changed to ac and the impedance is measured as

$$|\bar{Z}| = \sqrt{R_s^2 + \omega^2 L^2} = \frac{V}{I} \tag{7.44}$$

where V and I are the rms values of the voltage and current read from the meters. The experimentally measured value of the inductor may be determined from Eq. 7.44 from the measured values of V, I, and R_s and using the value of the frequency of the source.

There is inherent error is the measurement of circuit components using the voltmeter-ammeter technique due to meter loading. As shown in Fig. 7.17, the current that the ammeter actually measures is the current flowing through the parallel combination of the voltmeter and the unknown circuit component. However, if the impedance $|\bar{Z}_x|$ is small compared to the input impedance of the voltmeter, this technique provides accurate values.

7.4.2 Oscilloscope Measurement

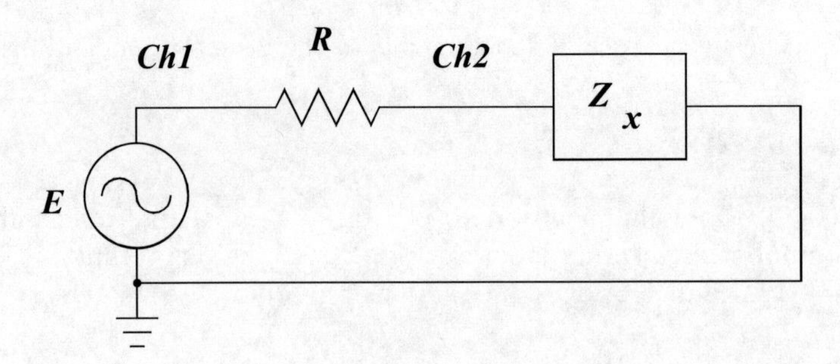

Figure 7.18: Oscilloscope measurement of impedance.

Measurement of voltage and current with meters provides no information about the phase relationship between these circuit component variables. If an oscilloscope is used to make these measurements as shown in Fig. 7.18, both the magnitude and angle of the impedance may be measured. The voltage on $Ch2$ is the impedance voltage. If the difference between the $Ch1$ and $Ch2$ voltage is displayed (for analog oscilloscopes the ADD function is used and $Ch2$ is inverted and with digital oscilloscopes the $MATH$ functions are used to display $Ch1 - Ch2$), this difference voltage is the current scaled by the resistor R. The technique is hampered by the same instrumentation loading problem as the voltmeter-ammeter technique.

7.4.3 Impedance Bridges

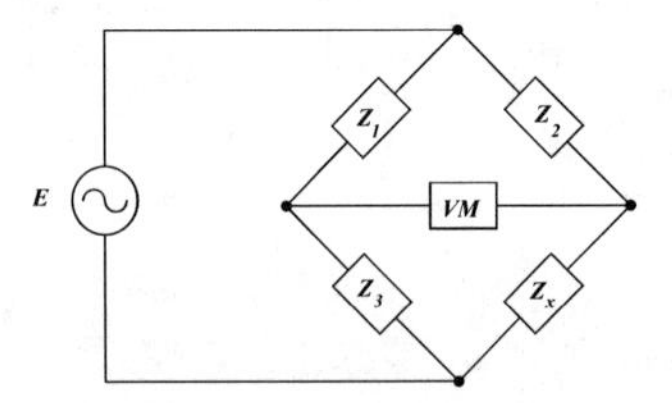

Figure 7.19: Impedance bridge.

Measurement of component values using an impedance bridge is an improvement over the voltmeter-ammeter technique in that there is no error due to meter loading. Shown in Fig. 7.19 is a basic diagram of an impedance bridge. The unknown impedance is $\bar{Z}_x$ and $\bar{Z}_1$, $\bar{Z}_2$, and $\bar{Z}_3$ are known impedances at least one of which can be varied. The measurement is made with a sinusoidal source for capacitors and inductors and with a DC source for resistors. One of the know impedance is varied until the VM reads zero. Once this occurs the unknown impedance is given by

$$\bar{Z}_x = \frac{\overline{Z}_2\overline{Z}_3}{\overline{Z}_1} \tag{7.45}$$

which is a measurement that is not corrupted by meter loading since the voltage drop across the voltmeter is zero. Numerous ingenious bridges have been developed for the measurement of R, C, and L and their dominant parasitic component. These instruments can also measure the dissipation factor for a capacitor and the quality factor for an inductor. However, except for specialized applications, these impedance bridges have been supplanted by digital instruments.

7.4.4 Digital Multimeter

The most common method of resistance measurement is the digital multimeter. The unknown resistor is connected to the instrument's input terminals and the voltage developed across it is precisely measured by the digital voltmeter section.

Most digital multimeters have both a "2 *Wire*" and "4 *Wire*" resistance measurement capability. The "2 *Wire*" resistance measurement is normally adequate. Only when the resistance to be measured is on the order of the resistance of the leads being used to make the connection should a "4 *Wire*" resistance measurement be necessary.

With "4 *Wire*" resistance measurement, two additional wires are used to sense the voltage across the unknown resistor. They are connected from the "4 *Wire*" inputs on the instrument to each side of the resistor. These additional two wires do not carry any current and, therefore, the voltage drop in the leads through which the current flows does not corrupt the measurement.

7.4.5 *RCL* Meters

An *RCL* meter is a digital instrument which automatically measures component value and displays the result. From the measurements that are made the instrument decides whether it is a resistor, capacitor, or inductor and displays this result along with component value.

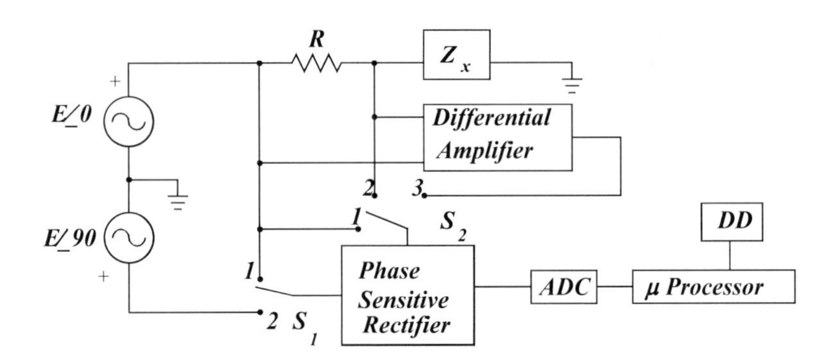

Figure 7.20: Block diagram of a digital *RCL* meter.

Shown in Fig. 7.20 is a block diagram of a digital *RCL* meter. The unknown impedance is $\bar{Z}_x$ and the resistor R is used to sense the current flowing through it. Two different oscillators are used that are 90° out of phase. The microprocessor controls the settings of the switches S_1 and S_2 (the control lines are not shown). The differential amplifier and phase sensitive receiver can be considered to have infinite input impedances.

The voltage at the output of the differential amplifier is proportional to the current flowing through $\bar{Z}_x$. There are two different inputs to the phase sensitive rectifier; they are determined by the settings of switches S_1 and S_2.

Five measurements are made. First both switches are in position 1 for which the output of the phase sensitive rectifier is proportional to $|\bar{E}|$ and this value is converted by the analog-to-digital converter (*ADC*) to a binary form and stored in the microprocessor. Next Switch 1 is set to position 1 and Switch 2 is set to position 2 and the output of the phase sensitive rectifier is proportional to the real part of the impedance voltage which is then digitized and stored in the microprocessor. Switch 1 is then set to position 2 and the imaginary portion of the impedance voltage is obtained. The same measurements are made with Switch 2 set to position 3 which yields the real and imaginary parts of the impedance current.

Once the real and imaginary parts of the impedance voltage and current have been measured, the microprocessor computes the component type and value and displays the result in the digital display (DD). Measurements are repetitively made and the results updated.

Both impedance bridges and RCL meters determine both a series and parallel representation for capacitors and inductors. These representations are valid only at the frequency at which the measurement is made and are not equivalent circuits. Both the series and parallel results are presented for computational convenience.

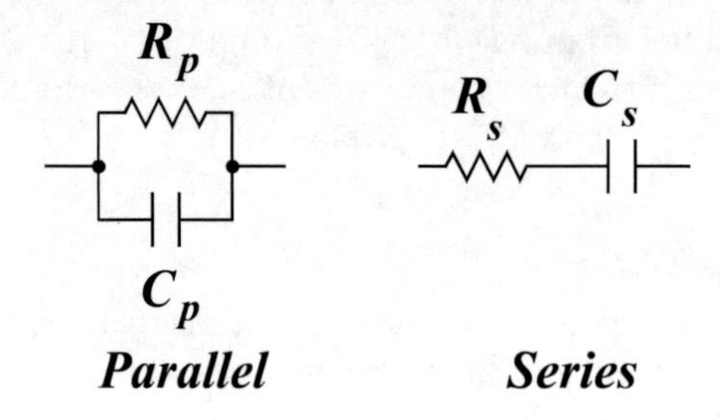

Figure 7.21: Series and parallel representation of a capacitor.

Shown in Fig. 7.21 is the series and parallel representation for a capacitor. Once one representation has been obtained, the other can be obtained from

$$C_s = (1 + D^2)C_p \tag{7.46}$$

$$R_s = \frac{D^2}{1 + D^2} R_p \tag{7.47}$$

$$D = \frac{1}{\omega R_p C_p} = \omega R_s C_s \tag{7.48}$$

which are obtained by equating the impedances of the two representations. The parallel representation is valid over a wide frequency range, but the series representation is valid only at the frequency of measurement, i.e. R_s and C_s will be functions of frequency.

Shown in Fig. 7.22 is the parallel and series representation for an inductor. Once one representation has been obtained, the other may be obtained from

$$R_p = (1 + Q^2)R_s \tag{7.49}$$

$$L_p = \frac{1 + Q^2}{Q^2} L_s \tag{7.50}$$

$$Q = \frac{\omega L_s}{R_s} = \frac{R_p}{\omega L_p} \tag{7.51}$$

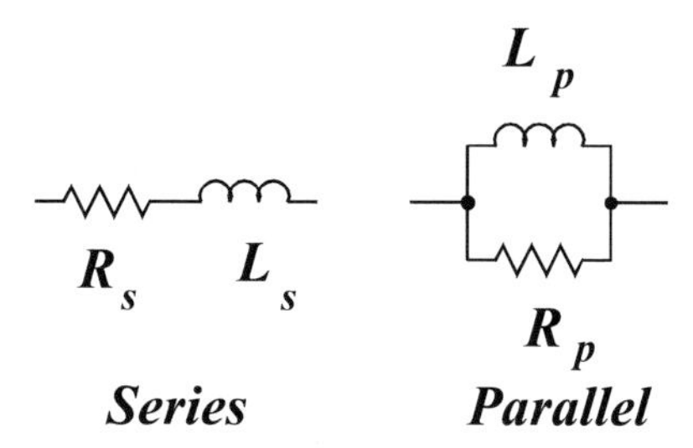

Figure 7.22: Series and parallel representation for an inductor.

which are obtained by equating the impedance of the series and parallel representations. The series representation is valid over a wide frequency range while the parallel representation is valid only at the measurement frequency, i.e. both R_p and L_p will be functions of frequency.

It should be borne in mind that the series and parallel representation for inductors and capacitors are not component models that are valid at all frequencies. The parameter values are valid at only the frequency at which the measurement is made.

7.5 Equipment

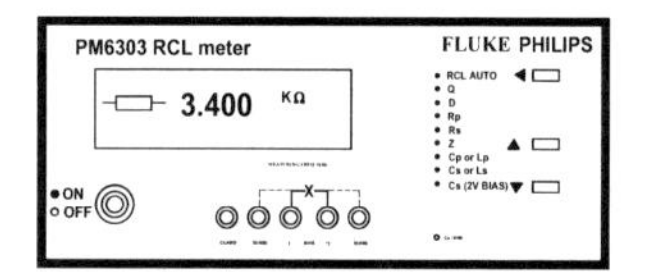

Figure 7.23: Fluke/Philips PM 6303 RCL meter.

Fluke/Philips PM 6303 RCL Meter. Shown in Fig. 7.23 is the **Fluke/Philips PM 6303 RCL** meter. This operation of this instrument was described in the previous section under *RCL* meters. It displays the type and component value in a *LCD* display. In the *AUTO* setting, the instrument automatically determined the component type and value. Quality factors for inductors and dissipation factors for capacitors can be determined and both series and parallel representations are available.

This instrument performs measurements at a frequency of 1 kHz. No other frequencies are permitted due to method that is used to produce the reference voltage. Namely, a crystal oscillator is used to set the frequency of the reference voltage and the frequency of operation of the microprocessor and analog-to-digital converter.

Sense terminal are provided on this instrument for the measurement of very low impedance values. They have the same function as those on the digital multimeters.

7.6 SPICE

SPICE can be used to perform the transient analysis of a first-order circuit. Only positive values of time are permitted and the initial value of the capacitor voltage or inductor current must be specified.

Digital computers solve differential equations by transforming them into difference equations for which the time variable becomes discrete. The accuracy of the solution may be increased by decreasing the time step size but this also increases the time required for the computer to solve the problem. Therefore, a compromise between accuracy and computation time must be made to perform a transient analysis.

7.6.1 Square Wave

The SPICE input deck code required to obtain the transient solution for a circuit energized by a 20 V peak-to-peak symmetric square wave with a dc level of 0 V and a frequency of 1 kHz is shown below. The `PULSE` function is used to specify the square wave. The arguments of `PULSE` are: initial voltage, peak voltage, initial delay time, rise time, fall time, pulse width, and pulse period.

The square wave is connected across the series combination of a 3 $k\Omega$ resistor and a 0.1 μF capacitor. The time constant $\tau = RC = 0.3\ ms$. The initial value of the capacitor voltage is obtained from Eq. 7.13

$$v_c(0) = -A \tanh\left(\frac{T}{4\tau}\right) = -6.8226\ V \tag{7.52}$$

where $A = 10\ V$ and $T = 1/f = 1\ ms$.

The arguments of the control line `.TRAN` are `print_step`, `final_time`, `results_delay`, `step_ceiling`, and `UIC`. The `print_step` is the time separation between which points would be printed to a file. The `final_time` is the final time for which the analysis is performed. The `results_delay` is the first time for which the results will be printed. The `step_ceiling` is a very important parameter that sets the time increments for the difference equation. Finally, `UIC` stands for Use the Initial Condition; if this parameter is omitted the initial condition will be ignored.

```
SQUARE WAVE EXCITATION
VI 1 0 PULSE(-10 10 0 1F 1 F 0.5M 1M)
R 1 2 3K
C 2 0 0.1U IC=-6.8226
.TRAN 0.1U 2M 0 0.02M UIC
.PROBE
.END
```

The capacitor voltage is the continuous trace while the capacitor current is the trace that is discontinuous every $T/2$ seconds in Fig. 7.24. The current is plotted in mA while the capacitor voltage is in Volts.

7.6.2 Triangular

The SPICE input deck code required to analyze the first-order RC circuit with triangular wave excitation is given below. The SPICE function PWL is used to specify two cycles of the triangular wave. The triangular wave has a peak-to-peak value of 20 V, a dc level of zero, and a frequency of 1 kHz. The initial value of the capacitor voltage is obtained as

$$v_c(0) = -A\left(1 - \frac{4\tau}{T}\tanh\left[\frac{T}{4\tau}\right]\right) = -1.8129\ V$$

from Eq. 7.20.

```
TRIANGULAR WAVE EXCITATION
VI 1 0 PWL(0 -10 0.5m 10 1m -10 1.5m 10 2m -10)
R 1 2 3K
C 2 0 0.1U IC=-1.8129
.TRAN 10U 2M 0 10U UIC
.PROBE
.END
```

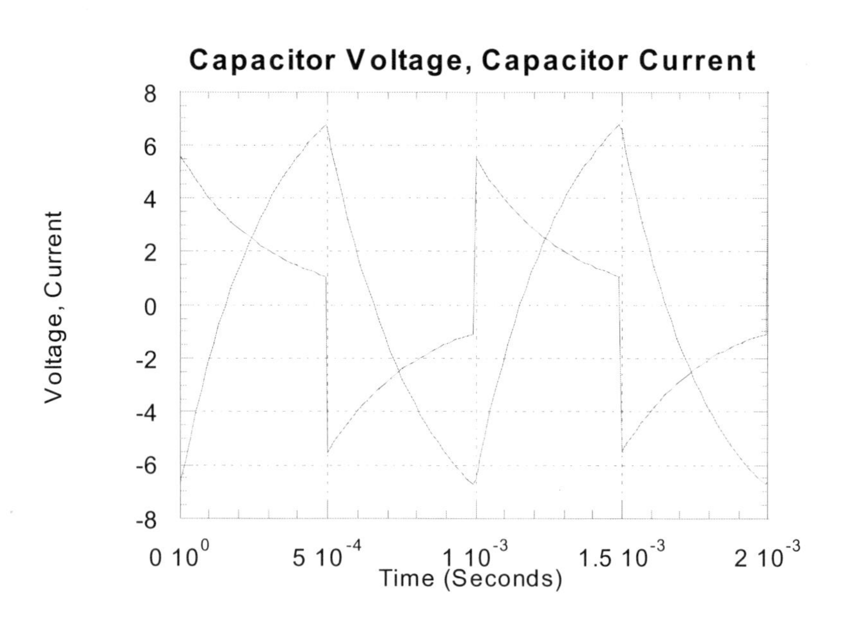

Figure 7.24: Transient response of first-order RC circuit.

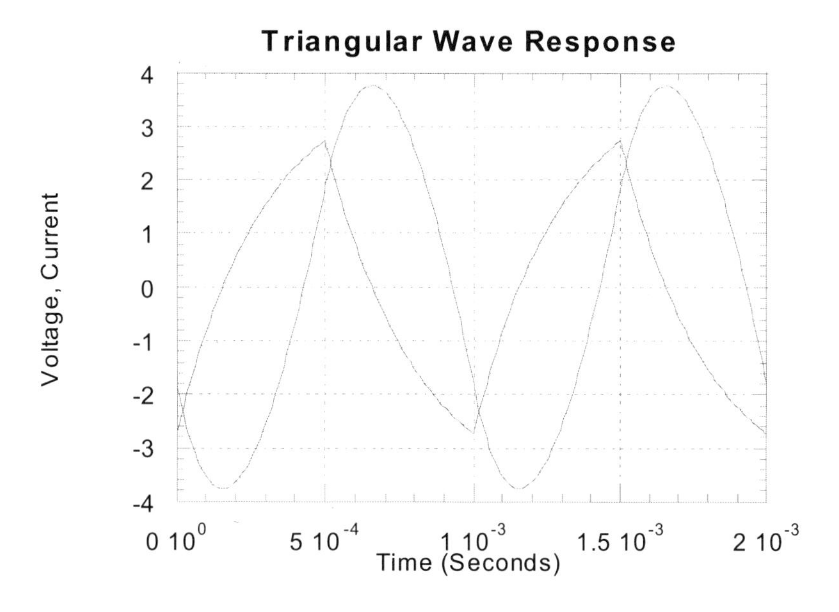

Figure 7.25: Transient response with triangular wave excitation.

The capacitor voltage is the one with the continuous first derivative or slope in Fig. 7.25. The capacitor voltage is plotted in volts and the capacitor current in mA.

7.6.3 Ramp

The SPICE input deck code required to analyze the first-order *RC* circuit with ramp wave excitation is given below. The ramp wave is a 20 Volt peak-to-peak wave with a dc level of zero and a frequency of 1 kHz. The initial value of the capacitor voltage is obtained from Eq. 7.23

$$v_c(0) = \frac{A}{\sinh(\frac{T}{2\tau})} - 2A\frac{\tau}{T} = -2.083\ V$$

```
RAMP WAVE EXCITATION
VI 1 0 PWL(0 0 0.5M 10 0.050001M -10 1.5M 10 1.500001M -10 2M 0)
R 1 2 3K
C 2 0 0.1U IC=-2.083
.TRAN 10U 2M 0 10U UIC
.PROBE
.END
```

The plot of the capacitor current and voltage are given in Fig. 7.26. The voltage is the trace that begins at −2.083 V.

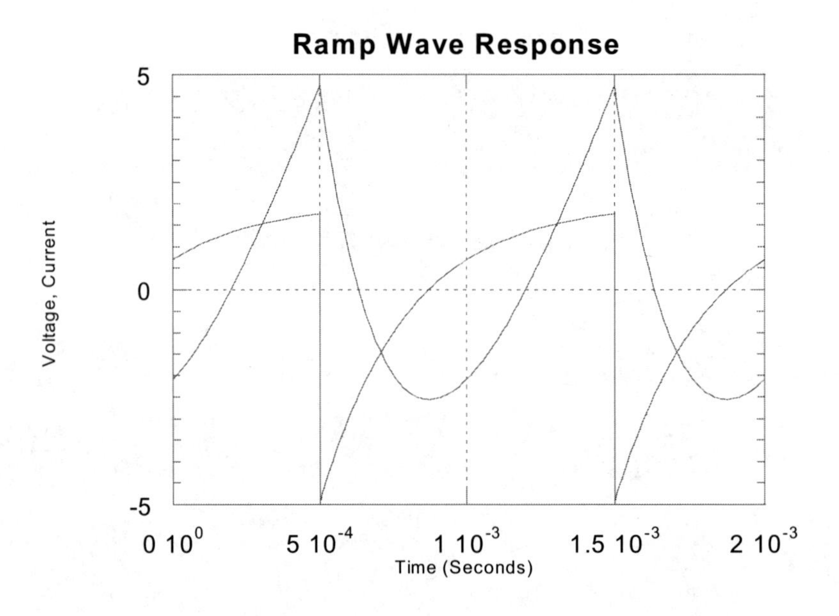

Figure 7.26: Ramp wave response.

Note that for each of the three excitations, once the voltage plot is obtained the current is given by $i(t) = Cdv_c(t)/dt$. This facilitates determines which trace is which.

7.7 Procedure

7.7.1 Parameter Measurement.

Resistance Measurement

The laboratory instructor will provide one capacitor, one inductor, and three resistors, 100 Ω, R_1, and R. Measure and record the actual value of the resistance of the 100 Ω resistor as well as the resistors R_1 and R with the **Agilent 34401A** digital multimeter set to "Ω 2 *W*" resistance measurement. Measure and record the resistance of the inductor and the capacitor using the "Ω 2*W*" mode. If any of the resistance values are less than 10 Ω, repeat the measurement using the "Ω 4*W*" mode. To make "Ω 4*W*" resistance measurements connect two additional wires from the banana jacks labeled "Ω 4*W Sense*" to the ends of the components in addition to the two wires connected to the "*HI Input*" and "*LO Input*" inputs and actuate the button labeled "Ω 4*W*" by pushing the blue *Shift* button and then blue "Ω 4*W*".

Repeat the resistance measurements for the resistors 100 Ω, R_1, and R using the **Fluke/Philips 6303 RCL** meter. Connect two wires from the banana jacks labeled "*X*" to the ends of the resistors. For resistances less than 10 Ω, connect two additional wires from the sense outputs of the meter to the ends of the component. Set the function to *RCL AUTO*.

Circuit Component	Nominal Resistance	Agilent 34401A	Fluke/Phillips RCL
100 Ω	100 Ω		
R			
R_1			
L	0		
C	∞		

Table 7.1: Resistance Measurement

Capacitance Measurement

Connect the capacitor to the inputs labeled "*X*" and set the function to "*RCL AUTO*" and record the capacitance value indicated. Set the function to "R_p" and record the resistance value indicated. Set the function to "R_s" and record the resistance value indicated. Set the function to C_p and record the capacitance value indicated. Set the function to C_s and record the capacitance value indicated. Set the function to D and record the value indicated.

Inductance Measurement

Connect the inductor to the inputs labeled "*X*". Set the function to *RCL AUTO* and record the inductance value indicated. Set the function to R_s and record the resistance value indicated. Set the function to R_p and record the resistance value indicated. Set the function to L_p and record the inductance value indicated. Set the function to L_s and record the inductance value indicated. Set the function to Q and record the value indicated.

Circuit Component	Measured Value
RCL Auto	
C_p	
R_p	
D	
C_s	
R_s	

Table 7.2: Capacitance Measurement

Circuit Parameter	Measured Value
RCL Auto	
L_s	
R_s	
Q	
L_p	
R_p	

Table 7.3: Inductance Measurement

7.7.2 Measurement of R_g for the Tektronix 3022B Function Generator

The purpose of this step is the measure the internal resistance of the source or function generator. The **Tektronix 3022B** function generator has specified internal, generator, or source impedance of $50\,\Omega$. This will be verified with a measurement in this section. It will be assumed that the impedance of both channels are the same so only measurements will be made on one.

The **Tektronix 3022B** function generator displays the amplitude of the voltage available at its output terminals. In order for the **Tektronix 3022 B** function generator to indicate the open circuit or Thévenin voltage in the display when amplitude is selected the output termination must be set to *HIGH Z*. The default or power-on setting is 50 Ω which would then result in a display one half of the Thévenin source voltage.

Terminating Impedance Adjustment for Function Generator

Turn on the **Tektronix 3022B** function generator and wait until it finishes its self-test power-on; this should only take a few seconds. Enable the CH1 output and set it for High Z. The function generator is now configured so that it will display the Thévenin source voltage when the amplitude display is selected.

Turn on the **Agilent 34401A** *DMM* and set it to measure an *AC* voltage. Set the function on the function generator to produce a sine with a *DC* offset of zero and a frequency of 1 kHz. Connect the output of the function generator to the input of the *DMM* using a *BNC* to banana plug lead. Set the frequency to 500 Hz and the amplitude of the output to approximately $5\,V_{rms}$ and record the value as V_g. Do not change the setting of the amplitude function for the rest of this step. Disconnect the leads from the function generator to the *DMM*.

Calculation of Internal Impedance of Function Generator

Assemble the circuit shown in Fig. 7.27. Use the *DMM* to measure the rms voltage, V, across the $100\,\Omega$ resistor. From this voltage measurement calculate the value of R_g. The value of R_g is given by

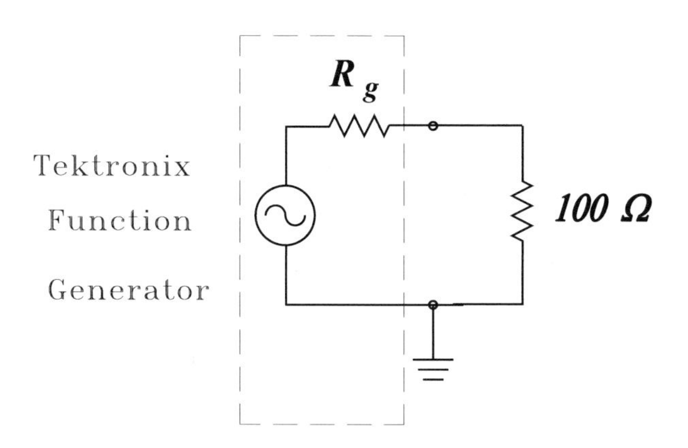

Figure 7.27: Measurement of R_g

$$R_g = 100 \times \left[\frac{V_g}{V} - 1\right] \tag{7.53}$$

and the actual resistance rather than the nominal resistance of 100 Ω resistor should be used in place of the "100" in this equation. This calculation will be used later in this experiment and should be made now.

$\mathbf{V}_g$	V	$\mathbf{R}_g$

Table 7.4: Tektronix Function Generator Resistance

7.7.3 Sine Wave Response

Break Frequency Calculation

(a) Calculate the break frequency

$$f_o = \frac{1}{2\pi\tau} = \frac{1}{2\pi RC} \tag{7.54}$$

using the actual component values measured in step 7.7.1.

Function Generator Adjustment

(b) Set the amplitude of the waveform produced by the **Tektronix 3022 B** function generator to 2 V peak-to-peak and the frequency to f_o.

Circuit Assembly

(c) Assemble the circuit shown in Fig. 7.28. $CH1+$ means that this is the point at which the red banana plug for the oscilloscope lead should be attached. The ground symbol (three parallel horizontal lines with decreasing width) means that this is the point at which the black banana plug for the scope lead should be attached (it is not necessary to connect both of the black banana plugs since they are connected to the shield on the BNC connectors which, when connected to the oscilloscope, are connected to the common power line ground on the AC cord which is also connected to the instrument chassis). The symbol $FG+$ means the red banana plug on the lead from $OUTPUT$ of the function generator.

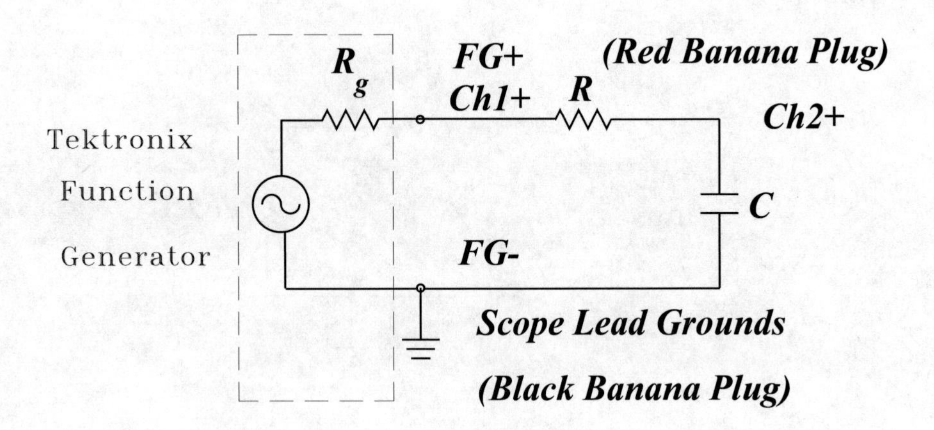

Figure 7.28: First Order Circuit

Voltage Peak-to-Peak	Volts/div Method	Measure Function Method
*Ch*1		
*Ch*2		

Table 7.5: Voltage Peak-to-Peak Measurement

Oscilloscope Adjustment

(d) Turn the **Tektronix 3012B** oscilloscope on and, after it boots, press *AUTOSET*. If the display is unstable, connect the Sync Output of the Function generator to the EXT TRIG input on the oscilloscope, press *TRIGGER Source* and switch the trigger source to EXT/10 and set the level to TTL

The waveform on $CH2$ is the capacitor voltage and the waveform on $CH1$ is the voltage across the output of the function generator which is also the voltage drop across the series combination of the resistor and capacitor.

Voltage Measurement

(e) Measure the peak-to-peak values of the sine waves on both channels. This is to be done in two ways. First, measure the number of vertical divisions that the waveform vertically spans and multiple by the $Volts/Div$ (the $Volts/Div$ are displayed in the upper left hand corner of the display). Second, press $MEASURE$ menu and scroll through the menus until peak-to-peak is found. The measurement is first selected for $CH1$ and then $CH2$.

Frequency Measurement

(f) Measure the frequency of the waveforms connected to the input to the oscilloscope. This will be done in two ways. First, measure the time interval corresponding to one period by measuring the number of major horizontals divisions that it spans and then multiple by $Time/div$ which is displayed in the upper right hand portion of the screen. Second, press MEASURE and then scroll through the menus until frequency is located.

Frequency	Time/Div	Measure Function Method
$Ch1$		

Table 7.6: Frequency Measurement

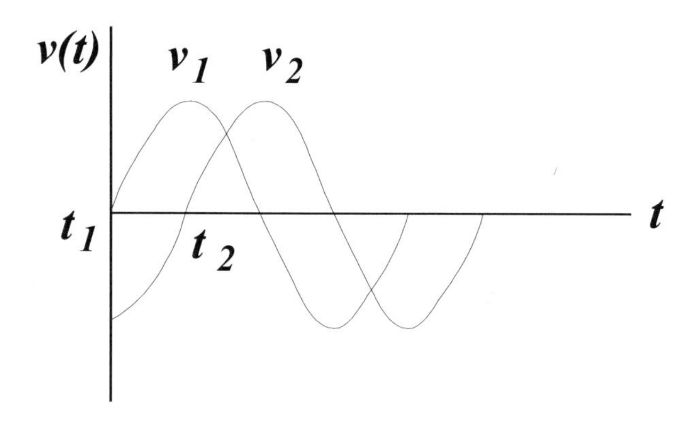

Figure 7.29: Phase shift measurement.

Phase Shift Measurement

(g) The phase shift between the waveforms connected to $CH1$ and $CH2$ will be measured in two different ways.

First, measure the time interval corresponding to the phase shift between the waveforms on $CH1$ and $CH2$ as shown in Fig. 7.29 . Adjust the vertical positions for $CH1$ and $CH2$ until the ground (small symbol on right hand portion of the screen) is at the same horizontal line such as the center. Turn on the cursors and select vertical bars. Position cursors $t1$ and $t2$ until they are aligned as shown in Fig. 7.29 with the rotary knob and SELECT button. [Remember the phase difference in degrees is given by $\phi = f \cdot (\Delta t) \cdot (360^o)$ where f is the frequency of the two waveforms.] Calculate and record the phase shift measured with the time cursors.

Print the display.

(h) Second, phase measurement is directly available as a measurement function. Press the *MEASURE* button, *Phase*, From $CH2$ to $CH1$, and finally OK Create Measurement. Compare with the result obtained with the previous methods.

(i) Change the frequency of the function generator to 10 f_o, press *AUTOSET* on the oscilloscope and repeat the measurement of the peak-to-peak voltages of the sine waves on $CH1$ and $CH2$ and the phase shift between them. Use either of the methods for the measurement.

(j) Change the frequency of the function generator to $f_o/2$ and repeat the measurement of the peak-to-peak voltages of the sine waves on $CH1$ and $CH2$ and the phase shift between them. Use either of the methods for the measurement.

Frequency	V_{pp1}	V_{pp2}	Phase Shift
f_o			
$f_o/2$			
$10f_o$			

Table 7.7: Voltage and Phase Shift Measurement

7.7.4 Square, Triangular, and Ramp Response of First Order *RC* Circuit

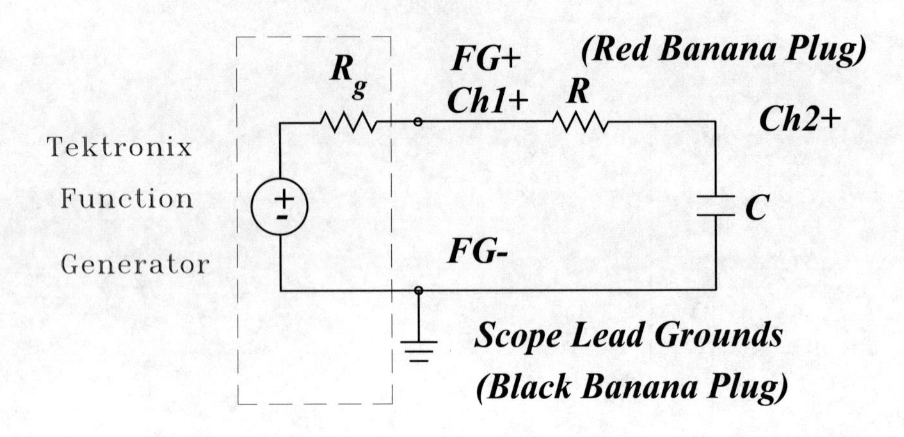

Figure 7.30: Square, Triangular, and Ramp Response

Square

(a) Assemble the circuit shown in Fig. 7.30 and use the **Tektronix 3022 B** function generator as the voltage source. Set the frequency of the function generator to f_o and the function to square. Connect the synch or TTL output of the function generator to the EXT TRG of the oscilloscope. Press *AUTOSET* and then switch the trigger source to *EXT*/10 and level to TTL(Press *TRIGGER*, *MENU*, *Source*, *Ext*/10,and then TTL).

Turn $CH1$ off. Do this by pressing the 1 button under the and then *OFF*.. Press MATH, Dual Wfm Math, and select the minus operator; this will display the difference of the waveforms connected to $CH1$ and $CH2$, i.e. the mathematical function $v_1(t) - v_2(t)$ where $v_1(t)$ is the voltage on $CH1$ and $v_2(t)$ is the voltage on $CH2$. Adjust the controls until a suitable display is obtained, i.e. nonoverlapping and a peak-to-peak value of several major divisions. When the MATH (red) button is displayed the SCALE and POSITION set this channel and when the $CH2$ button (blue) button is pushed these controls are for $CH2$..

The capacitor voltage is now being displayed on $CH2$ and the resistor voltage is the differential voltage $v_1(t) - v_2(t)$ which is, of course, proportional to the current.

Turn on the cursors. Switch the source to $CH2$ by pressing the blue $CH2$ button (switch between the cursors with the select button). Position the time and voltage cursors until they are as shown in Fig. 7.31. Print the display. Record any cursor settings that do not print.

Calculate the time constant from this experimental plot as

$$\tau = -\frac{t_2 - t_1}{\ln \dfrac{v_2 - A}{v_1 - A}} \tag{7.55}$$

where A is the peak value of the square wave (one-half of the peak-to-peak value of the square wave which is $2A$). Compare this with the theoretical time constant $\tau = RC$ and record both in the laboratory report.

It should be borne in mind that the time constant for this circuit is the total capacitance times the total resistance. The total resistance is the actual value of R plus the generator resistance R_g. The total capacitance with an oscilloscope lead connected across the circuit capacitor is given by the circuit capacitor's

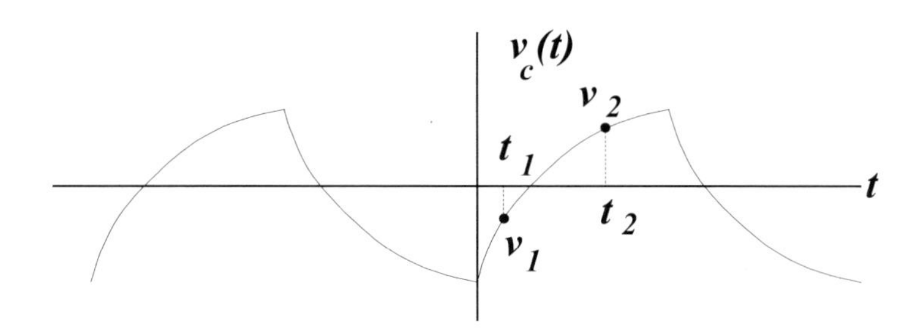

Figure 7.31: Capacitor voltage as a function of time.

capacitance plus the capacitance of the oscilloscope lead-input attenuator combination. The latter capacitance is normally on the order of picofarads and can be neglected if the circuit capacitance is sufficiently large. (It would be the capacitance of the input attenuator [13 pF for the **Tektronix 3012B** oscilloscope] plus the capacitance of the coaxial cable. The capacitance of RG-58AU coaxial cable is 28.5 pF/ ft. Thus the longer the lead the greater the capacitance.

The capacitance of the particular coaxial cable used may be obtained by measuring it with the **Philips RCL** meter. The banana plug leads are inserted into the inputs of the meter with the coaxial end not connected to anything. The capacitance can be directly obtained.

Change the frequency to $10f_o$ and print the function generator and capacitor voltage.

Triangular

(b) Set the frequency of the function generator to f_o. Turn the cursors off on the oscilloscope. (Cursors Off.) Switch the function on the function generator to triangular (a ramp with a symmetry of 50%) and print the display.

Change the frequency to $10f_o$ and print the function generator and capacitor voltage.

Ramp

(c) Set the frequency of the function generator to f_o. Switch the function on the function generator to ramp and print the display.

Change the frequency to $10f_o$ and print the function generator and capacitor voltage.

7.7.5 Square Response of First-Order GL Circuit

Assemble the first-order GL circuit shown in Fig. 7.32. Using the measured values of L and R_1 calculate the break frequency

$$f_o = \frac{1}{2\pi\tau} = \frac{1}{2\pi GL} = \frac{R}{2\pi L} \tag{7.56}$$

where R is the total resistance in the circuit $R = R_g + R_1 + R_s$.

Set the function generator to produce a square wave with a peak-to-peak value of 2 V, a DC level of 0 V (the default value), and a frequency of f_0.

Press *AUTOSET* on the oscilloscope and switch the trigger source to *EXT*. Turn $CH1$ off. Set the oscilloscope to display the difference of $CH1$ and $CH2$. Position the cursors as shown in Fig. 7.33 on $v_2(t)$

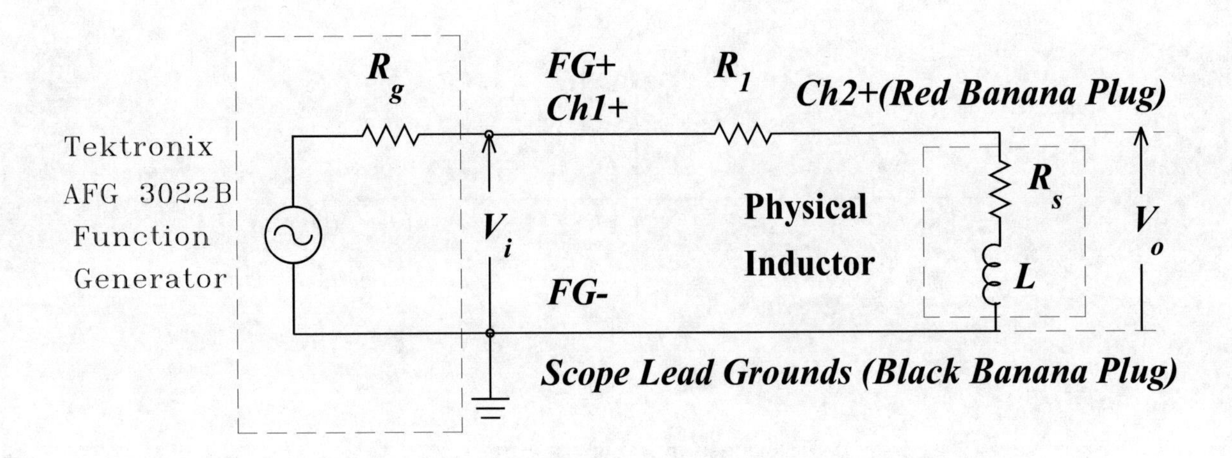

Figure 7.32: First Order GL Circuit

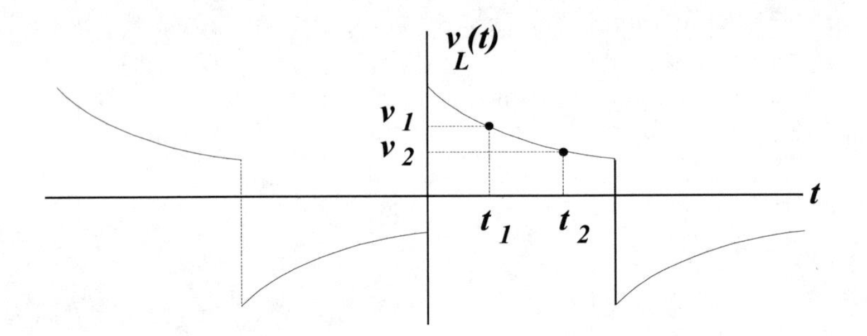

Figure 7.33: Inductor voltage.

which is the inductor voltage. Determine the time constant from the experimental plot as

$$\tau = \frac{t_2 - t_1}{\ln \frac{V_2}{V_1}} \tag{7.57}$$

Print the waveforms displayed on the screen. Compare this value of the time constant with the theoretical expression $\tau = L/R$ where R is the total resistance in the circuit.

7.8 Laboratory Report

1. Turn in all data that was taken and all the printouts obtained.
2. How do the values measured in step 7.7.1 for the inductor and capacitor parameters jibe with Eqs. 7.46 through 7.51 on page 128? Namely, using the measured values of C_p, R_p, and D as the theoretical values calculate the value that should be obtained for C_s and R_s. The frequency at which the measurement was made is 1 kHz. Similarly, using the measured values for L_s, R_s, and Q as the theoretical values calculate the values that should have been obtained for L_p and R_p. Also calculate the percentage error as

$$\% \, Error = 100 \times \frac{Measured - Theoretical}{Theoretical} \tag{7.58}$$

where the theoretical values for C_s and R_s for the capacitor are the calculated values and the theoretical values for L_p and R_p for the inductor are the calculated values.

Component Parameter	**Measured**	**Calculated**	**% Error**
C_p (*Capacitor*)		←	0
R_p (*Capacitor*)		←	0
D (*Capacitor*)		←	0
C_s (*Capacitor*)			
R_s (*Capacitor*)			
L_s (*Inductor*)		←	0
R_s (*Inductor*)		←	0
Q (*Inductor*)		←	0
R_p (*Inductor*)			
L_p (*Inductor*)			

Table 7.8: Calculation of Component Values

3. How well does the measured transient response for the capacitor voltage of the first-order RC circuit obtained in step 7.7.4 compare with the theoretical transient response? The theoretical time constant of the circuit is given by the total resistance (circuit resistance plus the internal resistance of the function generator) times the total capacitance (circuit capacitance plus the capacitance of the oscilloscope lead-input attenuator combination of the oscilloscope).

τ (*theoretical*) =____

τ (*measured*) =____

Use as the input impedance of the oscilloscope lead-input attenuator combination a $1\,\text{M}\Omega$ resistor in parallel with a $13\,\text{pF}$ capacitor.

4. Does the experimentally measured transient response of the first-order GL circuit agree with the theoretical response? The theoretical time constant is given by L/R where R is the total resistance of the circuit. The experimental time constant can be obtained by determining how long it takes for the inductor voltage to decay from its maximum value to $1/e$ of its maximum value.

τ (theoretical) =____

τ (experimental) =____

5. Address any questions posed in the procedure section. Discuss any anomalies that may have been observed during this experiment.

6. For value of $f \gg f_o$ the first-order RC circuit should function as an integrator. Namely, the output voltage should be

$$v_o(t) = \frac{1}{\tau} \int_{-\infty}^{t} v_i(\lambda) d\lambda \tag{7.59}$$

which is the integral of the input weighted by the factor $1/\tau$. Do the results obtained when $f = 10 f_o$ agree with this?

If the laboratory experiment and/or report cannot be completed by the end of the three hour lab session, turn in what has been completed to the laboratory instructor.

7.9 References

1. Banzhaf, W., *Computer-Aided Circuit Analysis Using PSpice*, 2nd edition, Prentice-Hall, 1992.
2. Cooper, W. D., *Electronic Instrumentation and Measurements Techniques*, 2nd edition, Prentice-Hall, 1978.
3. Dorf, R. C., *Introduction to Electric Circuits*, 2nd edition, Wiley, 1993.
4. Jones, L., and Chin, A. F., *Electronic Instruments and Measurements*, John Wiley, 1983.
5. Keown, J., *PSpice and Circuit Analysis*, Merrill, 1991.
6. Kerchner, R. M., and Corcoran, G. F., *Alternating Current Circuits*, Wiley, 1962.
7. Krenz, J. H., *Introduction to Electrical Circuits and Electronic Devices: A Laboratory Approach*, Prentice-Hall, 1987.
8. Kantrowitz, P., et. al., *Electronic Measurements*, Prentice-Hall, 1979.
9. Lago, G. V., and Waidelich, D. L., *Transients in Electrical Circuits*, Ronald, 1958.
10. LePage, W. R., *Complex Variables and the Laplace Transform for Engineers*, Dover, 1961.
11. MicroSim Corporation, *PSpice Circuit Analysis*, MicroSim, 1990.
12. Oliver, B. M., and Cage, J. M., *Electronic Measurements and Instrumentation*, McGraw-Hill, New York, 1971.
13. Rashid, *SPICE for Circuits and Electronics Using PSpice*, Prentice-Hall, 1990.
14. Reddick, H. W., and Miller, F. H., *Advanced Mathematics for Engineers*, Wiley, 1938.
15. Riaz, M., *Electrical Engineering Laboratory Manual*, McGraw-Hill, 1965.
16. Su, K. L., *Fundamentals of Circuit Analysis*, Waveland Press, 1993.
17. Tuinenga, P. W., *SPICE: A Guide to Circuit Simulation & Analysis Using PSpice*, Prentice-Hall, 1992.
18. Wolf, S., *Guide to Electronic Measurements and Laboratory Practices*, Prentice-Hall, 1973.
19. Wolf, S., *Guide to Electronic Measurements and Laboratory Practices*, 2nd edition, Prentice-Hall, 1983.
20. Wolf, S., and Smith, R. F. M., *Student Reference Manual for Electronic Instrumentation Laboratories*, Prentice-Hall, 1990.

Chapter 8

Passive RC Filter Circuits

8.1 Objective

Filters are among the most utilitarian and ubiquitous systems found in the lexicon of analog circuits. They permit analog signal processing by imparting an amplitude and phase change to a sinusoidal signal that is a function of the frequency of the signal. Most filters are designed to pass signals whose frequency content lies in certain bands while rejecting, to some extent, signals outside of these bands. Without these basic building blocks, electronic telecommunications, consumer electronics, control systems, biomedical instrumentation, digital signal processing, *ad infinitum,* would be impossible and civilization would revert back to the Dark Ages.

This experiment will examine filters implemented with resistors and capacitors, i.e. passive RC filters. Excluded will be filters which may also include inductors, transformers, op amps and other electronic amplifiers and/or nonlinear unilateral elements such as diodes; these filters are examined in subsequent experiments.

The classical filters types that will be examined are the low-pass, high-pass, shelving, all-pass, and band-pass. These filters will be designed, simulated, and experimentally examined. This permits a comparison of the theoretical, simulation, and experimental results.

8.2 Theory

Figure 8.1: Two-port network.

8.2.1 Network Definitions

Shown in Fig 8.1 is a linear two-port network. It will be assumed that the network contains no voltage or current sources and that all the voltages and currents found in the circuit are sinusoidal. Leads are extended from four nodes to form two ports. It is called a two-port network since each set of two nodes is called a port. There is an input port for which the input voltage is $\overline{V}_i$ and the input current is $\overline{I}_i$ and an output port for which voltage is $\overline{V}_o$and the output current is $\overline{I}_o$. Usually the word port is dropped and the two ports are simply referred to as the input and output. Since the input and output voltages and currents are sinusoidals, complex phasor notation is used.

Several network functions can be defined for the two-port network. These include the:

- complex transfer function

$$\overline{T}(s) = \left.\frac{\overline{V}_o}{\overline{V}_i}\right|_{\overline{I}_o=0} \tag{8.1}$$

- input impedance

$$\overline{Z}_i = \left.\frac{\overline{V}_i}{\overline{I}_i}\right|_{\overline{I}_o=0} \tag{8.2}$$

- output impedance

$$Z_o = \left.\frac{\overline{V}_o}{\overline{I}_o}\right|_{\overline{V}_i=0} \tag{8.3}$$

Other network functions can be defined but these three are normally the most useful. A filter is defined to be a network for which the complex transfer function is a function of frequency. This simply means that a purely resistive network is not a filter since the complex transfer function would be constant with respect to frequency.

If the elements contained inside the filter are limited to resistors, capacitors, inductors, transformers and op amps, the complex transfer function can be expressed as a ratio of polynomials in s, the complex frequency variable,

$$\overline{T}(s) = K\frac{s^m + b_{m-1}s^{m-1} + \cdots + b_1 s + b_0}{s^n + a_{n-1}s^{n-1} + \cdots + a_1 s + a_0} \tag{8.4}$$

where the coefficients of the polynomials, a_i's and b_i's, are real constants, K is a real constant, and s is the complex frequency variable. The order of the filter is the degree of the denominator polynomial, n, and is equal to the number of independent energy storage elements used in the filter.

If the numerator and denominator polynomials in Eq. 8.4 are factored, the complex transfer function may be expressed as

$$\overline{T}(s) = K\frac{(s-z_1)(s-z_2)\cdots(s-z_m)}{(s-p_1)(s-p_2)\cdots(s-p_n)} \tag{8.5}$$

where K is a real constant which may be positive or negative, the roots of the numerator, z_i, are known as the zeroes of the transfer function and the roots of the denominator, p_i, are known as the poles. When $s = z_i$ the transfer function is zero so this is known as a zero while when $s = p_i$ the transfer function is infinite so this is known as a pole (one with infinite height).

The magnitude of the complex transfer function, $\left|\overline{T}(j\omega)\right|$, is usually plotted on a log or decibel scale and the frequency on a log scale. This magnitude is referred to as the gain or attenuation of the filter. Normally, the term gain is used to imply that the linear magnitude is greater than one while attenuation is used to imply that the linear magnitude is less than one. Both the terms gain and attenuation are used where an attenuation is a negative gain (when expressed in terms of decibels) and vice versa.

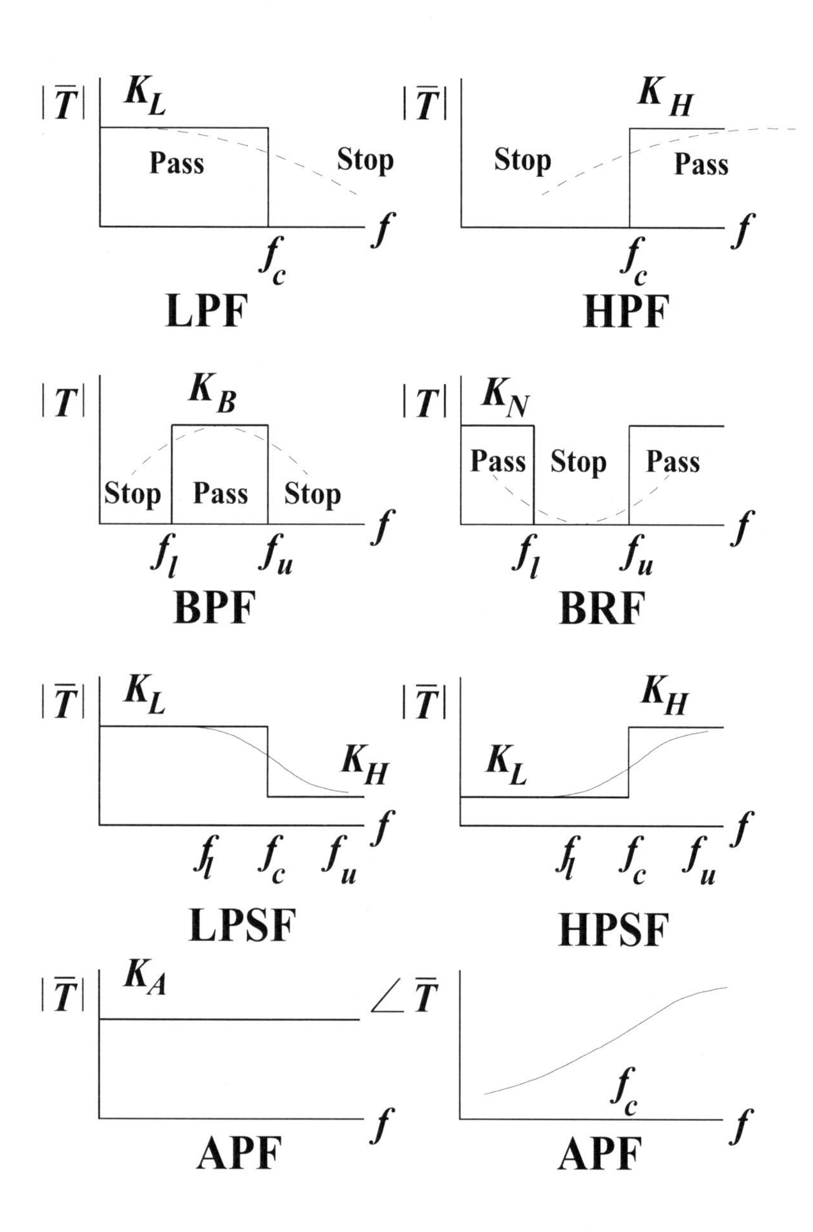

Figure 8.2: Filter types.

8.2.2 Filter Types

The filter type or classification is based upon the behavior of the magnitude of the complex transfer function as the frequency changes. Plots of the magnitude of the complex transfer function as functions of frequency, i.e. $s = j\omega$, are shown in Fig. 8.2. The solid lines are the ideal filter response and the dashed lines are the plots of a practical filter. An ideal filter would require that the degree of the denominator polynomial of the complex transfer function be infinite which would require an infinite number of energy storage elements inside the filter. The ideal filters are sometimes termed "brick wall" filters.

Plots of the magnitude and phase of the complex transfer function versus frequency are knows as Bode plots after the engineer who first made extensive use of them in analyzing signals, circuits, and systems (H. Bode). The magnitude is either plotted on a log scale or is expressed in terms of decibels. The phase is plotted in degrees or radians. The frequency is always plotted on a log scale. By examining the asymptotic behavior of the complex transfer function, it is possible to approximately plot the Bode plots using straight line segments if the poles and zeroes are real. (Some authors use the term Bode plot to refer to only the approximate plot of the frequency response.)

The basic classical filter classifications are:

- Low-Pass. For the ideal low-pass filter

$$\left|\overline{T}(j\omega)\right| = \begin{cases} K_L & f \leq f_c \\ 0 & f > f_c \end{cases} \tag{8.6}$$

  where K_L is a constant known as the dc gain of the filter. Obviously, it is known as a low-pass filter because it passes low frequencies and does not pass high frequencies. The band of frequencies $0 \leq f \leq f_c$ is known as the pass-band and the band of frequencies $f_c < f < \infty$ is known as the stop-band. The frequency f_c is known as the cut-off frequency. No circuit with a finite number of elements can perform the signal processing operation of Eq. 8.6 which requires a transition from the pass- to stop-bands in a band of zero width. Practical circuits require a more gradual transition from the pass- to stop-bands as shown with the dashed line. The frequency f_c for this practical filter is usually taken to be the frequency for which $\left|\overline{T}(j\omega)\right| = K_L/\sqrt{2} = 0.707K_L$ and the additional names half-power, $-3\ dB$, critical, corner, and break frequency are used. It should be noted that the low-pass filter is the only one of the classical filter types for which $\left|\overline{T}(j\omega)\right|$ is a constant at dc and zero at $f = \infty$ ($f = \infty$ will be used as a short hand notation for $\lim_{f\to\infty}$). The bandwidth of the filter is defined to be $BW = f_c$.

- High-Pass. This filter is the mirror image of the low-pass filter. It is defined by

$$\left|\overline{T}(j\omega)\right| = \begin{cases} K_H & f \geq f_c \\ 0 & f < f_c \end{cases} \tag{8.7}$$

  where K_H is the gain when $f = \infty$. All of the comments made about the low-pass filter are equally applicable to the high-pass filter with an interchange of the words high and low and the inequality signs. Since the upper frequency of the pass-band is infinity, there is no bandwidth associated with this filter. It should be noted that of the basic filter types that the high-pass filter is the only one for which the $\left|\overline{T}(j\omega)\right|$ is 0 at dc and a constant at $f = \infty$.

- Band-Pass. The ideal band-pass filter passes frequencies in a band $f_l \leq f \leq f_u$ and rejects frequencies outside of this band. It can be realized by cascading a low-pass with a high-pass filter where the cut-off frequency of the low-pass filter is f_u and the cut-off frequency of the high-pass filter is f_l. There are two $-3\ dB$ frequencies associated with the practical band-pass filter: f_l and f_u. It should be noted that of the basic filter types the band-pass filter is the only one for which the $\left|\overline{T}(j\omega)\right|$ is 0 at both dc and at $f = \infty$. The bandwidth may be taken as $BW = f_u - f_l$.

- Band-Reject. The band-reject filter is the mirror image of the band-pass filter. If the stop-band is narrow, the term notch filter is used.

- Low-Pass Shelving. The ideal low-pass shelving filter has a gain of K_L at low frequencies and a gain of K_H at high frequencies where $K_L > K_H$. At $f = f_c$ the gain abruptly changes from K_L to K_H. The gain is finite at both $f = 0$ and $f = \infty$. A real low-pass shelving filter requires a more gradual transition..

- High-Pass Shelving. The ideal high-pass shelving filter is the mirror image of the ideal low-pass shelving filter. Again, the gain changes abruptly at $f = f_c$ but, for this filter, $K_L < K_H$.

- All-Pass. These filters have an amplitude that is constant with respect to frequency but a phase shift that is a function of frequency.

8.2.3 Low-Pass Filter

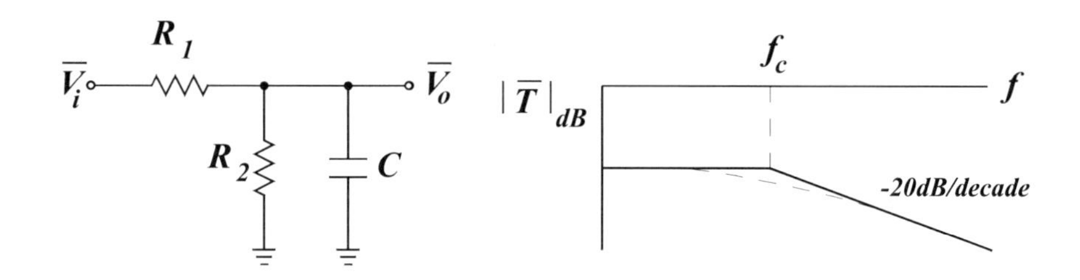

Figure 8.3: First-order low-pass filter.

A circuit diagram of a first-order low-pass filter and its Bode magnitude plot is shown in Fig. 8.3. The complex transfer function is given by

$$\overline{T}(s) = K_L \frac{1}{1 + s\tau} \tag{8.8}$$

where K_L is the dc gain

$$K_L = \frac{R_2}{R_1 + R_2} \tag{8.9}$$

and the time constant τ is given by

$$\tau = (R_1 \| R_2)\, C \tag{8.10}$$

This complex transfer function has a finite pole at $s = -1/\tau$ and no finite zero. The magnitude of the complex transfer function in decibels is

$$\left|\overline{T}(j\omega)\right|_{dB} = 20\log_{10}\left|\overline{T}(j\omega)\right| = 20\log_{10} K_L - 20\log_{10}\sqrt{1 + \omega^2\tau^2} = \tag{8.11}$$

$$20\log_{10} K_L - 20\log_{10}\sqrt{1 + \left(\frac{f}{f_c}\right)^2} \tag{8.12}$$

where f_c is the cut-off frequency

$$f_c = \frac{1}{2\pi\tau} \tag{8.13}$$

which is the frequency at which the response is down 3 dB from the dc value (actually it's 3.0103 dB). The magnitude of the complex transfer function is plotted as the dashed line in the figure to the right of the circuit diagram.

An approximate plot of the magnitude of the complex transfer function as a function of frequency may be obtained by considering the asymptotic behavior of Eq. 8.12. For frequencies $f \ll f_c$ the second term is negligible and the expression becomes

$$\left|\overline{T}(j\omega)\right|_{dB} = 20\log_{10} K_L \tag{8.14}$$

which is a straight line parallel to the frequency axis and located below it since K_L is a number less than 1 and the log of a number less than 1 is negative. For frequencies large compared to f_c Eq. 8.12 becomes

$$\left|\overline{T}(j\omega)\right|_{dB} = 20\log_{10} K_L - 20\log_{10}\frac{f}{f_c} \tag{8.15}$$

which, since the frequency is plotted on a log scale, is a straight line. The slope of this line is said to be $-20\ dB/decade$ since the function decreases by 20 dB when the frequency is increased by a factor of 10. (A change in the frequency by a factor of 10 is known as a decade.) This straight line intersects with the one given in Eq. 8.14 at $f = f_c$ for which the second term is zero.

Thus an approximate Bode plot of the magnitude of the complex transfer function as a function of frequency for the low-pass filter may be obtained by using two straight line segments. The first is drawn parallel to the frequency axis (which is plotted on a log scale) and located below the frequency axis by $20\log_{10} K_L$ where K_L is the low-frequency or dc gain. The second line is drawn through this line at $f = f_c$ and has a slope of $-20\ dB/decade$. This approximate plot is very close to the actual plot at all frequencies except near the break frequency f_c. Here the maximum error occurs which is 3 dB at $f = f_c$.

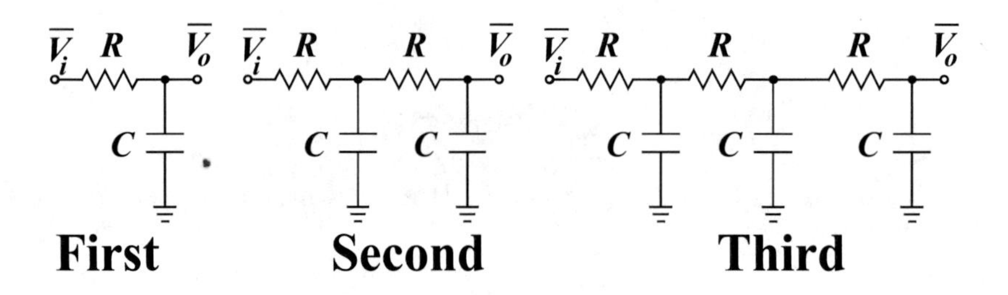

Figure 8.4: Unity-gain, low-pass filters.

Shown in Fig. 8.4 are a unity-gain first-, second-, and third-order low-pass filters. They are termed unity-gain because the gain at low-frequencies or dc is 1. The complex transfer functions and cut-off frequencies for these filters are:

Order	Complex Transfer Function	Cut-Off Frequency
First	$\dfrac{1}{1+s\tau}$	$\dfrac{1}{2\pi\tau}$
Second	$\dfrac{1}{s^2\tau^2+3s\tau+1}$	$\dfrac{0.374}{2\pi\tau}$
Third	$\dfrac{1}{s^3\tau^3+5s^2\tau^2+6s\tau+1}$	$\dfrac{0.194}{2\pi\tau}$

(8.16)

where $\tau = RC$. Although each resistor and capacitor have been shown with the same symbol for notational simplicity, it should be noted that if the object were to design filters with the same cut-off frequency the RC combinations would have to be different.

8.2.4 High-Pass Filter

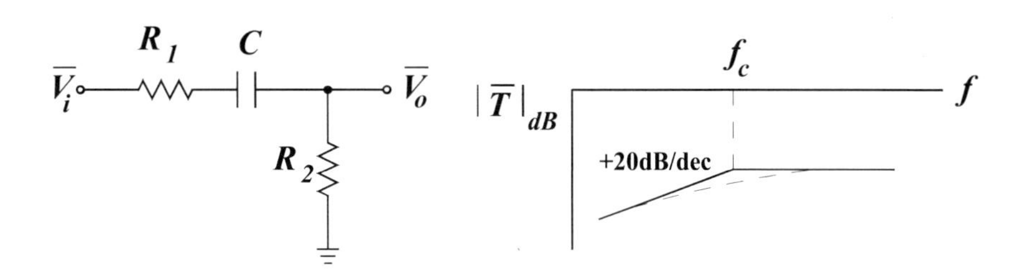

Figure 8.5: First-order high-pass filter.

A circuit diagram of a first-order high-pass filter and its Bode magnitude plot is shown in Fig. 8.5. The complex transfer function is given by

$$\overline{T}(s) = K_H \frac{s\tau}{1+s\tau} \tag{8.17}$$

where K_H is the high frequency gain or the gain at $f = \infty$

$$K_H = \frac{R_2}{R_1 + R_2} \tag{8.18}$$

and the time constant τ is given by

$$\tau = (R_1 + R_2)\, C \tag{8.19}$$

This transfer function has one finite pole at $s = -1/\tau$ and a finite zero at the origin. The magnitude of the complex transfer function in decibels is

$$\left|\overline{T}(j\omega)\right|_{dB} = 20\log_{10}\left|\overline{T}(j\omega)\right| = 20\log_{10} K_H + 20\log_{10}\omega\tau - 20\log_{10}\sqrt{1+\omega^2\tau^2} = \tag{8.20}$$

$$20\log_{10} K_L + 20\log_{10}\frac{f}{f_c} - 20\log_{10}\sqrt{1+\left(\frac{f}{f_c}\right)^2} \tag{8.21}$$

where f_c is the cut-off frequency

$$f_c = \frac{1}{2\pi\tau} \tag{8.22}$$

which is the frequency at which the response is down 3 dB from the $f = \infty$ value. The magnitude of the complex transfer function is plotted as the dashed line in the figure to the right of the circuit diagram.

An approximate plot of the magnitude of the complex transfer function as a function of frequency may be obtained by considering the asymptotic behavior of Eq. 8.21. For frequencies $f \gg f_c$ the second and third terms cancel and the expression becomes

$$\left|\overline{T}(j\omega)\right|_{dB} = 20\log_{10} K_H \tag{8.23}$$

which is a straight line parallel to the frequency axis and located below it since K_H is a number less than 1 and the log of a number less than 1 is negative. For frequencies small compared to f_c Eq. 8.21 becomes

$$\left|\overline{T}(j\omega)\right|_{dB} = 20\log_{10} K_L + 20\log_{10}\frac{f}{f_c} \tag{8.24}$$

which, since the frequency is plotted on a log scale, is a straight line. The slope of this line is said to be $+20$ $dB/decade$ since the function decreases by 20 dB when the frequency is increased by a factor of 10. (A change in the frequency by a factor of 10 is known as a decade.) This straight line intersects with the one given in Eq. 8.23 at $f = f_c$.

Thus an approximate Bode plot of the magnitude of the complex transfer function as a function of frequency for the high-pass filter may be obtained by using two straight line segments. The first is drawn parallel to the frequency axis (which is plotted on a log scale) and located be below the frequency axis by $20 \log_{10} K_H$ where K_H is the high-frequency or $f = \infty$ gain. The second line is drawn through this line at $f = f_c$ and has a slope of $+20\ dB/decade$. This approximate plot is very close to the actual plot at all frequencies except near the break frequency f_c. Here the maximum error occurs which is 3 dB.

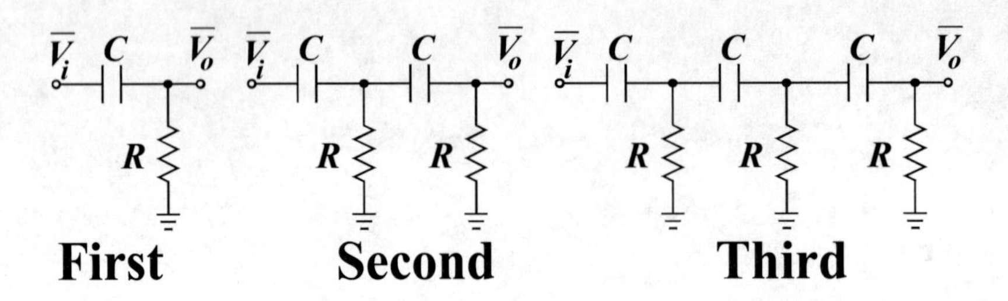

Figure 8.6: Unity-gain high-pass filters.

Shown in Fig. 8.6 are a unity-gain first-, second-, and third-order high-pass filters. They are termed unity-gain because the gain at high frequencies or $f = \infty$ is 1. The complex transfer functions and cut-off frequencies for these filters are:

Order	Complex Transfer Function	Cut-Off Frequency
First	$\dfrac{s\tau}{1 + s\tau}$	$\dfrac{1}{2\pi\tau}$
Second	$\dfrac{s^2\tau^2}{s^2\tau^2 + 3s\tau + 1}$	$\dfrac{2.67}{2\pi\tau}$
Third	$\dfrac{s^3\tau^3}{s^3\tau^3 + 6s^2\tau^2 + 5s\tau + 1}$	$\dfrac{5.15}{2\pi\tau}$

(8.25)

where $\tau = RC$. Although each resistor and capacitor have been shown with the same symbol for notational simplicity, it should be noted that if the object were to design filters with the same cut-off frequency the RC combinations would have to be different.

The transfer function for the high-pass filters may be obtained directly from that of the corresponding low-pass filters by invoking the low- to high-pass transformation which is

$$s\tau \longrightarrow \frac{1}{s\tau} \tag{8.26}$$

This simply means that all the $s\tau$ terms that appear in the transfer function for the low-pass filters are replaced with $1/s\tau$. Note that the coefficients in the numerators for the cut-off frequencies for the high-pass filters are the reciprocal of those appearing for the corresponding order low-pass filters.

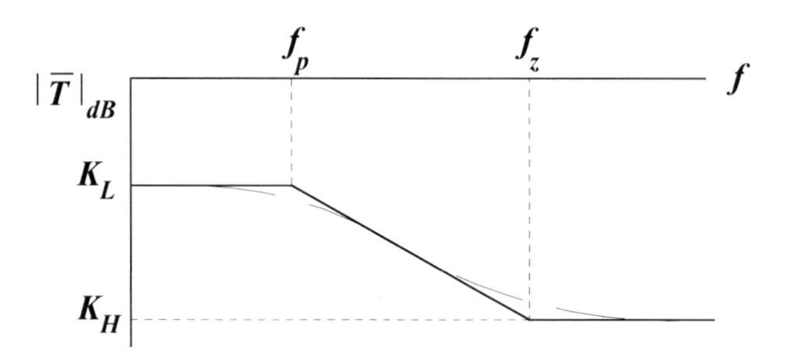

Figure 8.7: Magnitude Bode plot for first-order low-pass shelving filter.

8.2.5 First-Order Low-Pass Shelving Filter

The Bode plot for the magnitude of the complex transfer function for the first-order low-pass shelving filter is shown in Fig. 8.7. The complex transfer function is given as

$$\overline{T}(s) = K_L \frac{1 + s\tau_z}{1 + s\tau_p} \tag{8.27}$$

which has a finite zero at $s = -1/\tau_z$ and a finite pole at $s = -1/\tau_p$. The gain at low frequencies or dc is K_L and the gain at $f = \infty$ is $K_H = K_L\tau_z/\tau_p$. For this filter $\tau_p > \tau_z$ which means that $K_H < K_L$.

The magnitude of the complex transfer function in decibels is given by

$$\left|\overline{T}(j\omega)\right|_{dB} = 20\log_{10}\left|\overline{T}(j\omega)\right| = 20\log_{10} K_L + 20\log_{10}\sqrt{1+\omega^2\tau_z^2} - 20\log_{10}\sqrt{1+\omega^2\tau_p^2} = \tag{8.28}$$

$$20\log_{10} K_L + 20\log_{10}\sqrt{1+\left(\frac{f}{f_z}\right)^2} - 20\log_{10}\sqrt{1+\left(\frac{f}{f_p}\right)^2} \tag{8.29}$$

where

$$f_z = \frac{1}{2\pi\tau_z} \tag{8.30}$$

and

$$f_p = \frac{1}{2\pi\tau_p} \tag{8.31}$$

are the pole and zero frequencies. Since $\tau_p > \tau_z$,

$$f_p < f_z \tag{8.32}$$

which makes this a low-pass shelving filter with the shape shown in Fig. 8.7. The dashed line represents the exact frequency response.

An approximate plot of the magnitude of the complex transfer function in decibels as a function of frequency may be obtained by considering the asymptotic behavior of Eq. 8.29. For $f \ll f_p, f_z$ the imaginary parts for the second two terms are negligible and the complex transfer function is simply

$$\left|\overline{T}(j\omega)\right|_{dB} = 20\log_{10} K_L \tag{8.33}$$

which is a straight line parallel to the frequency axis. For $f \gg f_p, f_z$ the real parts of these two terms can be neglected and

$$\left|\overline{T}(j\omega)\right|_{dB} = 20\log_{10} K_L - 20\log_{10}\frac{f_z}{f_p} \tag{8.34}$$

which is a straight line parallel to the frequency axis located below the low frequency asymptote by $20\log_{10} f_z/f_p$ since $f_p < f_z$. For $f_p \ll f \ll f_z$ the asymptote is given by

$$\left|\overline{T}(j\omega)\right|_{dB} = 20\log_{10} K_L - 20\log_{10}\frac{f}{f_p} \tag{8.35}$$

which is a straight line with a slope of $-20\ dB/decade$ that intersects the two horizontal line at $f = f_p$ and $f = f_z$.

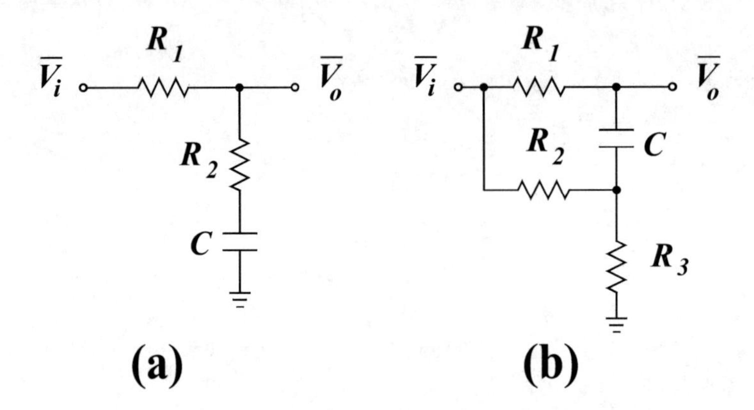

Figure 8.8: Circuit diagrams for two low-pass shelving filters.

Circuit diagrams for two low-pass shelving filters are shown in Fig. 8.8. The asymptotes for the Bode plots for the magnitude of the complex transfer function as a function of frequency can be obtained directly from the circuit diagrams. At $f = 0$ the capacitors are open circuits and so the networks are purely resistive. Similarly, at $f = \infty$ the capacitors are short circuits so the networks are purely resistive. To complete the asymptotic plots the time constants must be solved for so that the pole and zero frequencies may be obtained.

For the circuits shown in Fig. 8.8 the gains and time constants are:

Circuit	K_L	τ_z	τ_p
Figure 8.8 a	1	R_2C	$(R_1 + R_2)\,C$
Figure 8.8 b	1	$R_3\dfrac{R_1 + R_2}{R_2 + R_3}C$	$(R_1 + R_2\Vert R_3)\,C$

(8.36)

The low-pass shelving filters are known in the parlance of control systems as lag-lead filters. They are so-called because as the frequency is increased from dc the pole starts to introduce a lagging or negative phase. As the frequency approaches the zero it begins to introduce a phase lead which compensates, to some extent, the lagging phase introduced by the pole. At $f = 0$ and $f = \infty$ the phase is zero.

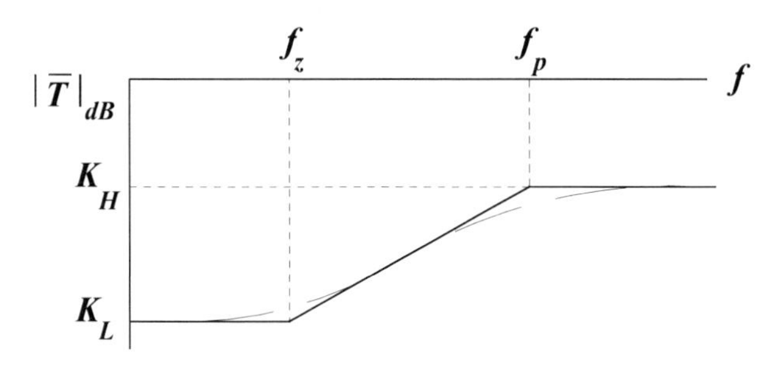

Figure 8.9: Magnitude Bode plot for first-order high-pass shelving filter.

8.2.6 First-Order High-Pass Shelving Filter

The Bode plot for the magnitude of the complex transfer function for the first-order high-pass shelving filter is shown in Fig. 8.9. The complex transfer function is given as

$$\overline{T}(s) = K_L \frac{1 + s\tau_z}{1 + s\tau_p} \tag{8.37}$$

which has a finite zero at $s = -1/\tau_z$ and a finite pole at $s = -1/\tau_p$. The gain at low frequencies or dc is K_L and the gain at $f = \infty$ is $K_H = K_L\tau_z/\tau_p$. For this filter $\tau_p < \tau_z$ which means that $K_H > K_L$.

The magnitude of the complex transfer function in decibels is given by

$$\left|\overline{T}(j\omega)\right|_{dB} = 20\log_{10}\left|\overline{T}(j\omega)\right| = 20\log_{10} K_L + 20\log_{10}\sqrt{1+\omega^2\tau_z^2} - 20\log_{10}\sqrt{1+\omega^2\tau_p^2} = \tag{8.38}$$

$$20\log_{10} K_L + 20\log_{10}\sqrt{1+\left(\frac{f}{f_z}\right)^2} - 20\log_{10}\sqrt{1+\left(\frac{f}{f_p}\right)^2} \tag{8.39}$$

where

$$f_z = \frac{1}{2\pi\tau_z} \tag{8.40}$$

and

$$f_p = \frac{1}{2\pi\tau_p} \tag{8.41}$$

are the pole and zero frequencies. Since $\tau_p < \tau_z$,

$$f_p > f_z \tag{8.42}$$

which makes this a high-pass shelving filter with the shape shown in Fig. 8.9. The dashed line represents the exact frequency response.

An approximate plot of the magnitude of the complex transfer function in decibels as a function of frequency may be obtained by considering the asymptotic behavior of Eq. 8.39. For $f \ll f_p, f_z$ the imaginary parts for the second two terms are negligible and the complex transfer function is simply

$$\left|\overline{T}(j\omega)\right|_{dB} = 20\log_{10} K_L \tag{8.43}$$

which is a straight line parallel to the frequency axis. For $f \gg f_p, f_z$ the real parts of these two terms can be neglected and

$$\left|\overline{T}(j\omega)\right|_{dB} = 20\log_{10} K_L + 20\log_{10}\frac{f_p}{f_z} \tag{8.44}$$

which is a straight line parallel to the frequency axis located above the low frequency asymptote by $20\log_{10} f_p/f_z$ since $f_p > f_z$. For $f_z \ll f \ll f_p$ the asymptote is given by

$$\left|\overline{T}(j\omega)\right|_{dB} = 20\log_{10} K_L + 20\log_{10}\frac{f}{f_z} \tag{8.45}$$

which is a straight line with a slope of $+20\ dB/decade$ that intersects the two horizontal lines at $f = f_p$ and $f = f_z$.

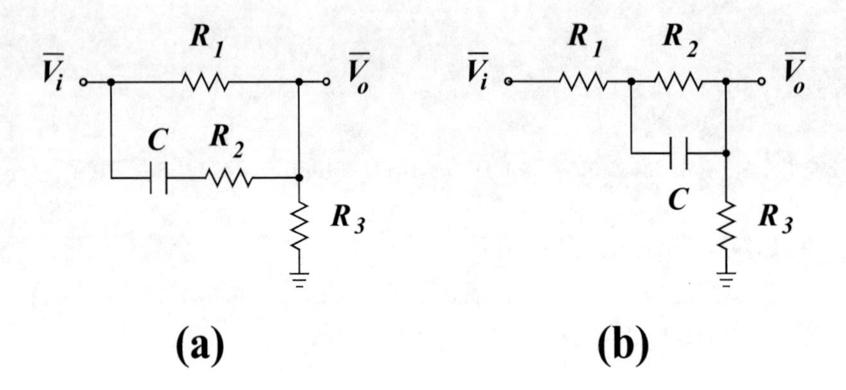

Figure 8.10: Circuit diagrams for two high-pass shelving filters.

Circuit diagrams for two high-pass shelving filters are shown in Fig. 8.10. The asymptotes for the Bode plots for the magnitude of the complex transfer function as a function of frequency can be obtained directly from the circuit diagrams. At $f = 0$ the capacitors are open circuits and so the networks are purely resistive. Similarly, at $f = \infty$ the capacitors are short circuits so the networks are purely resistive. To complete the asymptotic plots the time constants must be solved for so that the pole and zero frequencies may be obtained.

For the circuits shown in Fig. 8.10 the gains and time constants are:

Circuit	K_L	τ_z	τ_p
Figure 8.10 a	$\dfrac{R_3}{R_1+R_3}$	$(R_1+R_2)\,C$	$(R_2+R_1\Vert R_3)\,C$
Figure 8.10 b	$\dfrac{R_3}{R_1+R_2+R_3}$	R_2C	$(R_2\Vert\,[R_1+R_3])\,C$

(8.46)

The high-pass shelving filters are known in the parlance of control systems as lead-lag filters. They are so-called because as the frequency is increased from dc the zero starts to introduce a leading or positive phase. As the frequency approaches the pole it begins to introduce a phase lag which compensates, to some extent, the leading phase introduced by the zero. At $f = 0$ and $f = \infty$ the phase is zero.

8.2.7 All-Pass Filter

The circuit diagram for a first-order all-pass filter is shown in Fig. 8.11. The complex transfer function for this filter is given by

$$\overline{T}(s) = \frac{1}{2}\frac{1-s\tau}{1+s\tau} \tag{8.47}$$

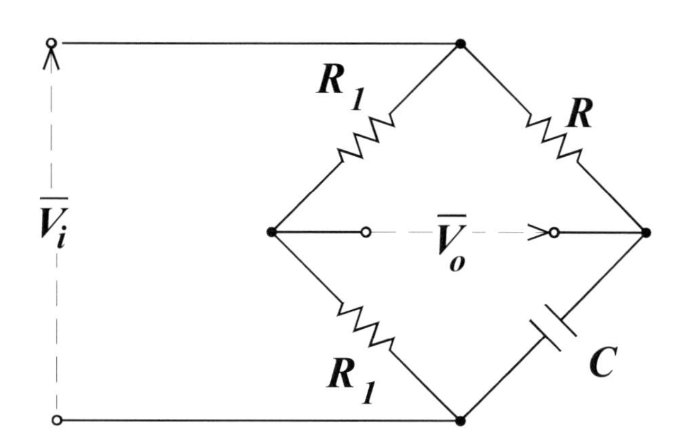

Figure 8.11: Circuit diagram for first-order all-pass filter.

where $\tau = RC$. It has a finite zero in the right half of the complex plane at $s = 1/\tau$ and a finite pole at $s = -1/\tau$. The magnitude of the complex transfer function equals 1/2 for all f. But the phase is given by

$$\angle\overline{T}(j\omega) = -2\tan^{-1}\omega\tau \tag{8.48}$$

which has a phase shift of 90° at $f = 1/(2\pi RC)$.

8.2.8 Band-Pass Filter

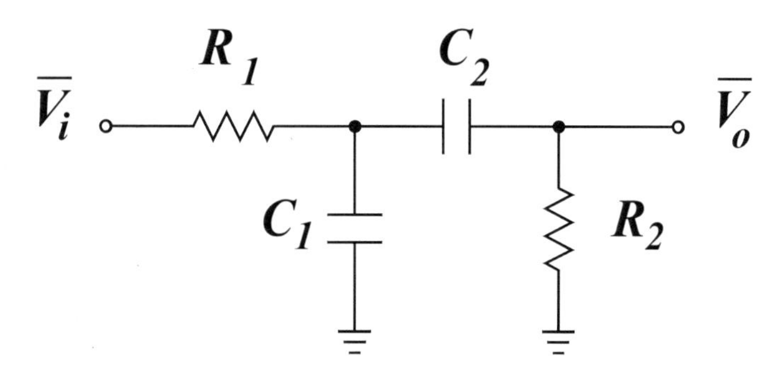

Figure 8.12: Circuit diagram of second-order band-pass filter.

The circuit diagram of a second-order band-pass filter is shown in Fig. 8.12. Were it not for the loading effect, this would be a first-order low-pass filter cascaded with a first-order high-pass filter. The complex transfer function is given by

$$\overline{T}(s) = \frac{sR_2C_2}{s^2R_1C_1R_2C_2 + [R_1C_1 + R_2C_2 + R_1C_2]\,s + 1} \tag{8.49}$$

The upper and lower $-3\ dB$ frequencies are relatively complicated functions of the component values.

Eq. 8.49 may be simplified for $f_u \gg f_l$ by assuming that

$$R_1C_1 + R_2C_2 \gg R_1C_2 \tag{8.50}$$

which then yields

$$\overline{T}(s) \cong \frac{sR_2C_2}{s^2R_1C_1R_2C_2 + [R_1C_1 + R_2C_2]\,s + 1} = \frac{sR_2C_2}{1 + sR_2C_2}\frac{1}{1 + sR_1C_1} \tag{8.51}$$

which is the cascade or product of a first-order high-pass filter with cut-off frequency

$$f_l = f_2 = \frac{1}{2\pi R_2C_2} \tag{8.52}$$

and a first-order low-pass filter with cut-off frequency

$$f_u = f_1 = \frac{1}{2\pi R_1C_1} \tag{8.53}$$

In order to use Eqs. 8.51, 8.52, and 8.53 the designer must assure that Eq. 8.50 is satisfied. Perhaps, the easiest way to achieve this is to pick $C_1 = C_2$ and $R_2 \gg R_1$.

8.2.9 Approximate Method of Determining Transfer Function for First-Order Circuits

The transfer function for all of the circuits considered in this experiment may be derived using elementary circuit analysis. Indeed, the simple voltage divider theorem is all that is required. However, for first-order circuits, containing a single capacitor and one or more resistors, the transfer function can be determined by inspection. This techniques is valid only if the circuit can be resolved into a single impedance from the input to the output node and another single impedance from the output node to ground.

The first step in this procedure is to determine what type of filter the two-port circuit is. For circuits with a single capacitor, there are only four types possible: low-pass, high-pass, shelving, and all-pass. This can be determined from the behavior of the circuit at $f = 0$ where the capacitors are open circuits and $f = \infty$ where the capacitors are short circuits. At these two frequency extremes the circuit is purely resistive and the gains may be easily determined.

Low-Pass Filter

A low-pass filter will have a gain at $f = \infty$ that is zero. None of the other three types will have an output that is zero at $f = \infty$. Once it has been determine that the circuit is a low-pass filter the next step is to write the transfer function

$$T(s) = K_L\frac{1}{1 + s\tau} \tag{8.54}$$

The dc gain, K_L, is determined by replacing the capacitor with an open circuit and using the voltage divider theorem to determine the gain.

The pole time constant, τ, is determined by placing a short circuit from the input node to ground and an open circuit from the output node to ground and determining the Thévenin equivalent resistance with respect to the capacitor, R_{th}. The pole time constant is then given by $\tau = R_{th}C$.

High-Pass Filter

A high-pass filter will have a gain at $f = 0$ that is zero. None of the other three types will have an output that is zero at $f = 0$. Once it has been determine that the circuit is a high-pass filter the next step is to write the transfer function

$$T(s) = K_H \frac{s\tau}{1 + s\tau} \tag{8.55}$$

The high-frequency gain, K_H, is determined by replacing the capacitor with an short circuit and using the voltage divider theorem to determine the gain.

The pole time constant, τ, is determined by placing a short circuit from the input node to ground and an open circuit from the output node to ground and determining the Thévenin equivalent resistance with respect to the capacitor, R_{th}. The pole time constant is then given by $\tau = R_{th}C$.

Shelving Filter

A shelving filter has a frequency response that is non zero at all frequencies including $f = 0$ and $f = \infty$. A low-pass shelving filter will have a gain at $f = 0$ that is larger than the gain at $f = \infty$ and, conversely, a high-pass shelving filter will have a gain at $f = 0$ that is smaller than that at $f = \infty$. (If the gains are the same at the two frequency extremes, it is an all-pass filter.)

Once it has been determined that the circuit is a shelving filter the next step is to write the transfer function

$$T(s) = K_L \frac{1 + s\tau_z}{1 + s\tau_p} \tag{8.56}$$

The dc gain, K_L, is determined by replacing the capacitors with open circuits and using the voltage divider theorem. The pole time constant is determined is the same manner used for the low- and high-pass filters.

The time constant for the zero, τ_z, is determined by placing an open circuit from the input to ground and a short circuit from the output to ground. The Thévenin equivalent resistance with respect to the capacitor is then determined, R_{th}. The time constant for the zero is then given by $\tau_z = R_{th}C$.

Example

To illustrate this technique Fig. 8.10b. The first step is to determine the gain at $f = 0$ and $f = \infty$. The gain at $f = 0$ is determined by replacing the capacitor with an open circuit; the circuit then consists of the three resistors and the gain is given by

$$K_L = \frac{R_3}{R_1 + R_2 + R_3} \tag{8.57}$$

The gain at $f = \infty$ is determined by replacing the capacitor with a short circuit; this then shorts the resistor R_2 and the gain is given by

$$K_H = \frac{R_3}{R_1 + R_3} \tag{8.58}$$

Since the gain at neither $f = 0$ nor $f = \infty$ is zero, the circuit is a shelving filter. And, since $K_H > K_L$ it is a high-pass shelving filter.

The next step is to write the transfer function

$$T(s) = K_L \frac{1 + s\tau_z}{1 + s\tau_p} \tag{8.59}$$

and determine the time constants. To determine τ_p the input is grounded and the output open circuited as shown on the left in Fig. 8.13. The Thévenin equivalent with respect to the capacitor is then

$$R_{th} = R_2 \| (R_1 + R_3) \tag{8.60}$$

since R_1 and R_3 are in series and the series combination is in parallel with R_2. The pole time constant is then $\tau_p = R_{th}C$.

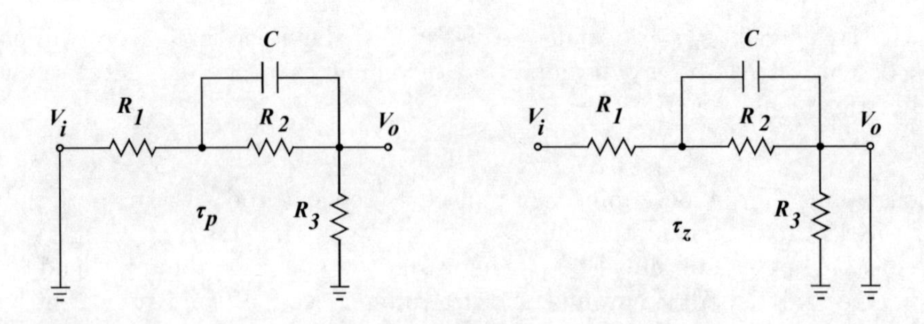

Figure 8.13: Circuits for determining time constants.

To determine the zero time constant the input is open circuited and the output short circuited as shown in the figure on the right in Fig. 8.13. The resistor R_3 is then shorted and the resistor R_1 is floating. So the Thévenin equivalent resistance with respect to the capacitor is simply R_2 and

$$\tau_z = R_2 C \tag{8.61}$$

which completes the determination of the transfer function.

The approximate magnitude Bode plot can now be plotted using three straight line segments as shown in Fig. 8.9. The low- and high-frequency asymptotes are straight lines parallel to the frequency axis given by Eqs. 8.57 and 8.58. The break frequencies are given by

$$f_p = \frac{1}{2\pi\tau_p} \tag{8.62}$$

and

$$f_z = \frac{1}{2\pi\tau_z} \tag{8.63}$$

from which a line is drawn from the low- to high-frequency asymptotes. This three line approximation is close to the actual response excepts at the break frequencies where the error is 3 dB.

The approximate technique to determine the transfer function of a circuit with a single capacitor can be used for each of the circuits in the previous sections with the sole exception of the Fig. 8.8b. It cannot be used on this circuit because the capacitor is situated so that it cannot be resolved into a single impedance from the input to the output node.

8.3 Design Assignment

A number of filter design assignments will have been made for this experiment by either the laboratory instructor or the homework problem assignment for the week. The filters should then be designed using the equations in the theory section and capacitor values that are available for the experimenter's use. The resistors values should then be calculated and the nearest 5% resistors should be used.

The following capacitor values are available for the experimenter's use: 10 pF to 0.47 μF with the following multipliers: 1, 1.5, 2.2, 3.3, 3.9, 4.7, and 6.8. These are 20% capacitors which means that the actual value of the capacitance will be found within 20% of the nominal value.

The following resistor values are available for the experimenter's use: 10 Ω to 1 $M\Omega$ with the following multipliers: 1, 1.1, 1.2, 1.3, 1.5, 1.6, 1.8, 2, 2.2, 2.4, 2.7, 3, 3.3, 3.6, 3.9, 4.3,

4.7, 5.1, 5.6, 6.2, 6.8, 7.5, 8.2, and 9.1. These are 5% resistors which means that the actual value of the resistance is guaranteed to lie within 5% of the nominal value.

The recommended range for practical circuit components are capacitances from 1 nF to 0.47 μF and resistances from 100 Ω to 100 kΩ.

8.4 SPICE

Each of the filter circuits that was designed for this experiment should be verified with a simulation using SPICE. The analysis type is AC. The input should be specified as AC with an amplitude of 1. After the SPICE simulation is performed the variable to be plotted should be `VDB(·)` of the output node.

The SPICE code to plot the magnitude of the complex transfer function of a first-order high-pass shelving filter shown in Fig. 8.10b is given below.

```
TITLE LINE
VI 1 0 AC 1
R1 1 2 10K
R2 2 3 470K
C 2 3 0.47N
R3 3 0 100
.AC DEC 256 10 1MEG
.PROBE
.END
```

The variable `VDB(3)` or `DB(V(3))` is then plotted. This is then the magnitude of the complex transfer function since the amplitude of the input is 1. If the phase were desired the variable `VP(3)` or `P(V(3))` would be plotted.

The magnitude Bode plot is shown in Fig. 8.14. The solid line is the actual plot and the dashed line is the approximate plot using the techniques described in previous sections.

8.5 Procedure

The **Tektronix 3022B** function generator displays the amplitude of the voltage available at its output terminals. In order for the **Tektronix 3022B** function generator to indicate the open circuit or Thévenin voltage in the display when amplitude is selected the output termination must be set to *HIGH Z*. The default or power-on setting is 50 Ω which would then result in a display one half of the Thévenin source voltage.

Terminating Impedance Adjustment for Function Generator

Turn on the **Tektronix 3022B** function generator and wait until it finishes its self-test on power-on; this should only take a few seconds. Enable the Output by pushing the button labeled Output. The function generator is now configured so that it will display the Thévenin source voltage when the amplitude display is selected.

8.5.1 Amplitude Adjustment

Set the amplitude of the waveform produced by the CH1 output of the **Tektronix 3022B** function generator to 2 V peak-to-peak. Set the frequency of the function generator to 10 kHz. Set the function to sine.

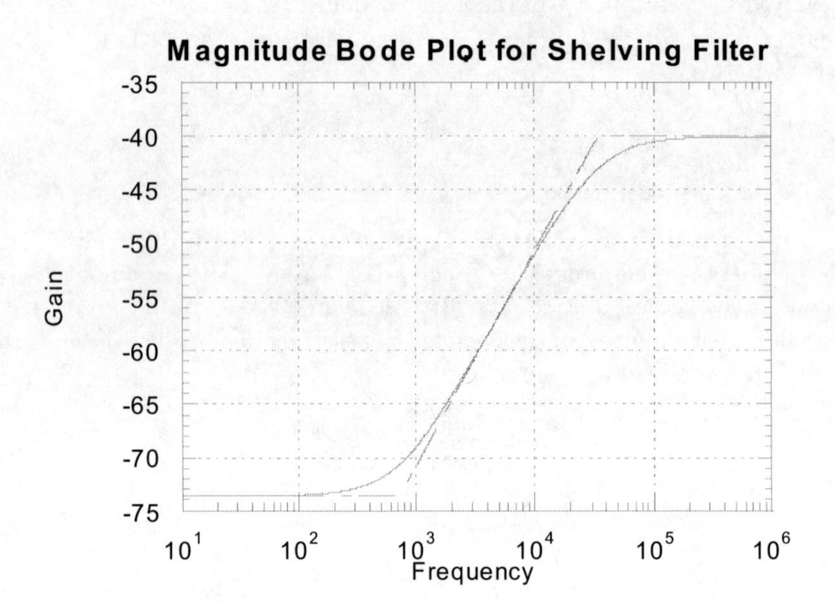

Figure 8.14: Magnitude Bode plot.

8.5.2 Frequency Response Measurement

Assemble the circuit shown in Fig. 8.15. Connect the Sync output of the FG to the external trigger on the oscilloscope. Set the trigger source of the scope to Ext/10 and the level to TTL. Frequency response data is to be obtained for each of the filters designed for this experiment. Data is to be obtained for at least one decade of the pass-band and two decades for the stop-band for each filter. It is left to the judgement of the experimenter to select the values of the frequency of the function generator for which data is taken.

The voltage at the output of the function generator will be constant unless the circuit significantly loads it. Thus the voltage measured here should be essentially constant.

The digital multimeter could be used to monitor the amplitude of the output of the filter instead of the oscilloscope. However, the oscilloscope must be used to measure the phase.

The actual values of the resistors and capacitors used should be measured and recorded. Use the **Fluke/Philips 6303 RCL** meter to measure the value of the capacitance of each capacitor. Use the digital multimeter to measure the resistance of each resistor.

The experimental value for the magnitude of the complex transfer function is given by

$$\left|\overline{T}(j\omega)\right|_{dB} = 20\log_{10}\left|\frac{\overline{V}_2}{\overline{V}_1}\right| \tag{8.64}$$

where either the peak, peak-to-peak, or *rms* value of the voltages at the input and output of the filter may be used.

Use **Agilent VEE, National Instruments LabVIEW,** or **National Instruments ELVIS** to plot the transfer function. Plot both the magnitude and phase of the transfer function. The lead connections for the function generator and oscilloscope are the same as previously but the output port of the filter must be connected to the input of the **Agilent 34401A DMM**.

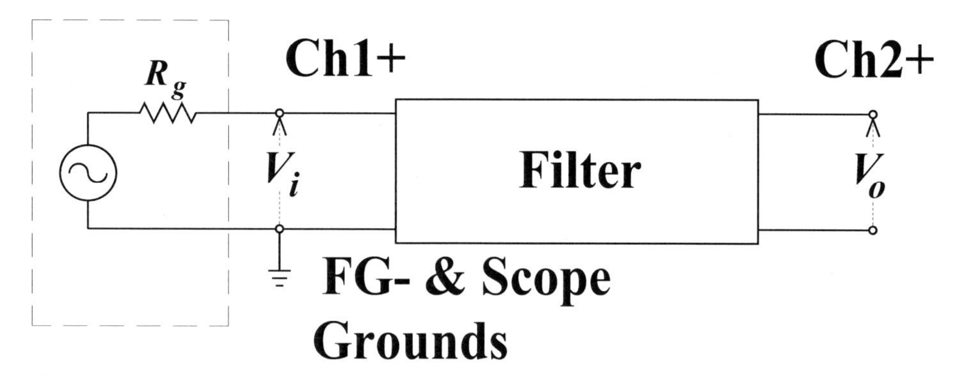

Figure 8.15: Frequency Response Measurement System

8.6 Laboratory Report

A comparison of the theoretical, simulation, and experimental results should be made. Any discrepancies should be explained. The simulation and experimental results should be presented as graphs or plots. From these it can be determined if the design specifications have been met.

8.7 References

1. Banzhaf, W., *Computer-Aided Circuit Analysis Using PSpice*, 2nd edition, Prentice-Hall, 1992.
2. Carlson, A. B., *Circuits*, Wiley, 1996.
3. Cooper, W. D., *Electronic Instrumentation and Measurements Techniques*, 2nd edition, Prentice-Hall, 1978.
4. D'Azzo, J. J., *Linear Control Systems*, McGraw-Hill, 1981.
5. Dorf, R. C., *Introduction to Electric Circuits*, 2nd edition, Wiley, 1993.
6. Jones, L., and Chin, A. F., *Electronic Instruments and Measurements*, John Wiley, 1983.
7. Keown, J., *PSpice and Circuit Analysis*, Merrill, 1991.
8. Kerchner, R. M., and Corcoran, G. F., *Alternating Current Circuits*, Wiley, 1962.
9. Krenz, J. H., *Introduction to Electrical Circuits and Electronic Devices: A Laboratory Approach*, Prentice-Hall, 1987.
10. Kantrowitz, P., et. al., *Electronic Measurements*, Prentice-Hall, 1979.
11. LePage, W. R., *Complex Variables and the Laplace Transform for Engineers*, Dover, 1961.
12. MicroSim Corporation, *PSpice Circuit Analysis*, MicroSim, 1990.
13. Oliver, B. M., and Cage, J. M., *Electronic Measurements and Instrumentation*, McGraw-Hill, New York, 1971.
14. Rashid, *SPICE for Circuits and Electronics Using PSpice*, Prentice-Hall, 1990.
15. Reddick, H. W., and Miller, F. H., *Advanced Mathematics for Engineers*, Wiley, 1938.
16. Riaz, M., *Electrical Engineering Laboratory Manual*, McGraw-Hill, 1965.
17. Sedra, A. S. and Smith, *K. C., Microelectronic Circuits*, 4th ed., Oxford, 1998.

18. Su, K. L., *Fundamentals of Circuit Analysis*, Waveland Press, 1993.
19. Tuinenga, P. W., *SPICE: A Guide to Circuit Simulation & Analysis Using PSpice*, Prentice-Hall, 1992.
20. Weinberg, L., *Network Analysis and Synthesis*, McGraw-Hill, 1962.
21. Wolf, S., *Guide to Electronic Measurements and Laboratory Practices*, 2nd edition, Prentice-Hall, 1983.
22. Wolf, S., and Smith, R. F. M., *Student Reference Manual for Electronic Instrumentation Laboratories*, Prentice-Hall, 1990.

Chapter 9

Second-Order Circuits

9.1 Objective

Circuits for which the relationship between the excitation and response is described by a second-order linear ordinary differential equation with constant coefficients are known as second-order circuits. Given the excitation and the circuit, the response may be obtained by solving this differential equation. Such circuits may have responses which differ radically from those of first-order circuits. Phenomenon such as resonance and damped oscillation may occur which do not occur in first-order circuits. Such circuits are widely used as electrical filters.

Two different energy storage devices are normally found in second order circuits: the inductor and the capacitor. Energy is stored in the magnetic field of the inductor and in the electric field of the capacitor. If the excitation does not contain impulses or higher order singularities, then the energy stored in the inductor and capacitor must be continuous functions of time. This requires that for nonimpulsive excitations the inductor current and capacitor voltage must be continuous functions of time. Since the resistor voltage is proportional to the current, all the circuit variables must be continuous except for the excitation and the inductor voltage.

This experiment will examine the response of two second-order circuits: the series RLC and the parallel GLC. The response will be obtained for step function excitation, square wave, and sinusoidal excitation. The principle of duality will be used to obtain the excitation for the parallel GLC circuit from the series RLC.

9.2 Theory

9.2.1 Series RLC

Shown in Fig. 9.1 is the series RLC circuit. The voltage source $e(t)$ is the excitation. The response may be either the current flowing through the circuit or the voltage across any of the three passive elements (RLC). If the capacitor voltage can be determined, then the current and the voltage drop across the other two passive elements can be determined from the constitutive equations for the voltage-current relations for the three passive elements.

An ordinary differential equation for the capacitor voltage can be obtained by applying Kirchoff's voltage law around the loop

$$v_L(t) + v_R(t) + v_c(t) = e(t) \tag{9.1}$$

where $v_L(t)$ is the voltage drop across the inductor, $v_R(t)$ is the voltage drop across the resistor, and $v_c(t)$ is the voltage drop across the capacitor. Since

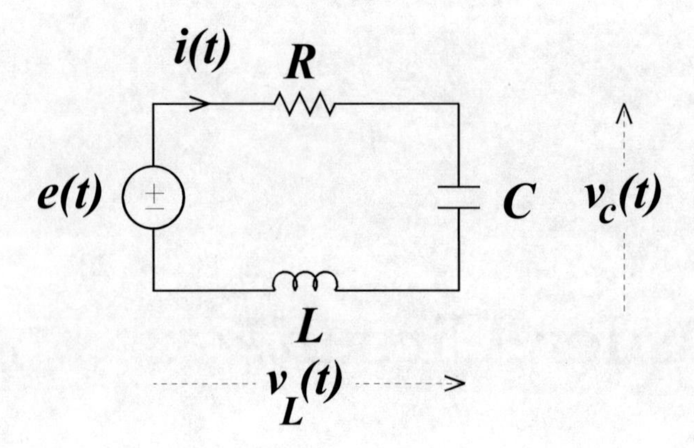

Figure 9.1: Series RLC circuit.

$$v_R(t) = R\, i(t) \text{ and } \quad i(t) = i_c(t) = C\frac{dv_c(t)}{dt} \tag{9.2}$$

the loop equation becomes

$$LC\frac{d^2v_c(t)}{dt^2} + RC\frac{dv_c(t)}{dt} + v_c(t) = e(t) \tag{9.3}$$

which is the desired second-order ordinary differential equation relating the capacitor voltage to the excitation. Thus, if the excitation is given along with two initial conditions, then the response, $v_c(t)$, may be obtained.

9.2.2 Step Function Response

The step function response is the solution for the current and element voltages when the excitation is a step function. Given $e(t) = E\ u(t)$ as shown in Fig. 9.2 the differential equation for $v_c(t)$ must be solved. Obviously, everything is zero prior to $t = 0$ because the voltage source is a short circuit across the series combination of the three passive components. To solve for $v_c(t)$ for $t > 0$ requires the initial conditions for $v_c(t)$ and $\frac{dv_c(t)}{dt}$ at $t = 0$. These initial conditions may be obtained from the required continuity for the inductor current and capacitor voltage. The continuity of capacitor voltage requires that $v_c(0) = 0$ and the continuity of inductor current requires that $\left.\frac{dv_c(t)}{dt}\right|_{t=0} = 0$.

The differential equation may be placed in a canonical form for second order systems by the introduction of the definitions:

$$\omega_o = \frac{1}{\sqrt{LC}} \equiv \text{the natural frequency } [radians/sec] \tag{9.4}$$

and

$$\zeta = \frac{R}{2}\sqrt{\frac{C}{L}} \equiv \text{the damping factor [dimensionless]} \tag{9.5}$$

which determines the form of the response. Namely, the form of the response is determined by whether ζ is less than, equal to, or greater than one.

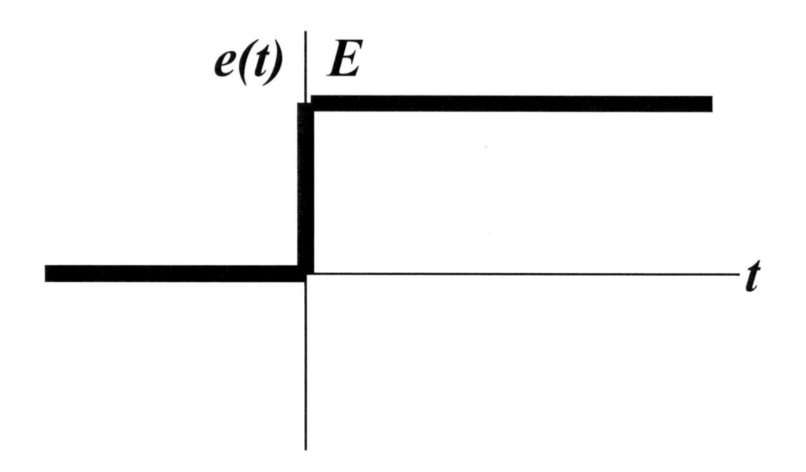

Figure 9.2: Step function $e(t) = Eu(t)$.

If $\zeta < 1$ the system or circuit is said to be underdamped and response consists of damped sinusoidal oscillations. The circuit is said to be critically damped if ζ is exactly equal to one which produces a response that is nonoscillatory but decays exponentially. Circuits with values of $\zeta > 1$ are overdamped and have responses that also decay exponentially.

Underdamped Response $(0 < \zeta < 1)$

The solution to the differential equation for $v_c(t)$ for the underdamped response $\zeta < 1$ is

$$i(t) = \frac{E}{R}\frac{2\zeta}{\sqrt{1-\zeta^2}}e^{-\alpha t}\sin\,\omega_d t \tag{9.6}$$

$$v_c(t) = E\left[1 - \frac{1}{\sqrt{1-\zeta^2}}e^{-\alpha t}\sin(\omega_d t + \phi)\right] \tag{9.7}$$

$$v_L(t) = -E\frac{1}{\sqrt{1-\zeta^2}}e^{-\alpha t}\sin(\omega_d t - \phi) \tag{9.8}$$

where $\alpha = \dfrac{R}{2L}$ is known as the decay constant of the envelope [$\sec^{-1}$], $\omega_d = \omega_o\sqrt{1-\varsigma^2}$ is the angular driven frequency [$radians/sec$], $\phi = \cos^{-1}\zeta$, and $\omega_o = 1\,/\sqrt{LC}$ is the resonant or natural frequency [$radians/sec$]. The damping factor $\zeta = \dfrac{R}{2}\sqrt{C/L}$ is a dimensionless quantity that lies in the range $0 < \zeta < 1$.

For the underdamped response the current and element voltages contain an oscillating term for which the amplitude of the oscillations decreases exponentially with time. This rate of decay is determined by α the decay constant of the envelope. Thus $1/\alpha$ is analogous to the time constants that are associated with first-order circuits. The frequency of the oscillations is given by $f_d = \omega_d/2\pi$ which is known as the driven frequency.

Critically Damped Response $(\zeta = 1)$

If the damping factor is exactly equal to one, the response is given by

$$i(t) = \frac{E}{R} 2\alpha t \; e^{-\alpha t} \tag{9.9}$$

$$v_L(t) = E[1 - \alpha t]e^{-\alpha t} \tag{9.10}$$

$$v_c(t) = E[1 - (1 + \alpha t)e^{-\alpha t}] \tag{9.11}$$

where $\alpha = \zeta\omega_o = R/2L$. For critically damping to occur R must equal $2\sqrt{L/C}$.

Overdamped Response ($\zeta > 1$)

The overdamped response is obtained when the damping factor $\zeta > 1$. The equations for this case may be obtained from the solution for the underdamped response by the introduction of the parameter $\beta = -j\omega_d$. The solution is then given by

$$i(t) = \frac{E}{R} \frac{2\zeta}{\sqrt{\zeta^2 - 1}} e^{-\alpha t} \; \sinh \beta t \tag{9.12}$$

$$v_c(t) = E\left[1 - \frac{1}{\sqrt{\zeta^2 - 1}} e^{-\alpha t} \; \sinh(\beta t + \phi)\right] \tag{9.13}$$

$$v_L(t) = -E\frac{1}{\sqrt{\zeta^2 - 1}} e^{-\alpha t} \; \sinh(\beta t - \phi) \tag{9.14}$$

where the constant $\alpha = R/2L$, $\phi = \cosh^{-1}\zeta$, and $\beta = \omega_o\sqrt{\varsigma^2 - 1}$.

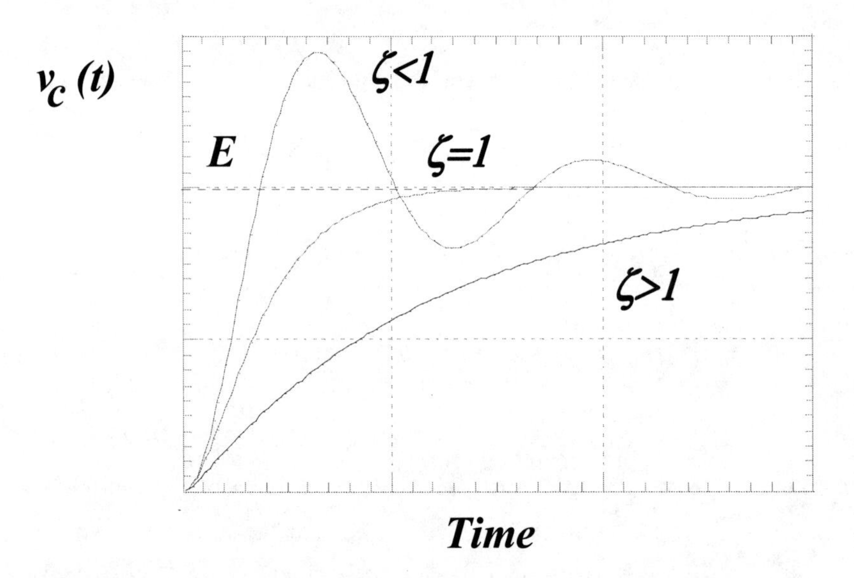

Figure 9.3: Capacitor voltage in series RLC circuit for step function excitation.

The capacitor voltage in the series RLC circuit is shown in Fig. 9.3 for various values of ζ. The value of ω_o has been held constant by holding the value of L and C constant while R is varied. When $R = 0$ the response would be undamped. Values of R in the range $0 < R < 2\sqrt{L/C}$ result in the underdamped response; $R = 2\sqrt{L/C}$ produces the critically damped response; $R > 2\sqrt{L/C}$ yields the overdamped response.

For the undamped response, $\zeta = R = 0$, the current is a sinusoidal function of time and the amplitude of the sinusoidal is constant. Energy is continuously exchanged between the magnetic field of the inductor and the electric field of the capacitor. The driven frequency, ω_d, is equal the resonant or natural frequency, ω_o, i.e. this is the frequency at which the circuit would naturally like to oscillate. As the resistance of the circuit is increased, the energy lost per cycle increases and the amplitude of the sinusoidal decreases on each cycle of the response. For the critically damped and overdamped responses, the energy loss is great enough to prevent oscillation.

9.2.3 Square Wave Excitation

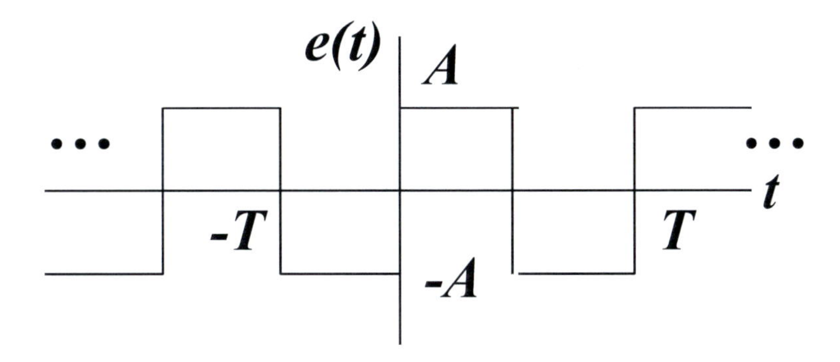

Figure 9.4: Symmetric square wave with a dc level of zero.

The square wave response is the solution for the current and element voltages when the excitation is a symmetric square wave with a DC level of zero. Given $e(t)$, as shown in Fig. 9.4, the differential equation for $v_c(t)$ must be solved. Since $e(t)$ is a periodic function with period T, the current and element voltages will also be periodic functions with period T; this means that, if a solution can be found for a time interval equal to the period, the solution has been obtained for all values of t. Additionally, because $e(t)$ has half-wave odd symmetry, i.e. $e(t) = -e(t + T/2)$, the current and element voltages will also have half-wave odd symmetry which means that a solution need only to be found for the time interval $0 \leq t \leq T/2$.

The differential equation for the time interval $0 < t < T/2$ is given by

$$\frac{d^2 v_c(t)}{dt^2} + 2\zeta\omega_o \frac{dv_c(t)}{dt} + \omega_o^2 v_c(t) = \omega_o^2 A \tag{9.15}$$

where A is the peak value of the square wave. If it is assumed that $\zeta < 1$, the solution to Eq. 9.15 is

$$v_c(t) = A + e^{-\alpha t}[\gamma_1 \cos(\omega_d t) + \gamma_2 \sin(\omega_d t)] \quad 0 \leq t \leq T/2 \tag{9.16}$$

where $\alpha = \zeta\omega_o = R/(2L)$, $\omega_d = \omega_o\sqrt{1-\zeta^2}$, $\zeta = \frac{R}{2}\sqrt{\frac{C}{L}}$, $\omega_o = 1/\sqrt{LC}$, and γ_1 and γ_2 are constants that will be determined from the initial conditions on the inductor current and capacitor voltage.

Since this is a series circuit, the capacitor current is equal to the series current and is given by

$$i(t) = i_R(t) = i_L(t) = i_c(t) = C\frac{dv_c(t)}{dt} \tag{9.17}$$

from which the current can be obtained by substituting Eq. 9.16 into Eq. 9.17 which yields

$$\begin{aligned} i(t) = Ce^{-\alpha t}\{&\omega_d[-\gamma_1 \sin(\omega_d t) + \gamma_2 \cos(\omega_d t)] \\ &-\alpha\left[\gamma_1 \cos(\omega_d t) + \gamma_2 \sin(\omega_d t)\right]\} \end{aligned} \tag{9.18}$$

which is also only valid in the time interval $0 \leq t \leq T/2$.

Substituting $t = 0$ into Eqs. 9.16 and 9.18 yields

$$v_c(0) = A + \gamma_1 \tag{9.19}$$

and

$$i(0) = C\omega_d\gamma_2 - \alpha C\gamma_1 \tag{9.20}$$

which can be used along with the trigonometric identity

$$\sin(A + B) = \sin(A)\cos(B) + \cos(A)\sin(B) \tag{9.21}$$

to express the capacitor voltage and the series current in the time interval $0 \leq t \leq T/2$ as

$$v_c(t) = A - \frac{e^{-\alpha t}}{\sqrt{1-\zeta^2}}\left[[A - v_c(0)]\sin(\omega_d t + \phi) - \frac{Ri(0)\sin(\omega_d t)}{2\zeta}\right] \tag{9.22}$$

and

$$i(t) = -\frac{e^{-\alpha t}}{\sqrt{1-\zeta^2}}\left[\frac{2\zeta}{R}[A - v_c(0)]\sin(\omega_d t) + i(0)\sin(\omega_d t - \phi)\right] \tag{9.23}$$

where $\phi = \cos^{-1}(\zeta)$. Thus, the capacitor current and voltage for the time interval $0 \leq t \leq T/2$ can be determined from Eqs. 9.22 and 9.23 if the initial values of the capacitor voltage and series current are known.

The capacitor voltage and series current can be determined for the time interval $T/2 \leq t \leq T$ by making use of the fact that because of the half-wave odd symmetry of $e(t)$, the current and all the voltages must also have this half-wave odd symmetry. The equations for the current and capacitor voltage in this time interval may be obtained from Eqs. 9.22 and 9.23 by making the substitution

$$t \longrightarrow t - T/2 \tag{9.24}$$

$$A \longrightarrow -A \tag{9.25}$$

$$t = 0 \longrightarrow t = T/2 \tag{9.26}$$

on the right hand side.

The initial values of the capacitor voltage and series current may be found by making use of the odd half-wave symmetry property which requires that $v_c(0) = -v_c(T/2)$ and $i(0) = -i(T/2)$. Substituting $t = T/2$ into Eqs. 9.22 and 9.23 then yields

$$-v_c(0) = A - \frac{e^{-\pi\zeta(T/T_o)}}{\sqrt{1-\zeta^2}}\left\{[A - v_c(0)]\sin\left[\pi\frac{T}{T_o}\sqrt{1-\zeta^2} + \phi\right] - \frac{Ri(0)}{2\zeta}\sin\left[\pi\frac{T}{T_o}\sqrt{1-\zeta^2}\right]\right\} \tag{9.27}$$

for $-v_c(0) = v_c(T/2)$ and

$$-i(0) = -\frac{e^{-\pi\zeta(T/T_o)}}{\sqrt{1-\zeta^2}}\left\{[A - v_c(0)]\frac{2\zeta}{R}\sin\left[\pi\frac{T}{T_o}\sqrt{1-\zeta^2}\right] + i(0)\sin\left[\pi\frac{T}{T_o}\sqrt{1-\zeta^2} - \phi\right]\right\} \tag{9.28}$$

for $-i(0) = i(T/2)$ where $T_o = 2\pi/\omega_o$. Eqs. 9.27 and 9.28 may be solved simultaneously for $i(0)$ and $v_c(0)$.

The solutions for the initial conditions obtained by solving Eqs. 9.27 and 9.28 becomes

$$\begin{bmatrix} v_c(0) \\ v_R(0) \end{bmatrix} = A \begin{bmatrix} a_{11} & a_{12} \\ a_{21} & a_{22} \end{bmatrix}^{-1} \begin{bmatrix} b_{11} \\ b_{21} \end{bmatrix} \tag{9.29}$$

where the coefficients in the matrices are given by:

$$a_{11} = -\left[1 + \frac{e^{-\pi\zeta(T/T_o)}}{\sqrt{1-\zeta^2}} \sin\left[\pi\frac{T}{T_o}\sqrt{1-\zeta^2} + \phi\right]\right] \tag{9.30}$$

$$a_{22} = \frac{a_{11}}{4\zeta^2} \tag{9.31}$$

$$a_{12} = a_{21} = b_{21} = -\frac{1}{2\zeta}\frac{e^{-\pi\zeta(T/T_o)}}{\sqrt{1-\zeta^2}} \sin\left[\pi\frac{T}{T_o}\sqrt{1-\zeta^2}\right] \tag{9.32}$$

$$b_{11} = 1 - \frac{e^{-\pi\zeta(T/T_o)}}{\sqrt{1-\zeta^2}} \sin\left[\pi\frac{T}{T_o}\sqrt{1-\zeta^2} + \phi\right] \tag{9.33}$$

where $v_R(0) = Ri(0)$ is the initial value of the voltage drop across the resistor.

If ζT is much greater than T_o, then

$$v_c(0) \simeq -A \tag{9.34}$$

and

$$i(0) \simeq 0 \tag{9.35}$$

which yields that the square wave response is identical to the step function response with a step amplitude of $2A$. Therefore, the step function response may be obtained by exciting the series RLC circuit with a symmetric square wave with a DC level of zero and a period that is long compared to T_o/ζ where T_o is the period associated with the frequency f_o which is the natural or resonant frequency of the circuit.

9.2.4 Sinusoidal Excitation

If the excitation is a sinusoidal function of time such as $e(t) = A\cos(\omega t)$, then the response may be obtained by solving the differential equation for $v_c(t)$ using classical techniques. However, for any problem with a sinusoidal excitation, it is far easier to use a complex phasor representation for the current and voltages. The real current and voltages may then be easily obtained from the corresponding phasors.

The complex phasor representation for $e(t) = A\cos\,\omega t$ is $\bar{E} = A/\sqrt{2}\angle 0^o$. This is known as a rms phasor since the magnitude of the complex quantity $\bar{E}$ is the peak value of the sinusoidal divided by the square root of 2 which is the rms value of the sinusoidal.

The complex phasor current $\bar{I}$ is given by $\bar{I} = \bar{E}/\bar{Z}$ where

$$\bar{Z} = R + j\omega L + 1/j\omega C = R + j[\omega L - 1/\omega C] \tag{9.36}$$

is the complex phasor impedance expressed in rectangular form. Expressed in polar form the complex impedance $\bar{Z}$ is

$$\bar{Z} = Z\angle\phi \tag{9.37}$$

where

$$Z = \sqrt{R^2 + [\omega L - 1/\omega C]^2} \tag{9.38}$$

is the magnitude of the impedance which is the ratio of the peak value of the voltage $e(t)$ to the peak value of the current $i(t)$ which is also equal to the ratio of the rms value of the voltage to the rms value of the current and

$$\phi = \tan^{-1}\left[\frac{\omega L - 1/\omega C}{R}\right] \tag{9.39}$$

is the phase angle by which the voltage $e(t)$ leads the current $i(t)$. The current as a function of time is then given by

$$i(t) = \frac{A}{Z}\cos(\omega t - \phi) \tag{9.40}$$

which is a sinusoidal with the same frequency as $e(t)$.

The complex admittance could be used to relate the complex phasor current to the complex phasor source voltage. The complex phasor admittance $\bar{Y}$ is just the reciprocal of the complex phasor impedance, i.e. $\bar{Y} = 1/\bar{Z}$. The magnitude of $\bar{Y}$ is just the reciprocal of the magnitude of $\bar{Z}$ and the angle of $\bar{Y}$ is the negative of the angle of $\bar{Z}$.

The time-average power supplied to the series combination of the resistor, inductor, and capacitor is given by

$$P = EI\ \cos(\phi) \tag{9.41}$$

where E is the magnitude of the rms phasor voltage $\bar{E}$ and I is the magnitude of the rms phasor current $\bar{I}$. Since no time-average power is dissipated in either the inductor or the capacitor, the time-average power supplied by the voltage source $e(t)$ is also given by

$$P = I^2R \tag{9.42}$$

which is the power dissipated in the resistor.

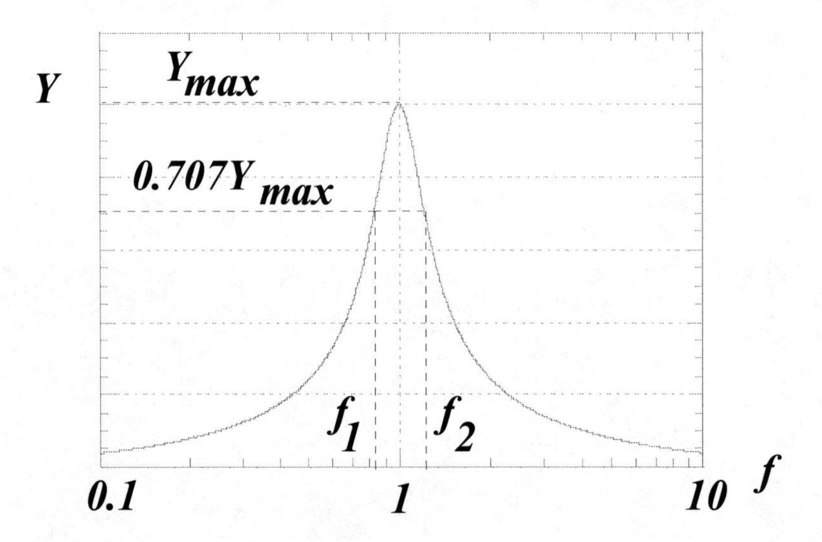

Figure 9.5: Magnitude of the admittance of a series RLC circuit versus frequency.

Shown in Fig. 9.5 is a typical plot of the magnitude of the admittance of a series RLC circuit as a function of the frequency of the source with the value of the circuit components held constant. The magnitude of the admittance achieves a maximum at a frequency f_o at which the inductive and capacitive reactance cancel. This frequency which is obtained by setting $\omega_o L = 1/\omega_o C$ and is given by

$$f_o = \frac{1}{2\pi\sqrt{LC}} \tag{9.43}$$

and is known as the resonant frequency of the circuit. At this frequency $\bar{Z} = R$ and the phase angle of $\bar{Z}$ is zero which means that there is no difference between the relationship between $e(t)$ and $i(t)$ and a circuit which simply has a resistor of R Ohms in series with a voltage source of $e(t)$ Volts.

The time-average power supplied by $e(t)$ to the series RLC circuit is

$$P = I^2 R = \frac{1}{2}\frac{A^2 R}{R^2 + [\omega L - 1/\omega C]^2} \tag{9.44}$$

which has its maximum value at resonance, i.e. $f = f_o$. The maximum power is $P = A^2/2R$. At two frequencies above and below resonance the power delivered to the circuit by the voltage source drops to one-half the maximum power. These two frequencies are known as the half-power frequencies and at these frequencies the difference between the inductive and capacitive reactance is $\pm R$. The lower half-power frequency, f_1, is given by

$$f_1 = \frac{R}{4\pi L}\left[\sqrt{1 + 4\left[\frac{\omega_o L}{R}\right]^2} - 1\right] \tag{9.45}$$

and the upper half-power, f_2, is given by

$$f_2 = \frac{R}{4\pi L}\left[\sqrt{1 + 4\left[\frac{\omega_o L}{R}\right]^2} + 1\right] \tag{9.46}$$

and the difference between these two frequencies $\Delta f = f_2 - f_1 = R/(2\pi L)$ is known as the half-power bandwidth. At the half-power frequencies the value of the magnitude of $\bar{Y}$ is $0.707\ G$ where $G = 1/R$ is the value of $\bar{Y}$ at resonance.

The quality factor of a series RLC circuit is defined to be the ratio of the resonant frequency to the half-power bandwidth. It determines the "sharpness" of the admittance plot or impedance plot and, therefore, the effectiveness of these circuits in filter applications. The quality factor for the series RLC circuit is

$$Q = \frac{f_o}{\Delta f} = \frac{f_o}{f_2 - f_1} = \frac{\omega_o L}{R} \tag{9.47}$$

which is a dimensionless quantity since it is the ratio of two frequencies.

9.2.5 Transfer Functions

The second order series RLC circuit can be considered as a system with the excitation as the input and either the resistor, inductor, or capacitor voltage as the output. Complex transfer functions can then be defined as the ratio of the complex phasor resistor, inductor, or capacitor voltage to the complex phasor source voltage.

The three possible transfer functions are then:

$$\bar{T}_R = \frac{\overline{V}_R}{\overline{E}} = \frac{1}{1 + jQ\left[\dfrac{f}{f_o} - \dfrac{f_o}{f}\right]} \tag{9.48}$$

$$\bar{T}_L = \frac{\overline{V}_L}{\overline{E}} = \frac{jQ\dfrac{f}{f_o}}{1 + jQ\left[\dfrac{f}{f_o} - \dfrac{f_o}{f}\right]} \tag{9.49}$$

$$\bar{T}_C = \frac{\overline{V}_C}{\overline{E}} = \frac{-jQ\,\dfrac{f}{f_o}}{1 + jQ\left[\dfrac{f}{f_o} - \dfrac{f_o}{f}\right]} \tag{9.50}$$

which express the relationships between the element voltages and the source voltage. These transfer functions show that at resonance the resistor voltage is equal to the source voltage, the inductor voltage has a value of Q times the source voltage and leads the source voltage by 90°, and that the capacitor voltage is equal in magnitude to the inductor voltage and lags the source voltage by $90°$. These three transfer functions are plotted in Fig. 9.6.

The maximum value of the magnitude of the resistor transfer function occurs at resonance. The maximum values of the capacitor and inductor transfer functions do not occur at resonance. However, for high Q circuits the peak values are near resonance.

9.3 Parallel GLC

The parallel GLC circuit is shown in Fig. 9.7. The current source $i(t)$ is the excitation and the voltage across the four parallel elements, $v(t)$, and the currents in the three passive circuit components are the response. The differential equation relating $i_L(t)$ to the excitation may be obtained and solved for a specified $i(t)$. Once $i_L(t)$ is known the capacitor current, parallel voltage, and resistor current may be found. Fortunately, this circuit is the dual of the series RLC circuit which means that the equations for the parallel RLC can be obtained from those for the series RLC.

The differential equation for $i_L(t)$ is

$$CL\frac{d^2 i_L(t)}{dt^2} + GL\frac{di_L(t)}{dt} + i_L(t) = i(t) \tag{9.51}$$

which is obtained by applying Kirchoff's current law to the parallel GLC circuit. This equation has exactly the same form as the differential equation for $v_c(t)$ for the series RLC circuit. This means that the equation for $i_L(t)$ for the parallel GLC circuit may be obtained from the equation for $v_c(t)$ for the series RLC circuit by replacing the "v" in the equation for the series RLC with an "i", changing the subscript from "c" to "L", replacing the "R" with a "G", replacing the "L" with a "C", and the "C" with an "L".

Any of the equations for the parallel GLC circuit may be obtained from the equations for the series RLC circuit by making the following exchanges in variable and parameters:

Series RLC	i	v	E	I	R	G	C	L	Z	Y	V	I
⇕	⇕	⇕	⇕	⇕	⇕	⇕	⇕	⇕	⇕	⇕	⇕	⇕
Parallel GLC	v	i	I	E	G	R	L	C	Y	Z	I	V

where the complex phasors are used only when the excitation is sinusoidal.

9.4 Circuit Component Models

The passive circuit components known as resistors, capacitors, and inductors, defined by the constitutive voltage current relationships

$$v(t) = Ri(t) \qquad i(t) = C\frac{dv(t)}{dt} \qquad v(t) = L\frac{di(t)}{dt} \tag{9.52}$$

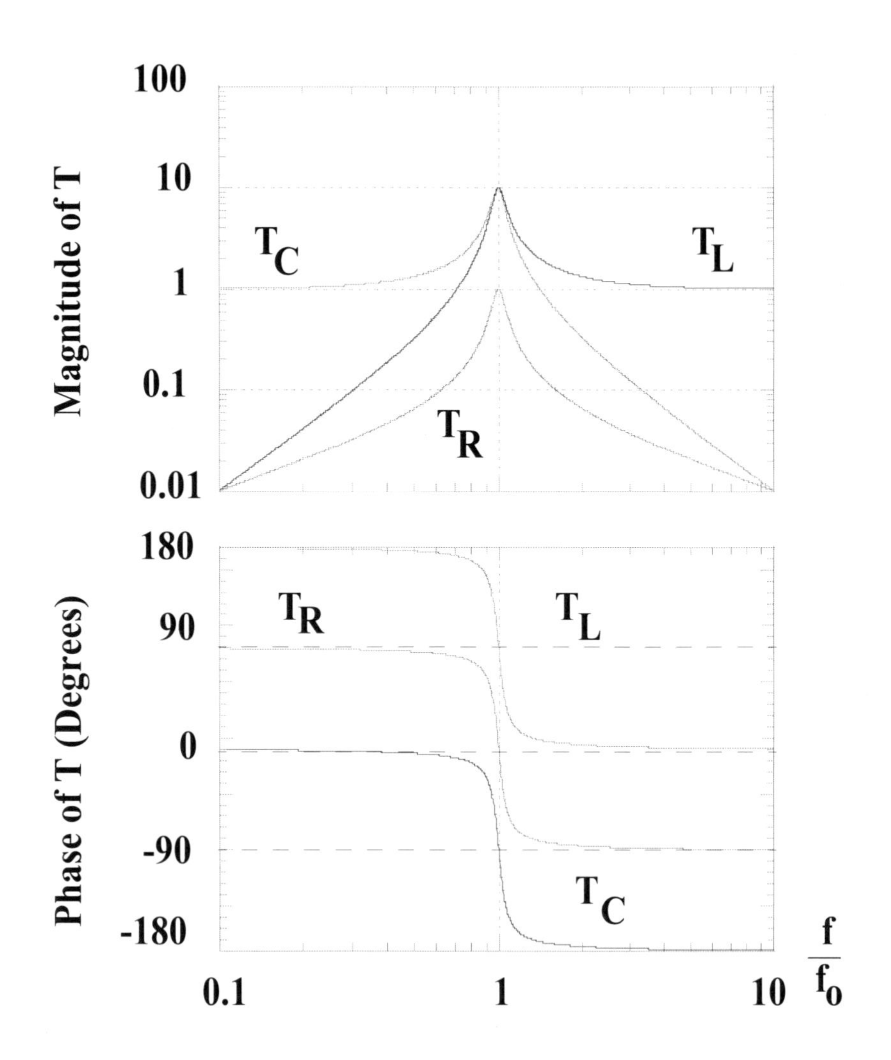

Figure 9.6: Transfer functions for second order series RLC circuit.

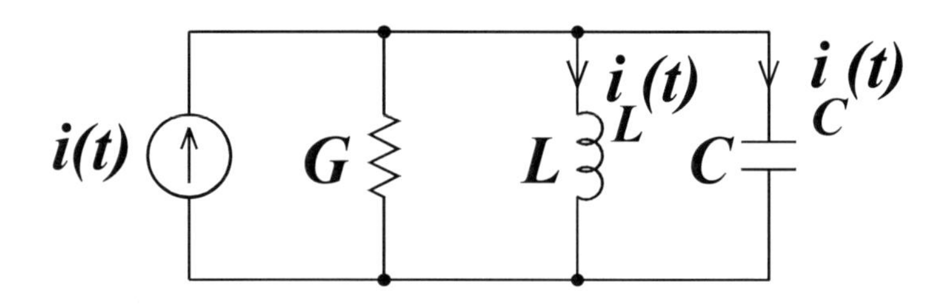

Figure 9.7: Parallel GLC circuit.

are mathematical idealizations. No physical circuit component can be modeled as purely resistive, inductive and capacitive over all ranges of frequency and voltage and current levels. Passive physical circuit elements are classified as resistors, inductors, or capacitors depending on which of these three mathematical idealizations they most closely resemble.

At frequencies below the VHF range, most physical resistors that are constructed of carbon can be accurately modeled as ideal resistors. As long as the power dissipated in a physical resistor is not excessive, at these low frequencies, the relationship $v(t) = Ri(t)$ is very accurate.

At frequencies below VHF, most physical capacitors are modeled as ideal capacitors in parallel with a resistance which represents leakage currents flowing between the plates of the capacitor. As long as the voltage drop across the plates of the capacitor is less than the voltage that produces breakdown of the dielectric material between the capacitor plates, this model is highly accurate. For many types of capacitors, such as ceramic disk and mylar, this parallel resistance is so large that it may be neglected as the physical capacitor can be accurately modeled as an ideal capacitor.

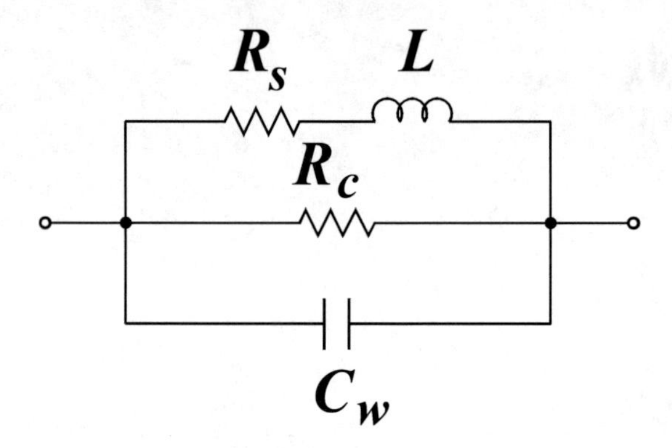

Figure 9.8: Equivalent circuit model for a physical inductor.

Physical inductors are not as well behaved as resistors and capacitors. Rarely can a physical inductor be accurately modeled as an ideal inductor. Shown in Fig. 9.8 is the circuit model that is usually used for physical inductors. The resistor R_s represents the "DC" resistance of the windings of wire that are used to construct the inductor. The ideal inductor L is in series with R_s and the series combination is in parallel with a resistor R_c, which represents losses in the core of the inductor, and a capacitor C_w which represents the capacitance of the windings of wire used to construct the inductor. If the frequency is sufficient low (below the VHF range), the capacitor C_w can be ignored and if the core is not particularly lossy the resistor R_c can be neglected. However, the resistor R_s will usually have to be included unless $\omega L >> R_s$. Inductors also act as "antennas" for picking up stray magnetic fields.

9.5 SPICE

The SPICE input deck code for the transient analysis of the under-damped second-order series RLC circuit shown in Fig. 9.3 is

```
SECOND-ORDER SERIES RLC
VI 1 0 PWL(0 1 30U 1)
R 1 2 190
L 2 3 2.53M IC=0
C 3 0 0.01U IC=0
```

```
.TRAN 1N 30U 0 1N UIC
.PROBE
.END
```

The parameters in the argument for PWL (piece wise linear) are ordered pairs of time and voltage. Only two points are required: t=0, VI=1; t=30U, VI=1 since SPICE begins at time t=0.

The initial conditions for the capacitor voltage and inductor current are zero. Unlike AC analyses the order of node specification is of the utmost importance in transient analyses. The first node for the capacitor specifies the assumed positive terminal for the voltage and the first node for the inductor is the assumed terminal into which a positive current flows.

The control line .TRAN invokes a transient analysis. The parameters for this control line are: print step interval (1N), final time (30U), results delay (0), step ceiling (1N), and use the initial condition (UIC). The parameter UIC is required if nonzero initial conditions are to be used in the analysis.

SPICE may also be used to obtain AC analyses of second-order circuits to obtain voltages, currents, and powers. To obtain the power the formula $P = VI\cos(\theta)$ must be specified in the argument of the probe function. Care must be exercised since the probe function P( ) returns the phase in degrees whereas the probe function COS( ) requires an argument in radians.

9.6 Procedure

9.6.1 Component Measurement

Measure and record the value of the resistor R_1, inductor L, and capacitor C provided with the **Fluke/Philips PM 6303** *RCL* meter set to *RCL AUTO* (the circuit component to be measured is placed across the terminals indicated as X on the *RCL* meter). The *RCL* meter will automatically determine whether the circuit component is an inductor, capacitor, or resistor and calculate and display the component value. Switch the function on the *RCL* meter to R_s and measure and record the series resistance of the inductor. Also measure the parallel resistance of the capacitor. These components will be used in a series *RLC* circuit with simulated step function excitation (simulated by using a low-frequency square wave), square wave excitation, and sinusoidal excitation. Calculate the value of α and ζ for the series *RLC* circuit using

$$\alpha = \frac{R}{2L} \quad \text{and} \quad \zeta = \frac{R}{2}\sqrt{\frac{C}{L}} \tag{9.53}$$

and record these values as the theoretical values of the attenuation constant of the envelope and the damping factor for the series *RLC* circuit. The value of R in the above equations is the total series resistance of the circuit to be assembled and is therefore the value of the resistance provided plus the internal resistance of the function generator as shown in Fig. 9.9 [$50\,\Omega$ for the **Tektronix 3022B**] plus the internal resistance of the inductor, R_s. If $\zeta \geq 1$ obtain a resistor with a smaller value of resistance and repeat the above. If $\zeta < 0.05$ obtain a larger value of R_1 and repeat the above.

Measure the values of these circuit components using **ELVIS**. Compare with the values measured with the **Fluke/Philips** meter.

9.6.2 Function Generator Setting

The **Tektronix 3022B** function generator displays the amplitude of the voltage available at its output terminals. In order for the **Tektronix 3022B** function generator to indicate the open circuit or Thévenin voltage in the display when amplitude is selected the output termination must be set to *HIGH Z*. The default or power-on setting is $50\,\Omega$ which would then result in a display one half of the Thévenin source voltage. Set the function generator for High Z.

9.6.3 Amplitude Adjustment

Set the amplitude of the waveform produced by the **Tektronix 3022B** function generator to 2 V peak-to-peak. Set the function to square.

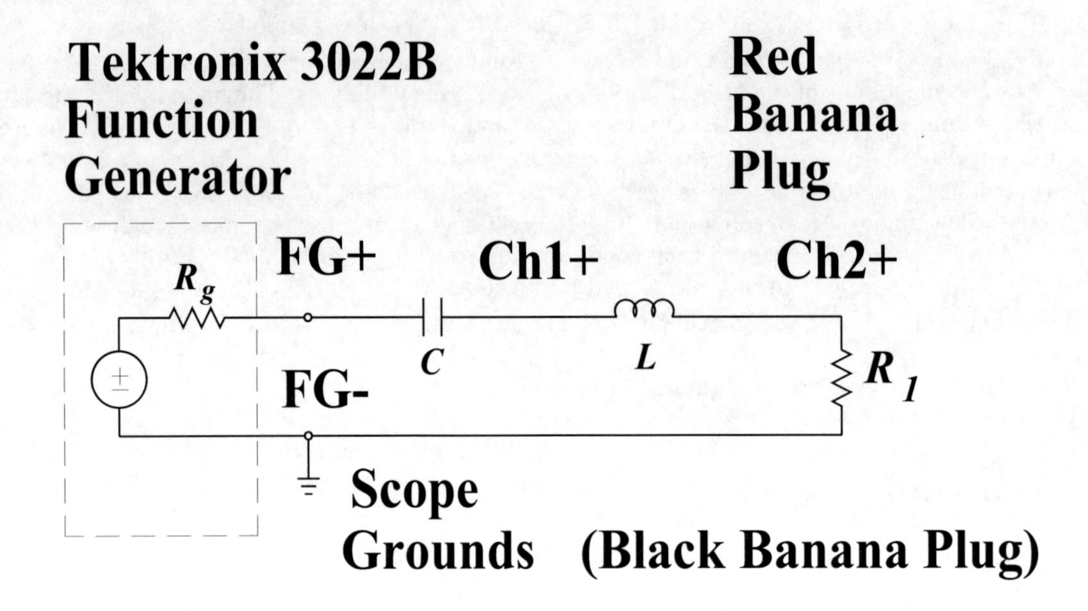

Figure 9.9: Series *RLC* Square Wave Excitation

9.6.4 Step Function Response of Series *RLC*

Assemble the circuit shown in Fig. 9.9 using the CH1 output of the **Tektronix 3022B** function generator as the excitation [the resistor R_s is not shown since it is the internal series resistance of the inductor]. Set the frequency of the function generator to $f_{sq} = 0.1\alpha$ where α is the theoretical value of the decay constant of the envelope calculated in step 9.6.1. Use a BNC to BNC connector to connect the *SYNC* or TTL output on the **Tektronix 3022B** function generator to *EXT TRIG* on the **Tektronix 3012B** oscilloscope.

Press *AUTOSET* on the **Tektronix 3012B** oscilloscope. Set the trigger source on the oscilloscope to Ext/10 and the trigger level to TTL Vary the vertical position knob for $CH2$ until the waveform is positioned in the lower half of the screen. Manually change the *Volts/div* knob for $CH2$ until the waveform occupies about 4 major vertical divisions (the lower half of the screen). Manually change the *Time/Div* knob until only one complete cycle is shown.

Press the *MATH* button followed by Dual Wfm Math, and set the operator to minus so that the difference of $CH1$ and $CH2$ will be displayed.. Turn the display for $CH1$ off by pressing the yellow $CH1$ button followed by *OFF*. Press *MATH* and then change the scale and vertical position so that it occupies the upper half of the screen.

The *MATH* function is now displaying the inductor voltage and $CH2$ the voltage across the resistor, R_1, which is, of course, proportional to the current. From this display the experimental values of f_d, the driven frequency, and α, the attenuation factor of the envelope, are determined. The measurement cursors on the oscilloscope can be used for these measurements. The driven frequency f_d is just the frequency of the damped oscillations and may be obtained from either the plot of the inductor or resistor voltage. The decay constant of the envelope α may be obtained from either the resistor or inductor voltage plots by picking two arbitrary times t_1 and t_2 and measuring the heights of the envelopes (V_1 and V_2) and determining α as

$$\alpha = \frac{1}{t_2 - t_1} \ln \left[\frac{V_1}{V_2} \right] \tag{9.54}$$

which has units of $s^{-1} [V_1 > V_2]$.

Turn on the cursors by pressing the *CURSORS* button. Select H Bars to measure voltages and V bars to measure times. The waveform being measured by the cursors is determined by pushing *CH*1, *CH*2, or *MATH*. The active cursor is determined with the SELECT button. Position the $V1$, $V2$, $t1$, and $t2$ cursors to correspond to two points on the envelope of the waveform as outlined above.

Print the display.

9.6.5 Square Wave Response of Series *RLC*

Set the frequency of the function generator to $\alpha/2$. Press *AUTOSET*. Manually vary the controls so that approximately two cycles are displayed. Print the display of the series current and inductor voltage

9.6.6 Component Measurement

Measure the value of the resistor R_2 with the **Fluke/Philips 6303** *RCL* meter set to *RCL AUTO*. Calculate the theoretical values of f_o and Q

$$f_o = \frac{1}{2\pi\sqrt{LC}} \qquad Q = \frac{\omega_o L}{R_s + R_2} \tag{9.55}$$

and record these values. If $Q > 10$ obtain a larger resistor for R_2 and repeat the above. If $Q < 5$ obtain a smaller resistor for R_2 and repeat the above.

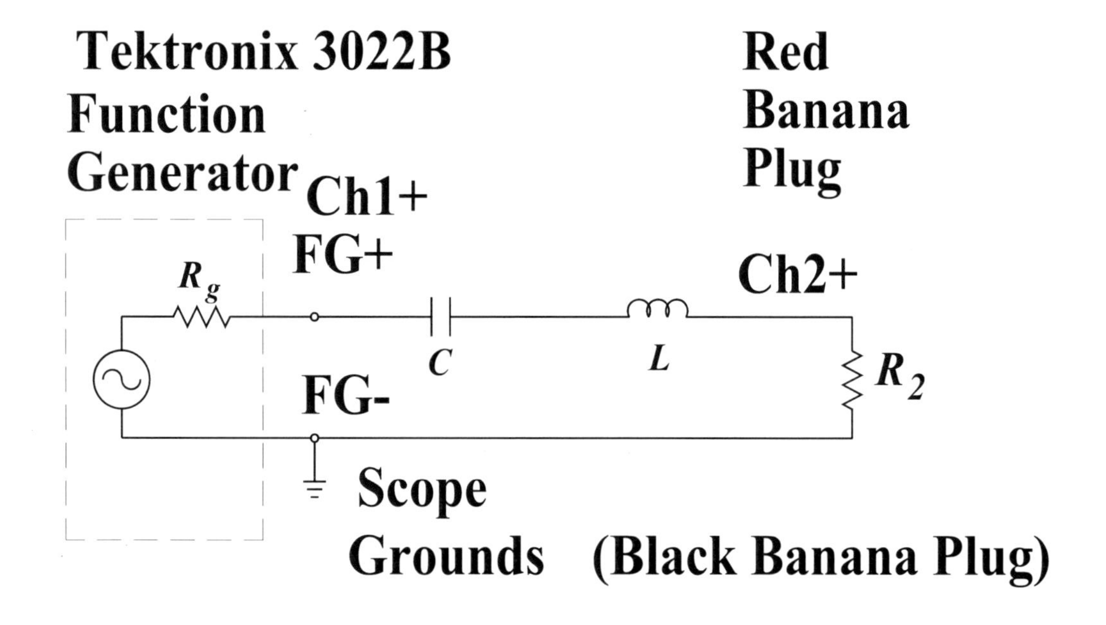

Figure 9.10: Sinusoidal *RLC* Response

9.6.7 Sinusoidal Response of Series *RLC* Circuit

Assemble the circuit shown in Fig. 9.10 [the resistor R_s is the internal series resistance of the inductor and is not shown on the diagram]. Leave the connection from the synch out of the function generator to external trigger of the oscilloscope.

Set the function on the function generator to sine and the frequency to the calculated value of the resonant frequency. Press *AUTOSET* and set the trigger source for external and select EXT/10 and set the level to TTL. Press DISPLAY, XY Display off, and then triggered xy. Slightly vary the frequency of the function generator until the ellipse becomes a straight line. This is the resonant frequency of the circuit. Simply record the number appearing in the display for the **Tektronix 3022B** function generator as the resonant frequency for the circuit. Print the display. Switch the display back to YT.

Data will be taken of the peak-to-peak voltage on $CH1$, V_{p-p1} (the voltage across the series combination of R_2, L, and C), the peak-to-peak voltage on $CH2$, V_{p-p2} (the voltage across the resistor R_2 which is proportional to the series current), and the phase shift between these two sinusoidal voltages. The voltage measurement menu may be used to measure these voltages. It is not necessary to print or sketch the display.

The amplitude of the function generator will be held constant and the frequency will be varied. (The 2 V_{p-p} used in the previous step is adequate.) Data should be taken for frequencies one decade above and below resonance. Convenient values to use are 1, 2, and 5 times the appropriate power of 10. Once the experimental value of resonance has been determined, several data points should be taken near resonance.

Measure the frequency response data and record it on the data sheet at the end of the experiment. The values of f, as stated above, are selected to cover the range one decade above and below resonance and the value of V_{p-p1}, V_{p-p2}, and $\Delta\phi$ are measured from the waveforms on the oscilloscope. The value of V_{p-p1} will remain approximately constant except near resonance while the value of V_{p-p2} will be small except for values of frequency near resonance. The phase shift is measured using $v_1(t)$ as the reference with the phase angle positive if $v_1(t)$ leads $v_2(t)$.

The phase shift may be measured using the time cursors. First, use the vertical position controls for $Ch1$ and $Ch2$ to adjust the reference or ground for each to the same horizontal line. Second, position the time cursors so that the interval of time Δt between which the two waveforms cross 0 V with a positive slope can be measured. The time interval Δt is measured and the phase shift $\Delta\phi$ is calculated as $\Delta\phi = 360° \times \Delta t f$.

Use **Agilent VEE, National Instruments LabVIEW**, or **National Instruments ELVIS** to obtain the frequency response of the circuit. Use the same connections for the function generator and oscilloscope but also connect the voltage across the resistor to the input of the **Agilent 34401A DMM**. Obtain only the amplitude of the frequency response. Obtain data for the same range as in the previous section using the same input amplitude for the function generator voltage. Print the results.

9.6.8 Component Measurement

Measure and record the value of R_3 and R_4 provided on the **Fluke/Philips 6303** *RCL* meter set to *RCL AUTOSET* and **ELVIS**. These will be used to construct the parallel *GLC* circuit shown in Fig. 9.11. Calculate and record the theoretical value of α, ζ, and ω_o for the parallel *GLC* circuit using

$$\alpha = \frac{G}{2C} \qquad \zeta = \frac{G}{2}\sqrt{\frac{L}{C}} \qquad \omega_o = \frac{1}{\sqrt{LC}} \tag{9.56}$$

where the value of G to be used is the reciprocal of the total parallel resistance in the circuit and C is the total parallel capacitance. The total parallel capacitance is the sum of the capacitance of the capacitor provided and the input capacitance of the oscilloscope that will be used to measure $v(t)$ [13 pf for the **Tektronix 3012B** plus the capacitance of the coaxial cable used to make the connection from the circuit to the oscilloscope]. The total resistance consists of the parallel combination of $(R_3 + 50)$, R_4, R_p, and R_{sp} where R_p is the equivalent parallel resistance of the physical inductor and R_{sp} is the input resistance of the oscilloscope that will be used to measure $v(t)$ [1 MΩ for the **Tektronix 3012B**]. The value that will be used for R_p is

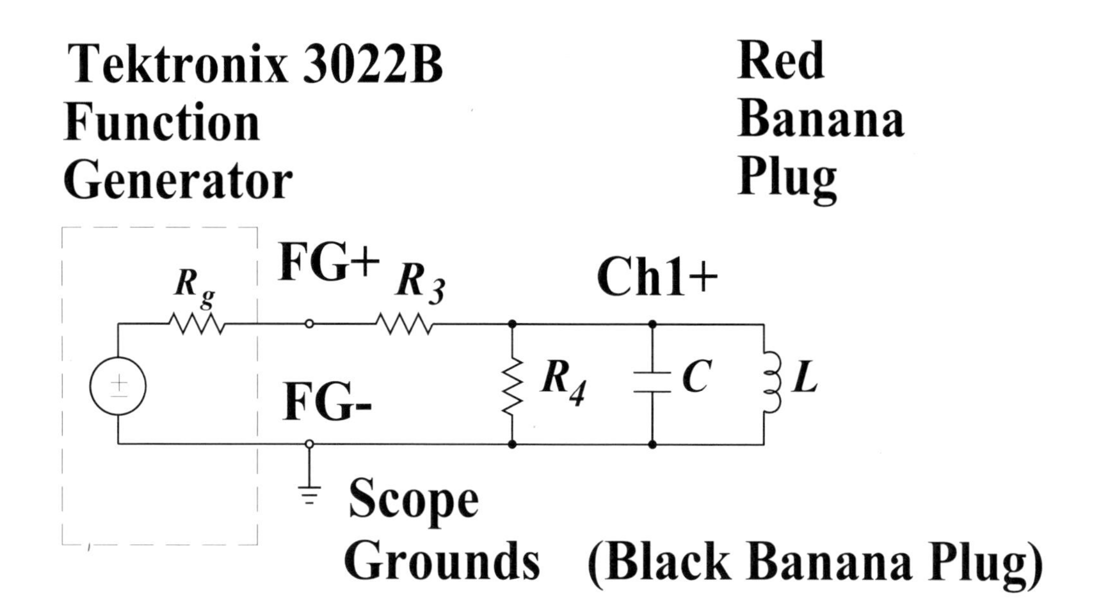

Figure 9.11: Parallel GLC Circuit

$$R_p = \frac{\omega_o^2 L^2}{R_s} \tag{9.57}$$

where R_s is the series resistance measured in step 9.6.1.

If $\zeta \geq 1$ obtain larger values for R_3 and R_4 and repeat the above. If $\zeta < 0.2$ obtain smaller value for R_3 and R_4 and repeat.

Follow the same procedure as in step 9.6.4, i.e. set the function generator to the same settings and to a frequency of $f_{sq} = 0.1\alpha$ where α is the theoretical value of the decay constant for the envelope of the parallel GLC circuit calculated above. Position cursors so that the attenuation factor of the envelope and driven frequency may be measured. Print or sketch $v(t)$ versus t. Determine from the sketch the experimental values of α and f_d.

9.7 Laboratory Report

The laboratory report should contain the following:

9.7.1 Parameter Measurement

A tabular summary of the measured value of the circuit components.

9.7.2 Step Function Excitation of Series RLC

1. Printouts of the current $i(t)$ and inductor voltage versus t that were measured on the oscilloscope. The current is the voltage that was measured on $Ch1$ divided by R_1. The inductor voltage $v_L(t)$ is the voltage that was obtained by taking the difference between the $Ch1$ and $Ch2$ voltages with the *Function* features of the oscilloscope.

Circuit Component	Measured Value
75 Ω R_1	
100Ω R_2	
1200Ω R_3	
1000Ω R_4	
3.3 mH L	
R_s	
.0068μF C	
R_p	

Table 9.1: Component Measurement

2. Theoretically calculated values of α, f_d, ζ, and f_o from the equations for these parameters in terms of the circuit components using the values of the circuit components that were measured by the **Fluke/Philips 6303** RCL meter. Experimentally measured values of α and f_d computed from Eq. 9.54 and the measured value of the driven frequency obtained in step 9.6.4. These parameters should be presented in tabular form and a comparison made.

Circuit Parameter	Theoretical Value	Measured Value	% Error
α			
f_d			
ζ		—	—
f_o		—	—

Table 9.2: Series RLC Step Parameters

9.7.3 Square Wave Excitation of Series RLC

The print outs taken.

9.7.4 Series RLC with Sinusoidal Excitation

1. The plots of the experimentally measured values of Y versus f.
The experimental values of Y are determined from the data by

$$Y = \frac{V_{p-p2}}{V_{p-p1}} \frac{1}{R_2} \tag{9.58}$$

because V_{p-p1} is measured across the series combination of the resistor, inductor, and capacitor while the current flowing through the series combination is the voltage V_{p-p2} divided by R_2.

The experimental values of Y versus f should be plotted on log-log graph paper. Three cycle by three cycle log-log graph paper should be used with the frequency (Hz) plotted one of the axis with three cycles and the magnitude of the admittance, Y (milli-mhos) plotted on the other. Plotted on the same sheet of graph paper should be the theoretical value of $Y = |\bar{Y}|$ where

$$\bar{Y} = \frac{1}{R_2} \frac{1}{1 + jQ\left[\frac{f}{f_o} - \frac{f_o}{f}\right]} \tag{9.59}$$

The plots may be made either with a spreadsheet or manually plotted on graph paper.
2. Experimental and theoretical values for f_o, Q, Δf.
3. The plot of the transfer function obtained with **VEE**. Compare this with the plot of the admittance obtained above. How do they differ?

Circuit Parameter	Theoretical	Experimental	% Error
f_o			
Q			
Δf			

Table 9.3: Series RLC Circuit Parameters

9.7.5 Parallel GLC with Step Function Excitation

1. Plot of $v(t)$ versus t.
2. Theoretically calculated values of α, f_d, ζ, and f_o. Experimentally measured values of α and f_d. These parameters should be presented in tabular form.

Circuit Parameter	Theoretical Value	Measured Value	% Error
α			
f_d			
ζ		—	—
f_o		—	—

Table 9.4: Parallel GLC Step Parameters

If the laboratory experiment and/or report cannot be completed by the end of the three hour lab session, turn in what has been completed to the laboratory instructor.

9.8 References

1. Banzhaf, W., *Computer-Aided Circuit Analysis Using PSpice*, 2nd ed, Prentice-Hall, 1992.
2. Carlson, A. B., *Circuits*, John Wiley, 1996.
3. Dorf, R. C., *Introduction to Electric Circuits*, 2nd edition, Wiley, 1993.
4. Irwin, J. D., *Basic Engineering Circuit Analysis*, MacMillan, 1993.
5. Keown, J., *PSpice and Circuit Analysis*, Merrill, 1991.
6. Kerchner, R. M., and Corcoran, G. F., *Alternating Current Circuits*, Wiley, 1962.
7. Krenz, J. H., *Introduction to Electrical Circuits and Electronic Devices: A Laboratory Approach*, Prentice-Hall, 1987.
8. Lago, G. V., and Waidelich, D. L., *Transients in Electrical Circuits*, Ronald, 1958.
9. LePage, W. R., *Complex Variables and the Laplace Transform for Engineers*, Dover, 1961.
10. MicroSim Corporation, *PSpice Circuit Analysis*, MicroSim, 1990.
11. Rashid, *SPICE for Circuits and Electronics Using PSpice*, Prentice-Hall, 1990.
12. Reddick, H. W., and Miller, F. H., *Advanced Mathematics for Engineers*, Wiley, 1938.
13. Riaz, M., *Electrical Engineering Laboratory Manual*, McGraw-Hill, 1965.
14. Su, K. L., *Fundamentals of Circuit Analysis*, Waveland Press, 1993.
15. Tuinenga, P. W., *SPICE: A Guide to Circuit Simulation & Analysis Using PSpice*, 3rd ed, Prentice-Hall, 1995.

Chapter 10

Transformers

10.1 Objective

The objective of this experiment is to examine an elementary broadband audio transformer. The equivalent circuit, transfer function, and nonlinear behavior will be examined. Applications that utilize the voltage, current, and impedance transformation properties will also be examined.

10.2 Theory

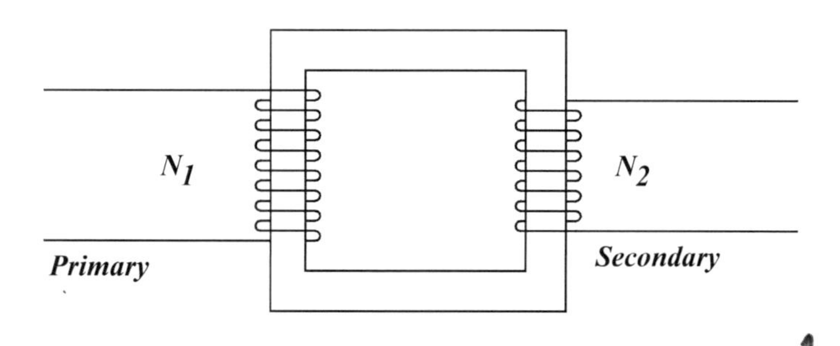

Figure 10.1: Elementary transformer.

An elementary transformer is shown in 10.1. It is a four-terminal, two-port circuit element that uses magnetic coupling to pass signals from the input port, known as the primary, to the output port, known as the secondary. Unless externally connected, complete electrical isolation exists between the primary and secondary, i. e., the dc resistance from the input to the output port is infinite. Normally, the primary consists of N_1 turns of wire and the secondary N_2 turns that are wound on a common core material, so that a significant portion of the flux produced by a current in one coil links the other. A time-varying current in one coil then induces a voltage in the other. The transformer may be used to step the voltage up or down.

Transformers, as circuit elements, offer the advantages of complete electrical isolation between the primary and secondary circuits, the ability to change voltage and current levels, and impedance transformation from the secondary to primary circuit. They are used at large power levels in the electric utility industry and at minute power levels in audio and radio-frequency applications. They are found in every electronic instrument or appliance which has an ac power cord.

10.2.1 Inductor

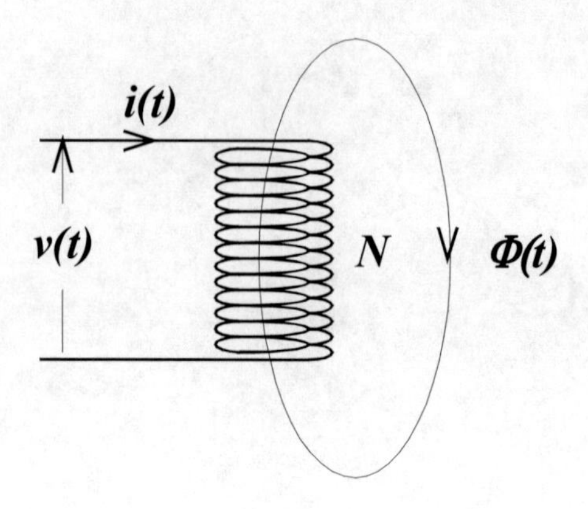

Figure 10.2: Elementary inductor.

Fundamental to the understanding of a transformer is the operation of an inductor. An inductor has only one coil, whereas a transformer has at least two coils. An inductor is shown in Fig. 10.2. It consists of N turns of wire that are wound so that the flux produced by each turn equally links all of the other turns. The current $i(t)$ passing through the wire produces a magnetic flux $\Phi(t)$. Assuming that this same magnetic flux, $\Phi(t)$, passes through each of the N turns, a quantity,

$$\lambda(t) = N\Phi(t) \tag{10.1}$$

known as the flux-linkages, may be defined. In the m. k. s. system of units, $\Phi(t)$ has units of webers.

Michael Faraday experimentally discovered that the voltage induced in a coil is given by

$$v(t) = \frac{d\lambda(t)}{dt} = N\frac{d\Phi(t)}{dt} \tag{10.2}$$

which is the basis for

$$\nabla \times \bar{E} = -\frac{\partial \bar{B}}{\partial t} \tag{10.3}$$

which is Maxwell's third equation, also known as Faraday's Law. Equation 10.2 is intuitively obvious since the N coils are in series, which means that the voltage induced is N times the voltage induced in only one turn.

The magnetic flux is a function of the current passing through the coil. Assuming that the current level is low enough so that the material through which the flux passes can be assumed to be linear, the flux is given by

$$\Phi(t) = \mathcal{P}Ni(t) \tag{10.4}$$

where $\mathcal{P}$ is a constant known as the permeance of the space occupied by the magnetic flux. The permeance is a function of the geometric shape of the inductor and the permeability of the media through which the magnetic flux passes. At large current levels, the permeance is a function of the level and history of the current that makes the equations non-linear and the analysis mathematically intractable by anything other than numerical means. The relationship for the flux as a function of the current in Eq. 10.4 is a direct consequence of Ampère's Law:

$$\nabla \times \bar{H} = \bar{J} + \frac{\partial \bar{D}}{\partial t} \tag{10.5}$$

which is Maxwell's fourth equation. The product $Ni(t)$ is known as the magnetomotive force and has units of ampere-turns.

By substituting Eq. 10.4 into Eq. 10.1 and the result into Eq. 10.2, yields

$$v(t) = \mathcal{P}N^2 \frac{di(t)}{dt} \tag{10.6}$$

from which the parameter

$$L = \mathcal{P}N^2 \tag{10.7}$$

known as the inductance or self-inductance of the inductor is defined. It has units of Henries in the meter-kilogram-second (m. k. s.) system. These equations disclose that the inductance is proportional to the number of turns squared and that induced voltage is equal to the product the inductance and the rate of change of the current. These equations are based on the assumption that the current level is low enough so that the equations may be assume to be linear.

10.2.2 Mutual Inductance

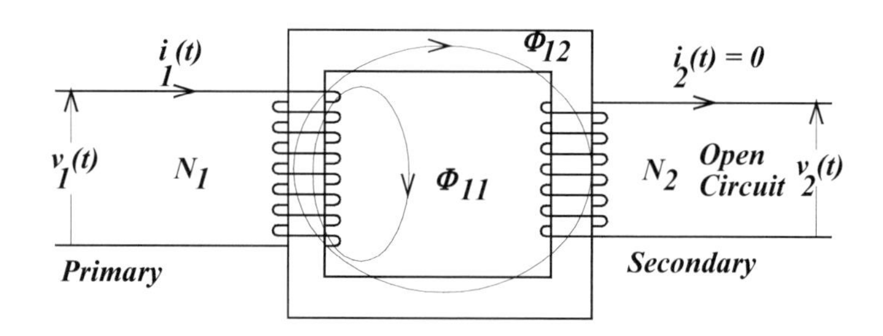

Figure 10.3: Mutually coupled coils with open secondary.

Two magnetically coupled coils are shown in Fig. 10.3. The secondary coil is open circuited, so that no current flows in it. There are N_1 turns of wire on the primary and N_2 turns of wire on the secondary. The coils are shown as being wound on a common closed structure. If this structure is composed of iron and has no breaks or air gaps, then virtually all of the flux produced by the current in the primary will link the secondary. If there are air gaps or if the structure is air, then only a portion of the flux produced in the primary will link the secondary.

The current $i_1(t)$ flowing in the primary coil produces the magnetic flux $\Phi_1(t)$. This flux can be resolved into two components as

$$\Phi_1(t) = \Phi_{11}(t) + \Phi_{12}(t) \tag{10.8}$$

where $\Phi_{11}(t)$ is the portion of the flux produced by the current in the primary that links only the primary and $\Phi_{12}(t)$ is the portion that links both coils.

The voltage induced in the secondary is given by Faraday's Law as

$$v_2(t) = \frac{d\lambda_2(t)}{dt} \tag{10.9}$$

where

$$\lambda_2(t) = N_2\Phi_{12}(t) \tag{10.10}$$

is the flux-linkages for the secondary. The flux $\Phi_{12}(t)$ is given by Ampere's Law as

$$\Phi_{12}(t) = \mathcal{P}_{12}N_1 i_1(t) \tag{10.11}$$

where $\mathcal{P}_{12}$ is the permeance of the path taken by the flux that is produced by the current in the primary that also links the secondary.

Combining Eqs. 10.9, 10.10, and 10.11 yields

$$v_2(t) = \mathcal{P}_{12} N_1 N_2 \frac{di_1(t)}{dt} \tag{10.12}$$

as the voltage induced in the secondary by a time varying current in the primary. Using Eq. 10.7, this can be expressed as

$$v_2(t) = \frac{\mathcal{P}_{12}}{\sqrt{\mathcal{P}_1 \mathcal{P}_2}} \sqrt{L_1 L_2} \frac{di_1(t)}{dt} \tag{10.13}$$

from which the mutual inductance M between the primary and secondary can be defined as

$$M = k\sqrt{L_1 L_2} \tag{10.14}$$

where

$$k = \frac{\mathcal{P}_{12}}{\sqrt{\mathcal{P}_1 \mathcal{P}_2}} \tag{10.15}$$

is known as the coefficient of coupling. The coefficient of coupling is a dimensionless quantity which satisfies the relationship

$$0 \leq k \leq 1 \tag{10.16}$$

which reflects the amount of flux produced by a current in one coil that links the second. If $k = 1$ there is unity coupling which means that all of the flux produced by the primary links the secondary. Conversely, if $k = 0$ the coils are uncoupled and none of the flux produced by the primary links the secondary. Finally, the induced voltage is given by

$$v_2(t) = M \frac{di_1(t)}{dt} \tag{10.17}$$

where M is the mutual inductance and has the same units as the self inductance, Henries in the m. k. s. system.

If the magnetic path consists of a closed ring of material with a high permeability such as iron or steel then k will be close to unity and essentially all of produced by the current in the primary will link the secondary. Conversely, if there are air gaps and the material on which the coils are wound has permeability near that of free space, then the coefficient of coupling will be small and only a minuscule amount of the flux produced by the primary will link the secondary.

10.2.3 Ideal Transformer

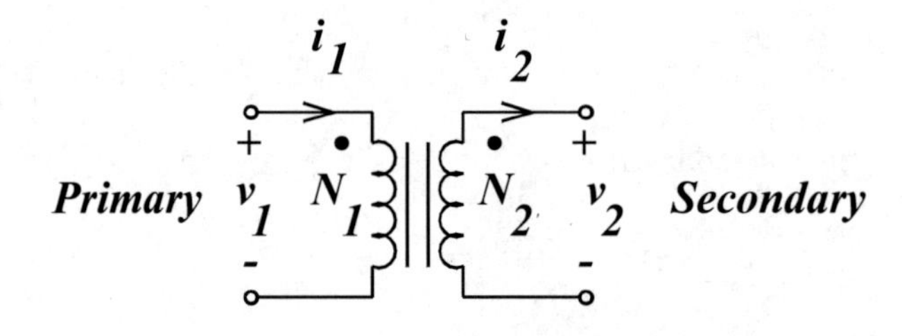

Figure 10.4: Ideal transformer.

The ideal transformer is shown in Fig. 10.4. It has N_1 turns of wire on the primary and N_2 turns on the secondary. The ideal transformer has a coefficient of coupling that is unity, i. e. $k = 1$. It is lossless which means that the electrical resistance of both coils is zero. The dots in the primary and secondary circuits are used to indicate the polarity of the voltages in these two windings. The two vertical lines mean that it has an iron core that yields the unity coupling.

The ideal transformer has the following properties

$$v_2 = nv_1 \tag{10.18}$$

and

$$i_2 = i_1/n \tag{10.19}$$

where

$$n = \frac{N_2}{N_1} \tag{10.20}$$

is known as the turns ratio. Thus, if $n > 1$, the voltage on the secondary is larger than the voltage on the primary, and it is known as a step-up transformer. For the converse, $n < 1$, the secondary voltage is smaller than the primary, and it is known as a step-down transformer. If $n = 1$, the only function of the transformer is to electrically isolate the circuits, and it is known as an isolation transformer.

Multiplying Eqs. 10.18 and 10.19 yields

$$p = v_1 i_1 = v_2 i_2 \tag{10.21}$$

which means that all of the power entering the primary leaves the secondary. Hence, the ideal transformer is lossless. It simply scales the voltage in one direction and the current in another, so that the product of the two remains constant.

10.2.4 Dot Convention

The dot convention is used with transformers to indicate the relative polarity of the voltage and currents in the primary and secondary circuits. The convention is such that the voltage at the dotted terminal in both the primary and secondary circuits has the same polarity. Thus, as current enters the dotted terminal in the primary in Fig. 10.4 ,this would make that terminal positive, this means that the voltage at the dotted terminal in the secondary is also positive, which would cause current to leave the dotted terminal in the secondary and make Eqs. 10.18 and 10.19 valid. If the positive direction for the current in the primary and secondary enters the dot, then the magnetic fluxes add. Since the direction of the flux is a function of both the direction of the current and the direction in which the wire is wound around the core, dots are required if current and voltage polarities are required. Equations applicable for other circuit variable and dot orientations are given in Fig. 10.5.

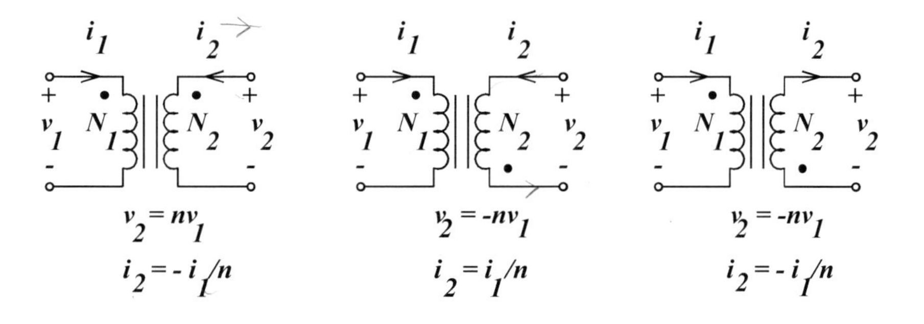

Figure 10.5: Dot convention.

Experimentally, dots can be established for a transformer by arbitrarily assigning a dot for the primary and observing the direction of the voltage induced in the secondary. For instance, a step function of voltage can be applied to the primary and the polarity induced in the secondary determines the placement of the dot.

The ideal transformer has some properties that, at first glance, seem peculiar. First, the inductance of both the primary and secondary windings is infinite as well as the mutual inductance. This is a consequence of Eq. 10.19. If the current in the secondary is zero, which occurs when the secondary is open circuited, then the current in the primary must also be zero. For this to occur when a voltage is applied to the primary, the inductance of the primary must be infinite. Second, the flux in the ideal transformer is zero. Since the voltage induced is proportional to the product of the inductance and the rate of change of flux, if the inductance is infinite, the flux must be zero if the voltage is to remain finite. Moreover, since the flux is zero, the energy stored in the ideal transformer is zero. That the flux stored is zero may also be seen from Eq. 10.19 which becomes

$$N_1 i_1 = N_2 i_2 \tag{10.22}$$

i.e., that the magnetomotive force applied to each coil is equal and in opposite directions which yields a flux of zero. The main utility of ideal transformers is modeling non-ideal transformers.

10.2.5 Controlled Source Model

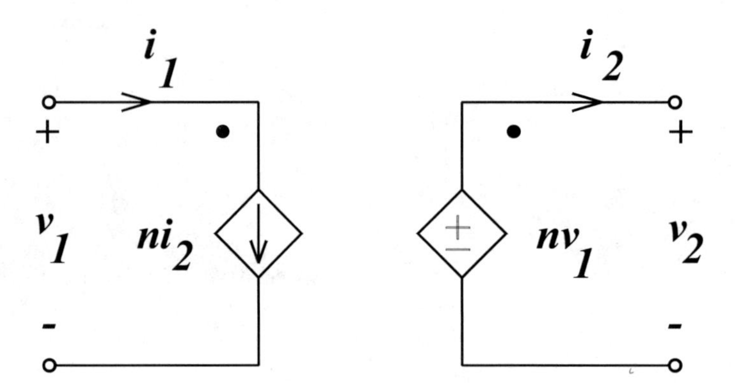

Figure 10.6: Controlled source model for ideal transformer.

A controlled source model for an ideal transformer is shown in Fig. 10.6. It follows directly from Eqs. 10.18 and 10.19. This model is useful in solving circuits problems with ideal transformers.

10.2.6 Impedance Transformation

In Fig. 10.7 an ideal transformer is shown with an impedance $\bar{Z}_L$ connected to the secondary. The input impedance is

$$\bar{Z}_i = \frac{\bar{V}_1}{\bar{I}_1} \tag{10.23}$$

where the circuit variables are being expressed using complex phasors. Eqs. 10.18 and 10.19 then yield

$$\bar{Z}_i = \frac{\bar{Z}_L}{n^2} \tag{10.24}$$

which is the impedance scaling property of an ideal transformer. So the impedance seen looking into the primary is equal to the load impedance on the secondary divided by the turns ratio squared. This is one

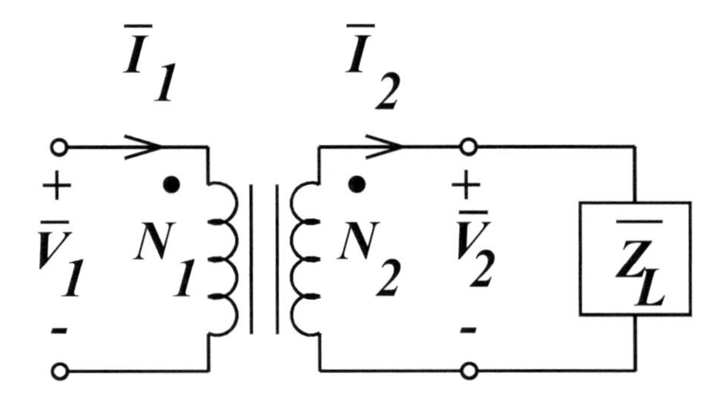

Figure 10.7: Impedance transformation.

of the most important and fundamental properties of a transformer. It permits the matching of source and load impedance for maximum power transfer.

10.2.7 Linear Non-Ideal Transformer

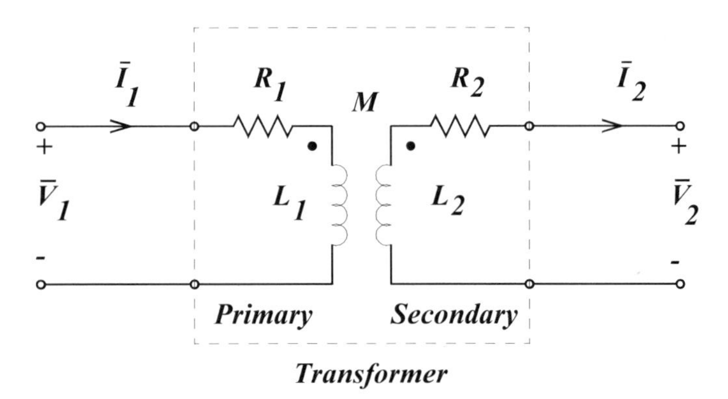

Figure 10.8: Lineary model for non-ideal transformer.

A circuit diagram for a linear model for a non-ideal transformer is shown in Fig. 10.8. Complex phasors are used to represent the voltages and currents. The inductance of the primary and secondary are L_1 and L_2, respectively, while M is the mutual inductance. The resistors R_1 and R_2 represent the dc resistance of the primary and secondary windings.

Applying Kirchoff's voltage law around the primary loop yields

$$\bar{V}_1 = (R_1 + j\omega L_1)\bar{I}_1 - j\omega M\bar{I}_2 \tag{10.25}$$

for the dot orientation shown in Fig. 10.8. The sign on the mutual inductance term in this equation is negative because the positive direction for the secondary current enters the un-dotted terminal which makes the un-dotted terminal in the primary positive in opposition to the primary current.

The application of Kirchoff's voltage law to the secondary loop yields

$$\bar{V}_2 = j\omega M\bar{I}_1 - (R_2 + j\omega L_2)\bar{I}_2 \tag{10.26}$$

for the output voltage. Here the sign on the mutual inductance term is positive since the primary current enters the dotted terminal which makes the induced voltage in the secondary positive at the dotted or top terminal.

Eqs. 10.25 and 10.26 become Eq. 10.18 if $k = 1$ and $R_1 = R_2 = 0$ which is the voltage relationship for an ideal transformer. But, to obtain the relationship for the currents in an ideal transformer, Eq. 10.19, the self and mutual inductances must be infinite.

10.2.8 Broadband Transformer

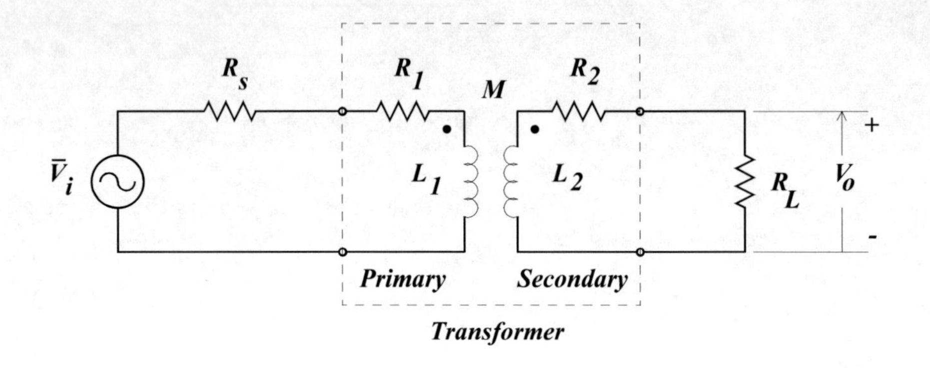

Figure 10.9: Broad-band, band-pass transformer.

The circuit shown in Fig. 10.9 is known as a broad-band, band-pass transformer if the primary and secondary are highly coupled, i. e., $k \approx 1$. This means that there is a relatively large band of frequencies for which the gain is approximately constant and the gain drops off at high and low frequencies. (The gain is the magnitude of the complex transfer function $\bar{T}(s) = \bar{V}_o(s)/\bar{V}_i(s)$ for the circuit shown.)

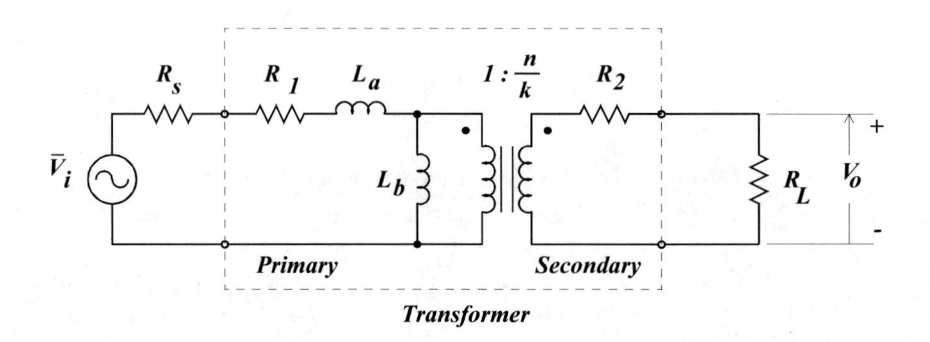

Figure 10.10: Equivalent circuit model for broad-band, band-pass transformer.

The analysis of the circuit shown in Fig. 10.9 is expedited by replacing it with the equivalent circuit shown in Fig. 10.10. In the equivalent circuit, the two coupled coils are replaced with two uncoupled inductors and an ideal transformer. It can be shown that these two circuits are equivalents by writing the terminal equations for the currents and voltages and showing that they are the same.

In the equivalent circuit the series inductor is given by

$$L_a = \left(1 - k^2\right) L_1 \tag{10.27}$$

and the shunt inductor

$$L_b = k^2 L_1 \tag{10.28}$$

where k is the coefficient of coupling and L_1 is the inductance of the primary of the transformer shown in Fig. 10.9. The ideal transformer has a turns ratio of n/k where n is the turns ratio as the transformer shown in Fig. 10.9. If $k \approx 1$, then $L_a \ll L_1$ and $L_b \approx L_1$.

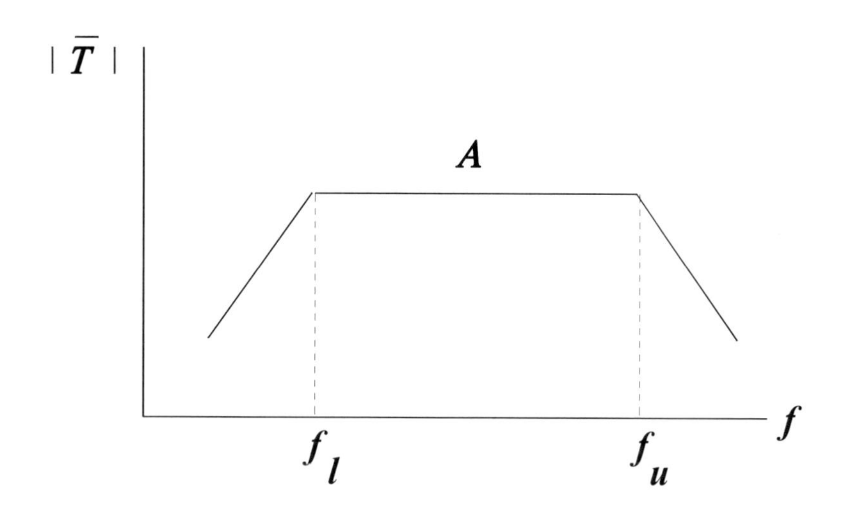

Figure 10.11: Transfer function for broad-band, band-pass transformer.

For $k \approx 1$ the magnitude of the complex transfer function of the circuit shown in Fig. 10.9 is approximately

$$\bar{T}(s) = \frac{\bar{V}_o(s)}{\bar{V}_i(s)} = \frac{As\tau_\ell}{(1+s\tau_\ell)(1+s\tau_u)} \tag{10.29}$$

which is sketched in Fig. 10.11. The mid-band gain is A and the upper and lower -3 dB frequencies are $f_\ell = 1/(2\pi\tau_\ell)$ and $f_u = 1/(2\pi\tau_u)$ respectively.

For frequencies in the mid-band $f_\ell \ll f \ll f_u$, the impedance of the series inductor, L_a, is small compared to the impedances in series with it (and can therefore be replaced with a short) and the impedance of the shunt inductor, L_b, is large compared to the impedance in parallel with it (and can therefore be replaced with an open circuit). In this frequency band the circuit elements are purely real and the midband gain is given by

$$A = \frac{k}{n}\frac{R_L}{R_a + R_b} \tag{10.30}$$

where

$$R_a = R_s + R_1 \tag{10.31}$$

is the total resistance in the primary circuit and

$$R_b = \frac{k^2}{n^2}(R_2 + R_L) \tag{10.32}$$

is the resistance coupled into the primary circuit by the ideal transformer.

The time constants for the poles may be determined by examining the high- and low-frequency behavior of the circuit shown in Fig. 10.10. At low frequencies, the inductor L_a is small compared to R_a, while the impedance of L_b is comparable to R_b. Therefore, the time constant τ_ℓ can be determined from the Thévenin equivalent circuit with respect to L_b as

$$\tau_\ell = \frac{L_b}{R_a \| R_b} \tag{10.33}$$

since at these frequencies L_a can be replaced with a short circuit. Similarly, the upper pole time constant is obtained by finding the Thévenin equivalent circuit with respect to L_a as

$$\tau_u = \frac{L_a}{R_a + R_b} \tag{10.34}$$

since at these frequencies L_b can be replaced by an open circuit.

An expression for k can be obtained using Eqs. 10.33 and 10.34 in terms of f_ℓ and f_u and the resistors in the circuit. The expression for k is

$$k = \frac{1}{\sqrt{1 + \dfrac{f_\ell}{f_u}\dfrac{R_a + R_b}{R_a \| R_b}}} \tag{10.35}$$

which is independent of L_1. This equation is useful in experimentally measuring k since the measurement of L_1 can be a Sisyphean task.

It should be borne in mind that Eq. 10.29 is only an approximate complex transfer function for the circuit shown in Fig. 10.9. Indeed, this expression is only valid when $k \approx 1$ which forces $f_u \gg f_\ell$. The general expression for complex transfer function for this circuit is

$$\bar{T}(s) = \frac{\bar{V}_o(s)}{\bar{V}_i(s)} = \frac{\dfrac{k}{n}\dfrac{R_L}{L_a} s}{s^2 + s\left(\dfrac{R_a}{L_a} + \dfrac{R_b}{L_a} + \dfrac{R_b}{L_b}\right) + \dfrac{R_a R_b}{L_a L_b}} \tag{10.36}$$

which reduces to Eq. 10.29 when k tends to unity.

10.2.9 Hysteresis

Inductors and transformers constructed with iron cores are nonlinear for sufficiently large magnetomotive forces because the core saturates at high current levels. Such inductors and transformers exhibit a behavior known as hysteresis, which is not possible with linear circuit elements. Linear circuit models such as those previously used do not exactly apply, although these circuits may be modified to approximate the nonlinear behavior.

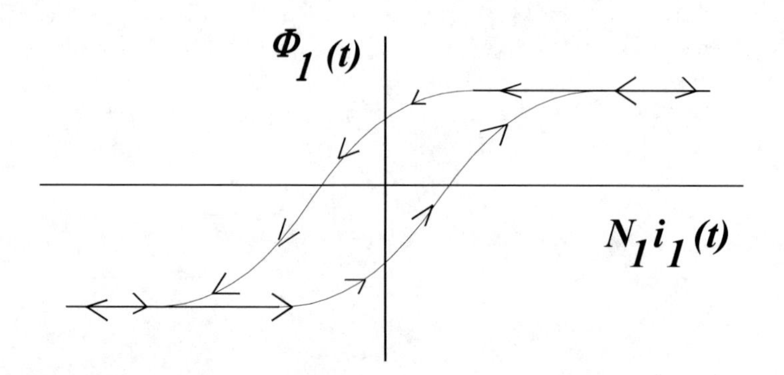

Figure 10.12: Hysteresis curve for iron core transformer.

If the secondary of a transformer is open circuited, a typical plot of the magnetic flux as a function of the primary current producing it for a iron core transformer is shown in Fig. 10.12. (For a linear transformer, this plot would be a straight line passing through the origin with a slope L_1/N_1.) This plot is known as a hysteresis curve for which the flux lags behind the current. (The term "hysteresis" comes from the Greek word "hysterein" which mean to lag.) The flux is not only a function of the magnitude of the current, but also whether the current is increasing or decreasing. The arrows indicate the portions of the plot that apply for increasing and decreasing current. (The upper portion is for decreasing current, and the lower portion is for increasing current.) For sufficiently large currents, the core saturates when $\Phi_1 = +\Phi_s$, i. e., increasing the current does not increase the flux once the core saturates. This hysteresis effect is a result of the magnetic domains in the core material being rotated in the presence of the time varying current flowing through the inductor.

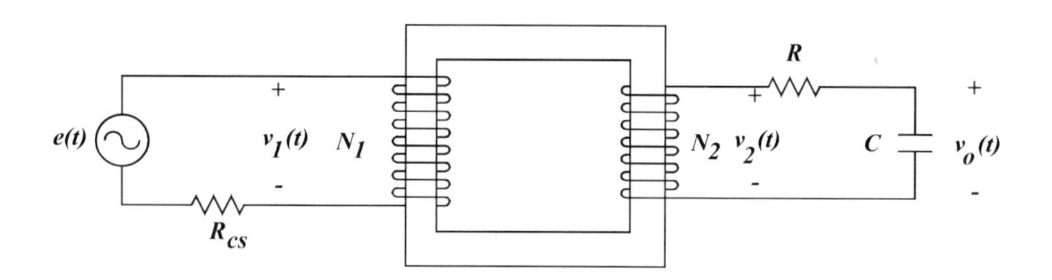

Figure 10.13: Circuit for obtaining hysteresis plot.

A plot of the hysteresis in an iron core transformer can be obtained by placing an integrator on the secondary. The voltage induced in the secondary is given by

$$v_2(t) = \frac{d\lambda_2(t)}{dt} \tag{10.37}$$

whether the circuit is linear or not. If this voltage is integrated, then the voltage at the output of the integrator is $\lambda_2(t)$. Such an arrangement is shown in Fig. 10.13.

If the excitation, $e(t)$, is a sinusoidal and the load is a capacitor, C, and the frequency of excitation $f >> 1/(2\pi RC)$, then the capacitor voltage, $v_o(t)$, is given by

$$v_o(t) = \frac{1}{RC}\int_{-\infty}^{t} v_2(u)du \quad = \frac{k}{RC}\frac{N_2}{N_1}\int_{-\infty}^{t} v_1(u)du \quad = \frac{k}{RC}N_2\Phi_1(t) \tag{10.38}$$

which means that the capacitor voltage is proportional to the magnetic flux linking coil 1. Since the voltage across resistor R_{cs} is proportional to the current flowing into the primary coil, this provides a method of obtaining the $\Phi - i$ curve for coil 1 with an instrument such as an oscilloscope. The capacitor voltage is displayed on the vertical axis of the oscilloscope and the voltage across resistor R_{cs} is displayed on the horizontal axis. This yields the desired hysteresis curve for the magnetic core material.

10.2.10 SPICE

Circuit analysis with SPICE requires that each node of the transformer be labeled. The coefficient of coupling and the inductance of the primary and the secondary are entered. The inductances must be specified with the dotted terminal entered first.

An example circuit in SPICE with the nodes labeled is shown in Fig. 10.14. The objective is to obtain a plot of the magnitude of the complex transfer function in dB as a function of the frequency in Hertz. The transformer is entered using the element K; the first two parameter for K are the primary and secondary inductors and the third is the coefficient of coupling. The PSpice input deck for the source code is

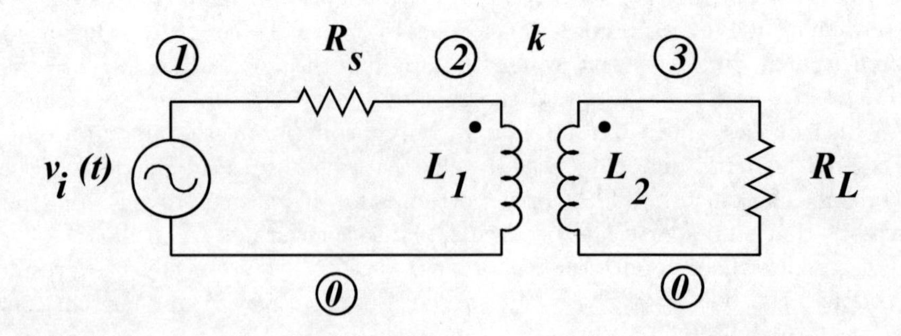

Figure 10.14: SPICE circuit.

```
BROAD-BAND BAND-PASS TRANSFORMER
VI 1 0 AC 1
RS 1 2 1K
L1 2 0 20
L2 3 0 5
K L1 L2 0.995
RL 3 0 3K
.AC DEC 100 1 100K
.PROBE
.END
```

Since a dc path to ground is required from each node to the ground node, the un-dotted node in both the primary and secondary was chosen as ground. If maintaining the electrical isolation between the primary and secondary is an important feature of the analysis, a large dummy resistor can be placed between the windings for the SPICE analysis.

The magnitude of the complex transfer function in dB (Gain) versus frequency is shown in Fig. 10.15. The input is the voltage source $v_i(t)$, and the output is the voltage across the load resistor R_L

10.3 Procedure

10.3.1 Specifications

The transformer examined in this experiment is an audio-frequency, broad-band, band-pass, two-coil device. One of its industrial applications is in telephone circuitry. It is specified to have a impedance transformation ratio of $4,000\,\Omega$ to $600\,\Omega$, which means that when a $600\,\Omega$ resistor is connected across the secondary the input impedance of the primary is $4,000\,\Omega$. When terminated with an impedance of $600\,\Omega$ and energized with a source with an impedance of $4,000\,\Omega$, it is specified to have a gain that is flat or constant to within $\pm$ 0.5 dB from 300 Hz to 3.5 kHz (the band of frequencies over which voice signals are transmitted over telephone lines). One of the goals of this experiment is to verify these specifications.

10.3.2 Resistance Measurement

One terminal should have a dot by it. This is for the primary. If there is no dot or two dots, the primary will be the winding with the largest resistance.

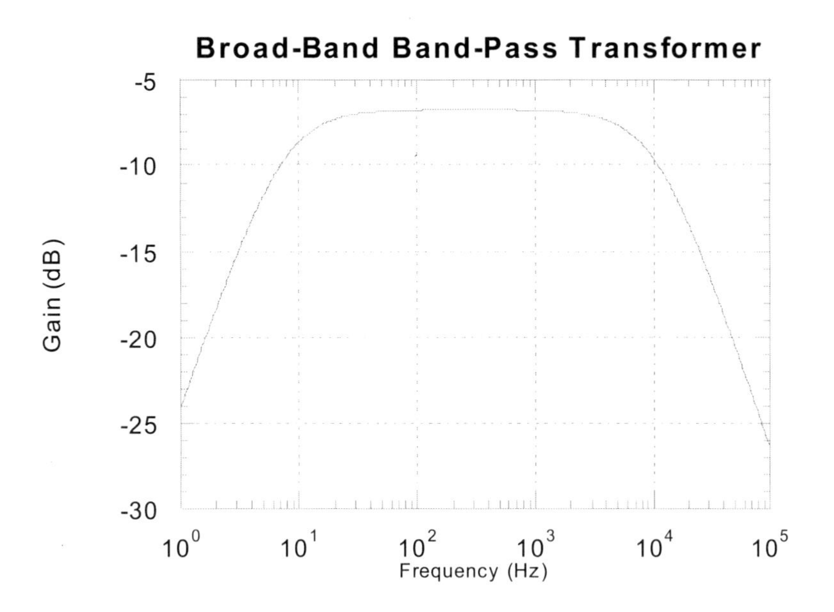

Figure 10.15: Gain versus frequency for broad-band, band-pass transformer.

Simpson Meter

Measure the dc resistance of the primary and secondary with a **Simpson Meter Model 260-7**. Do this by setting the function to one of the resistance scales, and connecting the two leads together and adjust the zero adjust knob until a resistance of zero is indicated. (If the meter is switched to another resistance scale, this procedure must be repeated.)

$Primary,\ R_1 =$ ______________________

$Secondary,\ R_2 =$ ______________________

Digital Multimeter

Turn on the **Agilent 34401A DMM**. Press the button "Ω $2W$" and measure the resistance of the primary and secondary.

$Primary, R_1 =$ ______________________

$Secondary, R_2 =$ ______________________

10.3.3 Dot Determination

Assemble the circuit shown in Fig. 10.16 (for simplicity, the resistances of the primary and secondary are omitted from this diagram since they are internal to the transformer). This assumes that the primary side of the transformer has a single dot indicated. If not, arbitrarily pick one side of the primary as the dotted terminal. Instead of the switch shown, the circuit will be completed at the appropriate time by inserting a hook-up wire which is equivalent to closing a switch.

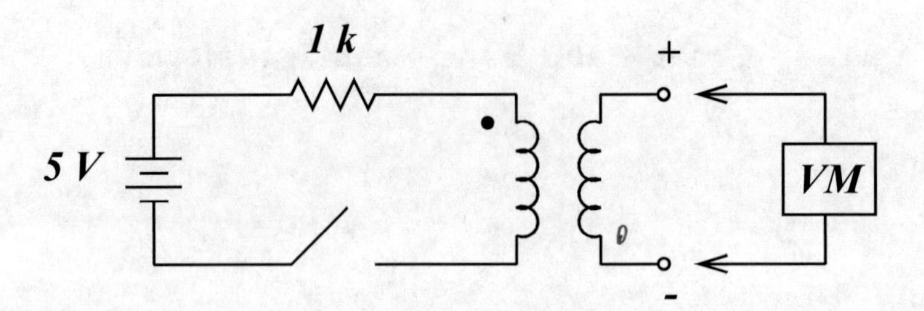

Figure 10.16: Dot determination circuit.

Use the **Simpson Model 260-7** as the dc voltmeter. Set the meter to dc volts and the 2.5 V range. Close the switch (complete the circuit) while watching the pointer on the meter. If the pointer deflects upscale, the positive lead of the voltmeter is on the dotted terminal, and if it deflects downscale, then vice versa is true.

Record which physical terminal in the secondary has the dot.

____

10.3.4 RCL Meter

Turn on the **Fluke/Philips 6303A RCL** meter. Make the following measurements:

Primary

$L_s =$ ____________________

$L_p =$ ____________________

$Q =$ ____________________

Secondary

$L_s =$ ____________________

$L_p =$ ____________________

$Q =$ ____________________

Sum

Connect the undotted terminal on the primary to the dotted terminal on the secondary. This should result in a single inductor $L_{sum} = L_1 + L_2 + 2M$. Use the two unconnected terminals to measure this inductance.

$L_s =$ ____________________

$L_p =$ ______________________

$Q =$ ______________________

Difference

Connect the two undotted terminals together. This should result in a single inductor $L_{diff} = L_1 + L_2 - 2M$. Use the two unconnected terminals to measure this inductance.

$L_s =$ ______________________

$L_p =$ ______________________

$Q =$ ______________________

M and k Calculations

Use the above measured values of the self-inductances to compute M and k using the L_s and L_p values.

L_s Measurements

$$M = \frac{L_{sum} - L_{diff}}{4} = \underline{\qquad\qquad\qquad\qquad} \tag{10.39}$$

$$k = \frac{M}{\sqrt{L_1 L_2}} = \underline{\qquad\qquad\qquad\qquad} \tag{10.40}$$

$$n = \sqrt{\frac{L_2}{L_1}} = \underline{\qquad\qquad\qquad\qquad} \tag{10.41}$$

L_p Measurements

$$M = \frac{L_{sum} - L_{diff}}{4} = \underline{\qquad\qquad\qquad\qquad} \tag{10.42}$$

$$k = \frac{M}{\sqrt{L_1 L_2}} = \underline{\qquad\qquad\qquad\qquad} \tag{10.43}$$

$$n = \sqrt{\frac{L_2}{L_1}} = \underline{\qquad\qquad\qquad\qquad} \tag{10.44}$$

10.3.5 Time-Constant Measurements

Assemble the circuit shown in Fig. 10.17 (for simplicity, the resistances of the primary and secondary are omitted from this diagram since they are internal to the transformer). Use 100 kΩ for the resistor R and place it in series with the primary winding of the transformer. Turn on the oscilloscope and place the leads as shown in Fig. 10.17.

Turn on the **Tektronix 3022B** function generator and wait until it finishes its self-test on power-on; this should only take a few seconds. Enable the CH1 output and configure it for High Z.

Set the frequency of the function generator to 1 kHz, the amplitude to 5 V peak-to-peak, and the function to square. Press *AUTOSET* on the oscilloscope. The waveform on $CH1$ should appear similar to that shown in Fig. 10.17. Position cursors at two times, t_1 and t_2, and the voltage cursors to measure the voltage at

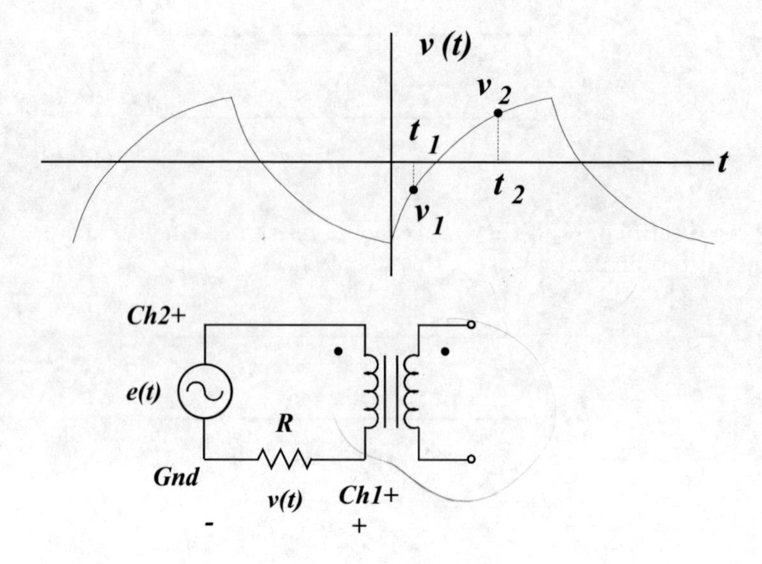

Figure 10.17: Circuit for time-constant measurements.

these two times, v_1 and v_2.Print the display. Be sure to record the values of the voltages and the times if they do not print. Compute the time constant

$$\tau = \frac{t_2 - t_1}{\ln\left(\dfrac{v_1 - A}{v_2 - A}\right)} \tag{10.45}$$

and then the inductance as

$$L_1 = \tau R \tag{10.46}$$

(the value of R that should be used is the total resistance but the 100 kΩ resistor is so large that the others may be neglected).

Repeat the above procedure for the other inductances

$L_1 =$ ____________________

$L_2 =$ ____________________

$L_{sum} =$ ____________________

$L_{diff} =$ ____________________

then calculate

$$M = \frac{L_{sum} - L_{diff}}{4} = \_\_\_\_\_\_\_\_\_\_ \tag{10.47}$$

$$k = \frac{M}{\sqrt{L_1 L_2}} = \_\_\_\_\_\_\_\_\_\_ \tag{10.48}$$

$$n = \sqrt{\frac{L_2}{L_1}} = \rule{4cm}{0.4pt} \tag{10.49}$$

10.3.6 Sinusoidal Measurements

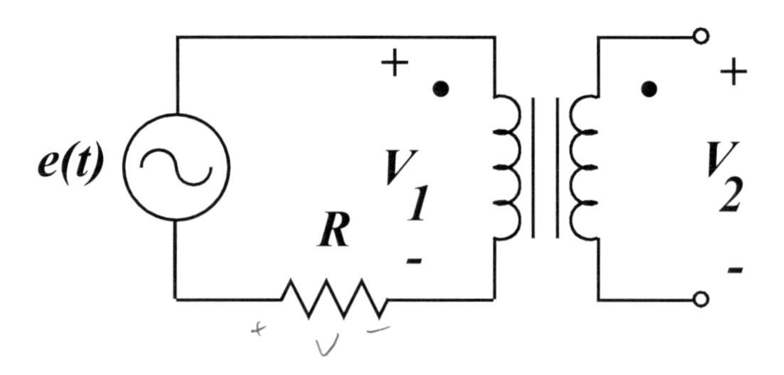

Figure 10.18: Sinusoidal measurments circuit.

Assemble the circuit shown in Fig. 10.18 (for simplicity, the resistances of the primary and secondary are omitted from this diagram since they are internal to the transformer). Use $R = 1\,\mathrm{k\Omega}$. Set the function generator to produce a 1 V rms sine wave with a frequency of 1 kHz. Use the DMM to measure the rms value of the primary and secondary voltages. Compute the turns ratio n as

$$n = \frac{V_2}{V_1} \tag{10.50}$$

which is a positive real quantity.

Measure V, the rms value of the voltage across R. Since this is proportional to the primary current, it may be used to compute the mutual inductance of the two coils and the self inductance of the primary. Compute

$$M = R\frac{V_2}{2\pi f V} \tag{10.51}$$

and

$$L_1 = R\frac{V_1}{2\pi f V} \tag{10.52}$$

where f is the frequency.

Use the above measured value for the turns ratio n and L_1; and

$$n = \sqrt{\frac{L_2}{L_1}} \tag{10.53}$$

to compute L_2; and then

$$k = \frac{M}{\sqrt{L_1 L_2}} \tag{10.54}$$

to compute the coefficient of coupling.

10.3.7 Impedance Measurement

Connect a 600 Ω resistor across the secondary terminals in the circuit shown in Fig. 10.18. (Use either two 300 Ω resistors in series or two 1.2 kΩ resistors in parallel to obtain the 600 Ω.) Measure V_1 and V and compute the input impedance as

$$Z_i = R\frac{V_1}{V} \tag{10.55}$$

10.3.8 Broad-Band Band-Pass Transformer

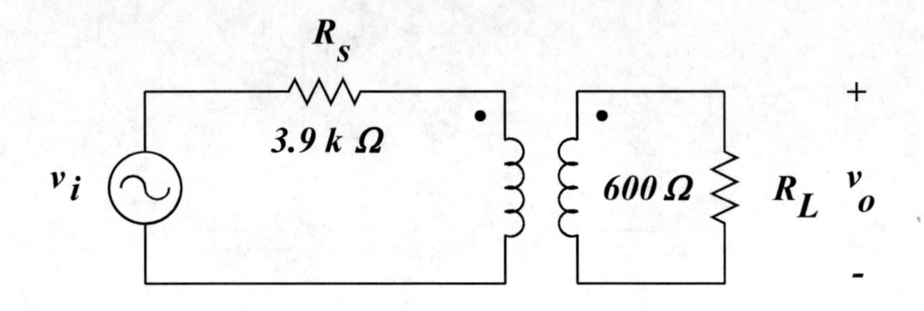

Figure 10.19: Broad-band, band-pass transformer.

Assemble the circuit shown in Fig. 10.19 (for simplicity the resistances of the primary and secondary are omitted from this diagram since they are internal to the transformer). Connect the DMM input to R_L.

Use **Agilent VEE**, **National Instruments LabVIEW**, or **National Instruments ELVIS** to measure and plot the magnitude of the complex transfer function $T = V_o/V_i$. Use an input amplitude of 1 V rms. Use the lowest starting frequency for which the program will work (1 Hz). Use a stop frequency of 100 kHz. The frequency range should be large enough so that the upper and lower −3 dB frequencies can be measured or, at least, estimated. Use the cursors to determine the mid-band gain and the upper and lower −3 dB frequencies.

Use Eq. 10.35 to determine k and either Eqs. 10.33 or 10.34 to determine L_1. In Eq. 10.32 assume that $k = 1$ and use the result in the above equations to obtain a more exact answer. Use the measured value of the mid-band gain and compare it with Eq. 10.30.

10.3.9 Hysteresis

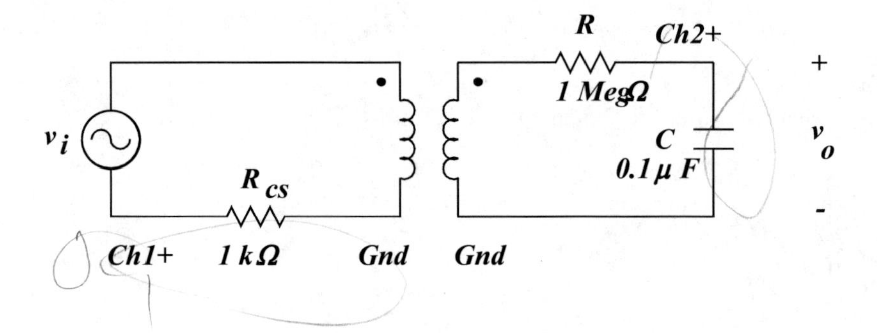

Figure 10.20: Circuit used to obtain hysteresis plot.

Assemble the circuit shown in Fig. 10.20. The function generator will be set to its maximum amplitude value so that the core of the transformer will be driven into saturation, and hysteresis can be observed. The resistor R (1 MΩ) and capacitor C (0.1 μF) has been selected to function as an integrator. The current sampling resistor has been picked to be relatively small (1 kΩ) so that a large current can be forced into the primary of the inductor. Set the function generator to produce a sine wave with a frequency of 50 Hz and an amplitude of 20 V peak-to-peak.

Place the oscilloscope in the XY mode. Manually adjust the horizontal and vertical Volts/div so that the display stays on the screen and takes up a significant portion of it. Print the display.

10.3.10 Automated Impedance Measurements

Modify the circuit shown in Fig. 10.19 and the **VEE** program to obtain a plot of the input impedance seen by the function generator as a function of frequency. Plot this impedance over the same range of frequencies as was used for the transfer function.

10.4 Laboratory Report

- Make a table of the resistances of the primary and secondary that were measured.
- Make a table of the values of L_1, L_2, M, k, and n that were measured.
- Compare the value of the input impedance obtained in step 10.3.7 with the specifications stated in step 10.3.1.
- Compare the amplitude data obtained in step 10.3.8 with the specifications in step 10.3.1.
- Turn in all data taken, calculations made, and plots obtained.

10.5 References

1. Banzhaf, W., *Computer-Aided Circuit Analysis Using PSpice*, 2nd edition, Prentice-Hall, 1992.
2. Carlson, A. B., *Circuits*, John Wiley, 1996.
3. Clark, K. K. and Hess, D. T., *Communications Circuits: Analysis and Design*, Addison-Wesley, 1971.
4. Keown, J., *PSpice and Circuit Analysis*, Merrill, 1991.
5. Kerchner, R. M. and Corcoran, G. F., *Alternating Current Circuits*, Wiley, 1962.
6. Krenz, J. H., *Introduction to Electrical Circuits and Electronic Devices: A Laboratory Approach*, Prentice-Hall, 1987.
7. Lago, G. V. and Waidelich, D. L., *Transients in Electrical Circuits*, Ronald, 1958.
8. LePage, W. R., *Complex Variables and the Laplace Transform for Engineers*, Dover, 1961.
9. MicroSim Corporation, *PSpice Circuit Analysis*, MicroSim, 1990.
10. Nilsson, J. W. and Riedel, S. A., *Electric Circuits*, 5th ed., Addison-Wesley, 1999.
11. Rashid, *SPICE for Circuits and Electronics Using PSpice*, Prentice-Hall, 1990.
12. Riaz, M., *Electrical Engineering Laboratory Manual*, McGraw-Hill, 1965.
13. Smith, J., *Modern Communications Circuits*, McGraw-Hill, 1986.
14. Terman, F. E., *Electronic and Radio Engineering*, 4th ed., McGraw-Hill, 1955.
15. Tuinenga, P. W., *SPICE: A Guide to Circuit Simulation & Analysis Using PSpice*, 3rd ed., Prentice-Hall, 1995.
16. Zahn, M., *Electromagnetic Field Theory*, John Wiley, 1979.

Chapter 11

Spectral Analysis

11.1 Objective

Physical signals such as voltage or current waveforms can oftentimes be expressed as a superposition of sinusoidal waveforms. This provides two ways of displaying the information contained in these signals: time plots and frequency plots. These two methods of displaying physical signals contain the same information and either can be obtained from the other with the appropriate mathematical operator. Which of these two equivalent representations is to be preferred is determined by the application.

The evaluation of the frequency content of a signal is known as spectral analysis. The basis of spectral analysis is the Fourier series and Fourier transform theory developed in the early Nineteenth century. Electronic instruments known as spectrum analyzers or waveform analyzers are used to obtain the spectral content of physical signals. Just as oscilloscopes are used to display time plots of signals, spectrum analyzers display frequency plots of signals.

Many modern digital oscilloscopes have the capacity to compute and display the discrete Fourier transform as well as plots signals as a function of time. The algorithm used in the computation of the spectra is known as the Fast Fourier Transform (FFT). Thus with a digital oscilloscope a spectrum analyzer is often an ancillary accoutrement that enables simultaneous displays of a signal in both the time and frequency domains. Albeit, due to the aliasing phenomenon that plagues digital instruments, the displays must be properly interpreted.

11.2 History

In 1807 a 39 year old French mathematician named Jean Baptiste Joseph Fourier (1768-1830) presented a paper at the Paris Academie that stated that a periodic function could be represented as a summation of sines and cosines. This hypothesis astonished Fourier's contemporaries; Lagrange, in particular, protested that such a representation was impossible. It was unfortunate for Lagrange that he had developed the Fourier series representation of periodic signals in 1759 while studying mechanical vibrations on a sting but did not realize the implications of his discovery. It is, therefore, Fourier who is accorded the credit for this landmark mathematical discovery since he had the insight to understand the implications of the equations. Although Fourier died well before the electronic era (1830), his name is prominently displayed on families of electronic instrumentation that are used to ascertain the spectral content of signals.

Fourier analysis was developed into a rigorous mathematical discipline by a mathematician named Dirichlet. He developed a set of conditions dealing with whether or not the function is piecewise differentiable over a bounded interval—appropriately known as the Dirichlet conditions—that a periodic signal must meet if it can be represented as a infinite series of sinusoidals. Fortunately, these conditions are met by practically all physical waveforms. This makes it unnecessary to consider the Dirichlet conditions if it is understood that pathological signals that are unlikely to be found in nature are excluded.

Fourier was a child prodigy and the last of nineteen children. Although once sentenced to the guillotine

Figure 11.1: Jean Baptiste Joseph Fourier.

during the French Revolution, he survived and performed mathematics research for Napoleon. He was sent to Egypt by Napoleon on a scientific expedition and wrote a 21 volume work which forms the basis of modern Egyptology.

11.3 Theory

11.3.1 Fourier Transform

The dual representation of a function or signal or waveform as either a function of time or frequency is established by the Fourier Transform. If $x(t)$ is a sufficiently well behaved function of time (i.e. it satisfies the Dirichlet conditions), the relationship

$$\mathfrak{F}\{x(t)\} = X(f) = \int_{-\infty}^{\infty} x(t)e^{-j2\pi ft}dt \tag{11.1}$$

and

$$x(t) = \mathfrak{F}^{-1}\{X(f)\} = \int_{-\infty}^{\infty} X(f)e^{j2\pi ft}df \tag{11.2}$$

exists. The function $X(f)$ is said to be the Fourier transform of the function $x(t)$. The two functions $x(t)$ and $X(f)$ are said to form a Fourier transform pair. The notation

$$x(t) \longleftrightarrow X(f) \tag{11.3}$$

is used to indicate that the two functions are Fourier transform pairs. The notation $\mathfrak{F}\{x(t)\}$ means the Fourier transform of $x(t)$ whereas the notation $\mathfrak{F}^{-1}\{X(f)\}$ means the inverse Fourier transform of $X(f)$.

Anent the Dirichlet conditions, essentially this requires that the function or signal or waveform $x(t)$ have a finite number of discontinuities in a finite time interval and satisfy

$$\int_{-\infty}^{\infty} |x(t)|^2\, dt < \infty \tag{11.4}$$

which means that $x(t)$ must have finite energy which is known as an energy signal. Signals with finite energy have zero power whereas signals with finite power have infinite energy; there are some signals which fall into neither class. The Fourier transform is the vehicle for energy signals (aperiodic or nonperiodic) while the Fourier series (to be developed) is the analytic tool for periodic waveforms which are power signals.

Physically the function $x(t)$ may represent a voltage or current or any other physical variable which may be expressed as a function of time. However, it must satisfy the Dirichlet conditions if it is to have a Fourier transform.

Rectangular Pulse

The rectangular pulse with amplitude A and duration τ is shown in Fig. 11.2 and defined as

$$x(t) = \begin{cases} A & |t| \leqq \tau/2 \\ 0 & |t| > \tau/2 \end{cases} \tag{11.5}$$

which obviously satisfies Eq. 11.4. The expression for this $x(t)$ may be expressed as $x(t) = Au(\tau/2 - |t|)$ where $u(t)$ is unit step function. The Fourier transform is

$$X(f) = A\int_{-\tau/2}^{\tau/2} e^{-j2\pi ft}dt = A\frac{e^{-j2\pi ft}}{-j2\pi f}\Bigg|_{-\tau/2}^{\tau/2} = A\frac{e^{j\pi f\tau} - e^{-j\pi f\tau}}{j2\pi f\tau} = A\tau sinc(f\tau) \tag{11.6}$$

where the mathematical identity

$$\sin\theta = \frac{e^{j\theta} - e^{-j\theta}}{2j} \tag{11.7}$$

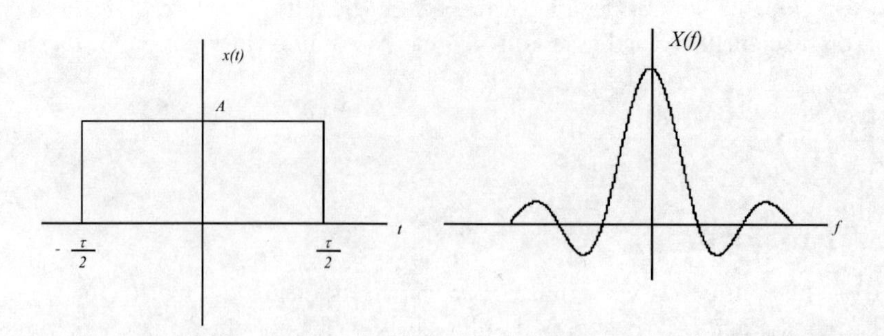

Figure 11.2: Rectangular pulse.

and the function $sinc(u)$ is defined as

$$sinc(u) = \frac{\sin \pi u}{\pi u} \tag{11.8}$$

or

$$Au(\tau/2 - |t|) \longleftrightarrow A\tau sinc(f\tau) \tag{11.9}$$

are a Fourier transform pair. The magnitude of the transform is plotted in Fig. 11.2 next to the square pulse. This Fourier transform pair illustrates that a function which is time limited must extend infinitely in frequency. Due to the symmetry of Eqs. 11.1 and 11.2, the converse is also true; namely, a function which has a finite bandwidth, or is limited in frequency, must extend infinitely in time.

The symmetry of the Fourier transform pair establishes a principle of duality. Namely, if

$$x(t) \longleftrightarrow X(f) \tag{11.10}$$

then

$$X(t) \longleftrightarrow x(-f) \tag{11.11}$$

which results in a twofold reduction in the effort required to evaluate Fourier transforms. Employing the results for the Fourier transform of the rectangular pulse yields

$$AW sinc(tW) \longleftrightarrow Au(W/2 - |f|) \tag{11.12}$$

as a Fourier transform pair.

11.3.2 Fourier Series

The complex Fourier series representation of a periodic waveform, $x(t)$, is given by

$$x(t) = \sum_{n=-\infty}^{\infty} \bar{c}_n e^{jn\omega_p t} \tag{11.13}$$

where $\bar{c}_n$ are the complex Fourier expansion coefficients, n is a summation index, $\omega_p = 2\pi f_p = 2\pi/T$, $f_p = 1/T$ is known as the fundamental frequency of the periodic wave, and T is the period of $x(t)$. Physically, $x(t)$ may be a voltage or current or any other process which is periodic. The complex expansion coefficients $\bar{c}_n$ can be determined from $x(t)$ by

$$\bar{c}_n = \frac{1}{T} \int_o^T x(t) e^{-jn\omega_p t}\, dt \tag{11.14}$$

where the range of integration indicated as from 0 to T can be generalized to any time interval having a length of T. Thus given a periodic function $x(t)$ the complex Fourier series representation can be obtained if the above integral can be evaluated. Although the expansion coefficients, $\bar{c}_n$, are complex, the indicated infinite summation of $\bar{c}_n$ times the complex exponentials $e^{jn\omega_p t}$ results in a real function of time, $x(t)$.

The complex Fourier series can be expressed directly as a real infinite series of trigonometric functions with the substitution

$$\bar{c}_n = \frac{a_n - jb_n}{2} \tag{11.15}$$

where a_n and b_n are real. The Fourier series representation becomes

$$x(t) = \frac{a_o}{2} + \sum_{n=1}^{\infty}[a_n \cos(n\omega_p t) + b_n \sin(n\omega_p t)] \tag{11.16}$$

where the expressions for the real expansion coefficients are given by

$$a_n = \frac{2}{T}\int_o^T x(t)\cos(n\omega_p t)\,dt \tag{11.17}$$

and

$$b_n = \frac{2}{T}\int_o^T x(t)\sin(n\omega_p t)\,dt \tag{11.18}$$

where the integration may be performed over any time interval of length T. The above representation of the Fourier series is known as the sine-cosine representation or real trigonometric representation. Both the complex and real trigonometric representation result in equally valid expansions for $x(t)$.

The real trigonometric Fourier series may be expressed as

$$x(t) = \frac{a_o}{2} + \sum_{n=1}^{\infty} \sqrt{a_n^2 + b_n^2}\,\sin(n\omega_p t + \phi_n)] \tag{11.19}$$

by applying the trigonometric identity

$$\sin(A + B) = \sin A \cos B + \cos A \sin B \tag{11.20}$$

to Eq. 11.19 where

$$\phi_n = \tan^{-1}\frac{a_n}{b_n} \tag{11.21}$$

The expression for $x(t)$ can be placed in the form

$$x(t) = \bar{c}_0 + 2\sum_{n=1}^{\infty} \mid \bar{c}_n \mid \cos\left(n\omega_p t + \angle\bar{c}_n\right) \tag{11.22}$$

by virtue of Eq. 11.15 where

$$\angle\bar{c}_n = \phi_n - \frac{\pi}{2} \tag{11.23}$$

is the angle of the complex Fourier expansion coefficients.

The above equations disclose that a periodic function of time can be expressed as a sum of weighted sinusoidals plus a constant term which represents the DC level of the waveform ($\bar{c}_o$ or $a_o/2$). The lowest frequency sinusoidal in the expansion has the same frequency as the periodic function ($n = 1$) and is known as the fundamental. The spectral components with higher frequencies are known as harmonics. Thus if a frequency plot were made of a time periodic function it would consist of points at frequencies DC, f_p, $2f_p$, $3f_p$, $\cdots$ with the height of the points above the frequency axis determined by the Fourier expansion coefficients.

Strictly speaking the Fourier series does not have a Fourier transform since it does not satisfy the Dirichlet conditions. A periodic function of time is known as a power signal—it has finite power rather than finite energy. However, the Fourier transform of the Fourier series can be taken with the use of the Dirac delta function, $\delta(t)$, defined as

$$\begin{array}{ll} \delta(t) = 0 & t \neq 0 \\ \int_{-\infty}^{\infty} \delta(t) dt = 1 & \end{array} \tag{11.24}$$

which means that it has a singularity at the origin. The Dirac delta function is loathed by mathematicians and revered by engineers and physicists. Mathematicians are not fond of the Dirac delta function since it is not an analytic function. Engineers and physicists employ the Dirac delta function to analyze systems which are concentrated or highly localized in time or space. The Dirac delta function is the well known impulse function from elementary circuit theory (the derivative of the unit step function).

The Fourier transform of the Dirac delta function is given by

$$\mathfrak{F}\{\delta(t)\} = \int_{-\infty}^{\infty} \delta(t) e^{-j2\pi ft} dt = 1 \tag{11.25}$$

which means that

$$\delta(t) \longleftrightarrow 1 \tag{11.26}$$

form a Fourier transform pair. This means that the inverse Fourier transform of 1 must be the Dirac delta function which provides the mathematical identity for the Dirac delta function

$$\int_{-\infty}^{\infty} e^{j2\pi ft} df = \delta(t) \tag{11.27}$$

which also means

$$\int_{-\infty}^{\infty} e^{j2\pi ft} dt = \int_{-\infty}^{\infty} e^{-j2\pi ft} dt = \delta(f) \tag{11.28}$$

by exchanging the variable of integration.

The Fourier transform of $x(t)$ may now be evaluated where $x(t)$ is a periodic function which has a Fourier series representation. The Fourier transform of $x(t)$ is given by

$$X(f) = \mathfrak{F}\{x(t)\} = \int_{-\infty}^{\infty} x(t) e^{-j2\pi ft} dt = \int_{-\infty}^{\infty} \left[\sum_{n=-\infty}^{\infty} \bar{c}_n e^{jn2\pi f_p t} \right] e^{-j2\pi ft} dt \tag{11.29}$$

which becomes

$$X(f) = \sum_{n=-\infty}^{\infty} \bar{c}_n \int_{-\infty}^{\infty} e^{-j2\pi(f-nf_p)t} dt = \sum_{n=-\infty}^{\infty} \bar{c}_n \delta(f - nf_p) \tag{11.30}$$

upon exchanging the order of integration and using Eq. 11.28 for the Dirac delta function. Thus the spectra or Fourier transform of a periodic function of time, $x(t)$, with period T and fundamental frequency $f_p = 1/T$

$$x(t) = \sum_{n=-\infty}^{\infty} \bar{c}_n e^{jn2\pi f_p t} \tag{11.31}$$

is

$$X(f) = \sum_{n=-\infty}^{\infty} \bar{c}_n \delta(f - nf_p) \tag{11.32}$$

which is a line spectra. It is called a line spectra because $X(f) = 0 \forall f \exists f \neq nf_p$where n is an integer. The power in this signal is concentrated at frequencies that are integer multiples of the fundamental frequency f_p of the periodic waveform. A plot of the magnitude of $X(f)$ as a function of f would be lines (actually Dirac delta functions) located at $f = nf_p$ with amplitudes $|\bar{c}_n|$.

The spectra of $X(f)$ is symmetric with respect to the frequency axis, i.e. $X(-f) = -X(f)$. For periodic signals this means that $|\bar{c}_{-n}| = |\bar{c}_n|$ and $\angle\bar{c}_{-n} = -\angle\bar{c}_n$. Therefore, if a plot is to be made of the magnitude of the spectra, it would be pointless to make the plot for both positive and negative frequencies since they are the same. For periodic signals what is often plotted is the rms power spectra which is given by

$$X_{rms}(f) = |\bar{c}_0| + \sum_{n=1}^{\infty} \frac{2\,|\bar{c}_n|}{\sqrt{2}}\delta(f - nf_p) = |\bar{c}_0| + \sum_{n=1}^{\infty} \sqrt{2}\,|\bar{c}_n|\,\delta(f - nf_p) \tag{11.33}$$

where $|\bar{c}_0| = a_0/2$ is the dc component and $\sqrt{2}\,|\bar{c}_n|$ is the rms value of the sinusoidal components at frequencies $f = nf_p$. This rms spectra is often what is plotted by spectrum analyzers. The rms magnitude may be plotted on a linear or log scale (usually in decibels). If the plot is linear the vertical scale is $\sqrt{2}\,|\bar{c}_n|$ whereas it is $20\log_{10}\left(\sqrt{2}\,|\bar{c}_n|\right)$ if a logarithmic plot is made in decibels.

The value of the Fourier expansion coefficients is determined by the location of the time origin. Thus a judicious choice of the time origin may minimize the effort required to evaluate the Fourier expansion coefficients. The value of the Fourier expansion coefficients is independent of the frequency of $x(t)$. Thus, the frequency of the periodic waveform determines the spacing of the spectral components but not their height.

Square Wave

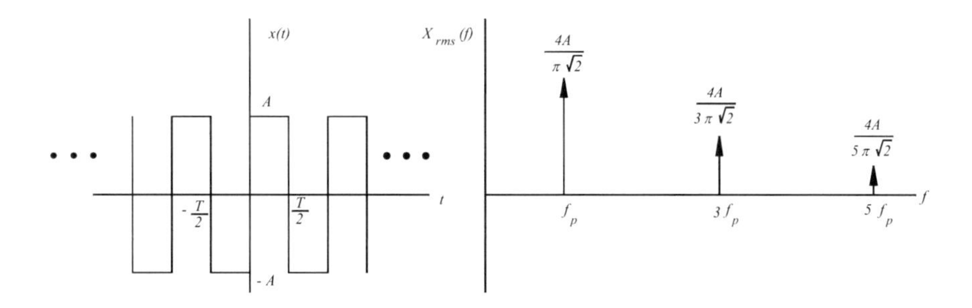

Figure 11.3: Symmetric square wave.

Shown in Fig. 11.3 is a symmetric square wave with a dc level of zero, a period T, a frequency $f_p = 1/T$, and a peak value of A. For this wave the expansion coefficients are most easily evaluated by choosing the time origin to coincide with an instant of time when the square wave is switching from its lower to its upper level. The expansion coefficients are given by

$$\begin{aligned} \bar{c}_n &= \frac{1}{T}\int_0^T x(t)e^{-jn\omega_p t}\,dt = \frac{1}{T}\int_{-T/2}^{T/2} x(t)e^{-jn\omega_p t}\,dt \\ &= \frac{1}{T}\int_{-T/2}^{0} (-A)e^{-jn\omega_p t}\,dt + \frac{1}{T}\int_0^{T/2} (A)e^{-jn\omega_p t}\,dt \end{aligned} \tag{11.34}$$

which can be simplified by making the change in the variable of integration in the first integral $u = -t$ and $u = t$ in the second integral which then yields

$$\bar{c}_n = \frac{A}{T}\int_0^{T/2} \left[-e^{jn\omega_p u} + e^{-jn\omega_p u}\right] du = \frac{-2jA}{T}\int_0^{T/2} \sin(n\omega_p u)\,du \tag{11.35}$$

because of the mathematical identity

$$\sin\,\theta = \frac{e^{j\theta} - e^{-j\theta}}{2j} \tag{11.36}$$

which yields

$$\bar{c}_n = \frac{2jA}{n\omega_p T}\cos(n\omega_p u)\Bigg|_0^{T/2} = \frac{2jA}{2n\pi}[\cos(n\pi) - 1] = \begin{cases} -\dfrac{2jA}{n\pi} & n \; odd \\ 0 & n \; even \end{cases} \tag{11.37}$$

where $\omega_p T = 2\pi$ was used to simplify the expression for $\bar{c}_n$. Eq. 11.37 states that $\bar{c}_n = 0$ for all even value of n and $\bar{c}_n = \dfrac{-2jA}{n\pi}$ for n odd. Thus this periodic waveform can be expressed as

$$f(t) = -\frac{j2A}{\pi}\sum_{\substack{n=-\infty \\ n\ odd}}^{\infty}\frac{1}{n}e^{jn\omega_p t} = \frac{4A}{\pi}\sum_{\substack{n=1 \\ n\ odd}}^{\infty}\frac{1}{n}\sin(n\omega_p t) \tag{11.38}$$

where the summation extends over only odd integers. Note that the peak value of the fundamental has a value that is larger than the peak value of the square wave. If the spectral components are normalized with respect to the fundamental, the expressions are particularly simple and given by

$$\bar{c}_n/\bar{c}_1 = \begin{cases} 0 & n \text{ even} \\ \dfrac{1}{n} & n \text{ odd} \end{cases} \tag{11.39}$$

which discloses that the even harmonics are zero and the odd decrease as $1/n$.

Triangular Wave

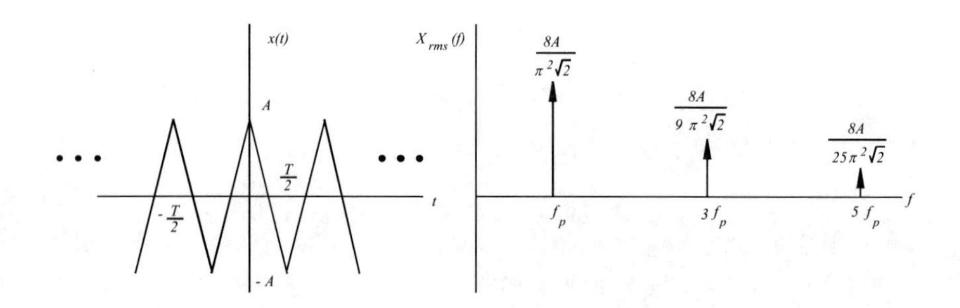

Figure 11.4: Symmetric triangular wave.

Shown in Fig. 11.4 is the symmetric triangular wave with a dc level of zero, a period T, a frequency of $f_p = 1/T$, and a peak-to-peak value of $2A$. For this wave, the most judicious choice of the time origin is that instant of time when the waveform is at its peak. For this choice, the expression for the complex Fourier expansion coefficient is

$$\bar{c}_n = \frac{1}{T}\int_0^T x(t)e^{-jn\omega_p t}\,dt = \frac{1}{T}\int_{-T/2}^{T/2} x(t)e^{-jn\omega_p t}\,dt \tag{11.40}$$

$$= \frac{1}{T}\int_{-T/2}^{0} x(t)e^{-jn\omega_p t}\,dt + \frac{1}{T}\int_{0}^{T/2} x(t)e^{-jn\omega_p t}\,dt = \tag{11.41}$$

$$= \frac{1}{T}\int_{-T/2}^{0}\left(\frac{4A}{T}t + A\right)e^{-jn\omega_p t}\,dt + \frac{1}{T}\int_{0}^{T/2}\left(-\frac{4A}{T}t + A\right)e^{-jn\omega_p t}\,dt \tag{11.42}$$

which could be evaluated as integrals of exponentials and polynomials times exponentials. However it is more convenient to make a change in the variable of integration in the first integral in Eq. 11.42 by letting $u = -t$ and then replace the dummy variable of integration u with t. The expression for the complex Fourier expansion coefficients then becomes

$$\bar{c}_n = \frac{1}{T}\int_{0}^{T/2}\left[-\frac{4A}{T}t\left(e^{jn\omega_p t} + e^{-jn\omega_p t}\right) + A\left(e^{jn\omega_p t} + e^{-jn\omega_p t}\right)\right]dt \tag{11.43}$$

which becomes

$$\bar{c}_n = \frac{1}{T}\int_{0}^{T/2}\left[-\frac{8A}{T}t\cos n\omega_p t + 2A\cos n\omega_p t\right]dt \tag{11.44}$$

by virtue of the trigonometric identity

$$\cos\theta = \frac{e^{j\theta} + e^{-j\theta}}{2} \tag{11.45}$$

which greatly facilitates the evaluation of this integral. Eq. 11.44 can be evaluated by employing the standard integral

$$\int x\cos ax\,dx = \frac{x\sin ax}{a} + \frac{\cos ax}{a^2} \tag{11.46}$$

which yields

$$\bar{c}_n = \frac{1}{T}\left[-\frac{8A}{T}\left(\frac{t\sin n\omega_p t}{n\omega_p} + \frac{\cos n\omega_p t}{n^2\omega_p^2}\right) + 2A\frac{\sin n\omega_p t}{n\omega_p}\right|_0^{T/2} \Longrightarrow \tag{11.47}$$

$$\bar{c}_n = -\frac{2A}{n^2\pi^2}\left(\cos n\pi - 1\right)\ \forall n \tag{11.48}$$

as the general expression for the complex Fourier expansion coefficients since $\omega_p T = 2\pi$. The expression becomes

$$\bar{c}_n = \begin{cases} \frac{4A}{\pi^2}\frac{1}{n^2} & n \text{ odd} \\ 0 & n \text{ even} \end{cases} \tag{11.49}$$

for n even or odd. Therefore, for the symmetric triangular wave, the ratio of the harmonics to the fundamental is given by $1/n^2$ for n odd and zero for even n. The triangular wave is then given by

$$f(t) = \frac{8A}{\pi^2}\sum_{\substack{n=1 \\ n\ odd}}^{\infty}\frac{\cos n\omega_p t}{n^2} \tag{11.50}$$

where the sum is over odd integers since the even harmonics are zero. Note that the peak amplitude of the fundamental has a value which is slightly less that the peak value of the triangular wave.

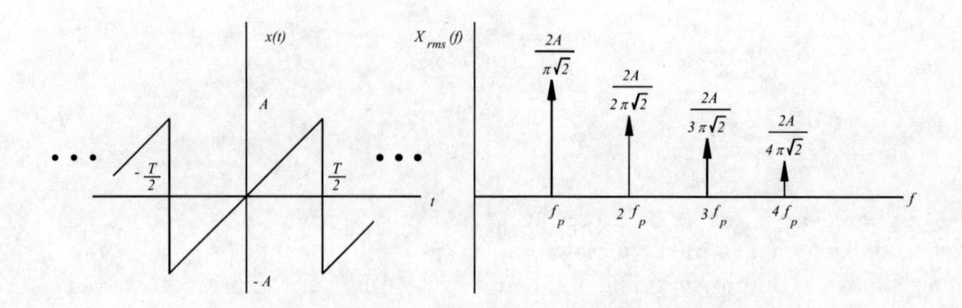

Figure 11.5: Symmetric ramp wave.

Symmetric Ramp Wave

Shown in Fig. 11.5 is the symmetric ramp wave with a peak value of A, DC level of zero, frequency $f_p = 1/T$, and a period T. For this wave the complex Fourier series coefficients are given by

$$\overline{c}_n = \frac{1}{T}\int_{-T/2}^{T/2} x(t)e^{-jn\omega_p t}\,dt = \frac{1}{T}\int_{-T/2}^{T/2} \frac{2A}{T} te^{-jn\omega_p t}\,dt \tag{11.51}$$

$$= \frac{2A}{T^2}\left[\frac{te^{-jn\omega_p t}}{-jn\omega_p} - \frac{e^{-jn\omega_p t}}{(-jn\omega_p)^2}\right|_{-T/2}^{T/2} = 2A\left[e^{-j\pi n}\left(\frac{j}{4\pi n} + \frac{1}{4\pi^2 n^2}\right) - e^{j\pi n}\left(-\frac{j}{4\pi n} + \frac{1}{4\pi^2 n^2}\right)\right] = \tag{11.52}$$

$$jA\frac{\cos n\pi}{n\pi} - jA\frac{\sin n\pi}{n^2\pi^2} = j\frac{A}{n\pi}(-1)^n \tag{11.53}$$

since $e^{j\theta} = \cos\theta + j\sin\theta$ and $\sin n\pi = 0\ \forall\ n$, $\cos n\pi = (-1)^n$, and $\omega_p T = 2\pi$. Therefore, the complex Fourier expansion coefficients for the ramp wave are given by

$$\overline{c}_n = \frac{jA}{\pi n}(-1)^n \ \ \forall n \neq 0 \tag{11.54}$$

and $\overline{c}_0 = 0$, viz.

$$\overline{c}_n = \begin{cases} 0 & n = 0 \\ j\dfrac{A}{\pi n}(-1)^n & n \neq 0 \end{cases} \tag{11.55}$$

for all n. Therefore, all the spectral components are present for the ramp wave and the real trigonometric Fourier series representation is given by

$$x(t) = \frac{2A}{\pi}\sum_{n=1}^{\infty}(-1)^{n+1}\frac{\sin(n\omega_p t)}{n} \tag{11.56}$$

where the sum is over all integers. Note that the peak value of the fundamental has a value that is less than two thirds the peak value of the ramp wave.

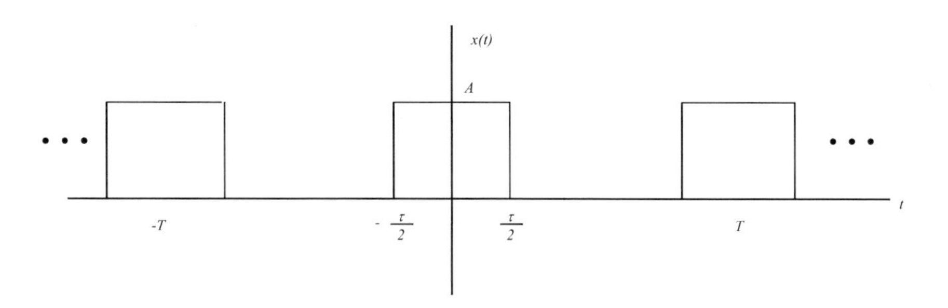

Figure 11.6: Rectangular pulse train.

Rectangular Pulse Train

Shown in Fig. 11.6 is the rectangular pulse train with peak value A, duration τ, period T, frequency $f_p = 1/T$, and dc level dA where $d = \tau/T$ is known as the duty cycle. The duty cycle is simple the fraction of the time the waveform is non zero. The complex Fourier expansion coefficients are given by

$$\overline{c}_n = \frac{1}{T}\int_{-\tau/2}^{\tau/2} Ae^{-jn\omega_p t}dt = \frac{1}{-jn\omega_p T}e^{-jn\omega_p t}\bigg|_{-\tau/2}^{\tau/2} = dAsinc(nd) \tag{11.57}$$

where $sinc(u)$ is defined in Eq. 11.8.

If the duty cycle d is one half, the rectangular pulse train becomes a symmetric square with DC level $A/2$. The expansion coefficients have a slightly different form than that for the square wave considered previously due to the choice of the time origin, the shift in the DC level, and the differing peak-to-peak levels. But, the rms spectra are, of course, identical when properly normalized.

RF Pulse Train

The waveform shown in Fig. 11.7 is known as a cosine pulse train. The duration of the pulse is τ, the frequency $f_p = 1/T$, the amplitude A, and, for simplicity, it will be assumed that the frequency f_c is an integral multiple of f_p which makes the DC level zero and the waveform continuous. If the frequency f_c is large compared to $f_p = 1/T$ it is known as a radio frequency pulse train.

The complex Fourier expansion coefficients are

$$c_n = \frac{1}{T}\int_{-\tau/2}^{\tau/2} A\cos\omega_c t\; e^{-jn\omega_p t}dt = \frac{1}{T}\int_{-\tau/2}^{\tau/2} A\cos\omega_c t\; \cos n\omega_p t\; dt \tag{11.58}$$

by virtue of Euler's relation

$$e^{j\theta} = \cos\theta + j\sin\theta \tag{11.59}$$

and using the fact that cos is an even function while sin is an odd function and that the limits of the integral are symmetric. Using the mathematical identity

$$\cos A\cos B = \frac{1}{2}\left[\cos\left(A+B\right) - \cos\left(A-B\right)\right] \tag{11.60}$$

permits the evaluation of the expansion coefficients as

$$c_n = \frac{A\tau}{2T}\left[sinc\left(f_c - nf_p\right)\tau + sinc\left(f_c + nf_p\right)\tau\right] \tag{11.61}$$

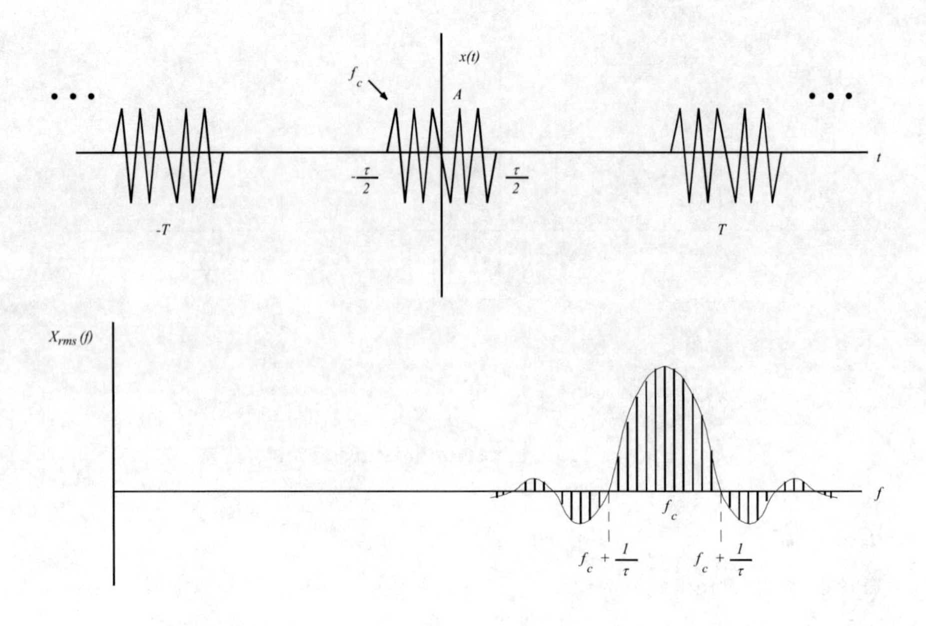

Figure 11.7: RF pulse train.

which is a line spectra centered about $\pm f_c$. The envelope of the spectra (a smooth curve drawn through the tip or top of the lines) is a *sinc* function. The envelope first crosses zero at frequencies $f = f_c + 1/\tau$ and $f = f_c - 1/\tau$ above and below f_c respectively. The difference of these two frequencies $2/\tau$ is known as the width of the main lobe.

This type of waveform is used by radar systems. The duty cycle $d = \tau/T$ is picked to be small. The frequency f_c specifies the band of the radar, e.g. X band, Ka band, etc. The waveform $x(t)$ is transmitted and the received waveform $r(t)$ (which has bounced off a target) has a shape similar to $x(t)$, i.e. the amplitude is reduced, there is a time delay between the two, and the frequency of the RF pulse of the received signal has, in general, slightly shifted. The transmitted and received waveforms are compared and the time delay between the two measured. The delay Δt multiplied by c (the speed of light) divided by 2 is the range or distance of the target. The shift in the frequency Δf_c for the received signal is caused by the Doppler shift and determines the radial velocity of the target.

Root-Mean-Square (rms) of Periodic Waveforms

The root-mean-square (rms) value of a voltage or current waveform is the value of a dc source which would produce the same heating effect in a resistor. This means that the time average power of the signal is equal to the value of a dc source that produces the same power dissipation in a resistor.

If $x(t)$ is a periodic waveform with period T, the *rms* value of $x(t)$, X_{rms}, is given by

$$X_{rms} = \sqrt{\frac{1}{T}\int_o^T x^2(t)\,dt} \tag{11.62}$$

where the integration may be performed over any time interval of length T. The familiar result that is obtained from the above expression for a sine wave with a DC level of zero and peak value A is $X_{rms} = A/\sqrt{2}$.

The above expression for the rms value of a waveform may be used to obtain the rms value of a periodic waveform in terms of the Fourier expansion coefficients. Namely,

$$X_{rms} = \sqrt{\sum_{n=-\infty}^{\infty} |\, \bar{c}_n|^2} = \sqrt{\frac{a_o^2}{4} + \sum_{n=1}^{\infty}\left[\frac{a_n^2 + b_n^2}{2}\right]} \tag{11.63}$$

which discloses that the rms value of a periodic waveform is determined by the amplitude of the individual spectral components.

Many digital multimeters (DMM) can measure the rms value of a waveform. Some of these can perform the measurement correctly only if the waveform is a sinusoidal with a dc level of zero and are called averaging meters. Others can measure the rms value of an arbitrary waveform and are called true rms meters.

Alternative Method of Computing Fourier Expansion Coefficients

An alternative method of evaluating Fourier expansion coefficients other than the straightforward computation of Eq. 11.14 exists. This consists of defining a nonperiodic function which consists of one cycle of the periodic function, determining the Fourier transform of this nonperiodic function, and evaluating the resulting transform at the frequency $f = nf_p$.

If $x(t)$ is a periodic function with period T, define the nonperiodic function $x_p(t)$ as

$$x_p(t) = \begin{cases} x(t) & |t| < T/2 \\ 0 & \text{otherwise} \end{cases} \tag{11.64}$$

which is $x(t)$ multiplied by a rectangular pulse with amplitude 1 and duration T. The periodic function $x(t)$ can then be represented as

$$x(t) = \sum_{n=-\infty}^{\infty} x_p(t - nT) = x_p(t) * \sum_{n=-\infty}^{\infty} \delta(t - nT) \tag{11.65}$$

where $*$ is the convolution operator. Taking the Fourier transform of both sides and utilizing the result that the Fourier transform of the convolution of two functions is given by the product of their Fourier transforms yields

$$X(f) = \frac{X_p(f)}{T} \sum_{n=-\infty}^{\infty} \delta(f - nf_p) \tag{11.66}$$

since

$$\sum_{n=-\infty}^{\infty} \delta(t - nT) \longleftrightarrow \frac{1}{T} \sum_{n=-\infty}^{\infty} \delta(f - nf_p) \tag{11.67}$$

form a Fourier transform pair (the infinite sum of equally spaced Dirac delta functions in either time or frequency is known as a comb function). Eq. 11.66 can be expressed as

$$X(f) = \sum_{n=-\infty}^{\infty} \frac{X_p(nf_p)}{T} \delta(f - nf_p) \tag{11.68}$$

which, compared with Eq. 11.30 discloses

$$\bar{c}_n = \frac{X_p(nf_p)}{T} \tag{11.69}$$

as the alternative expression for the complex Fourier expansion coefficients.

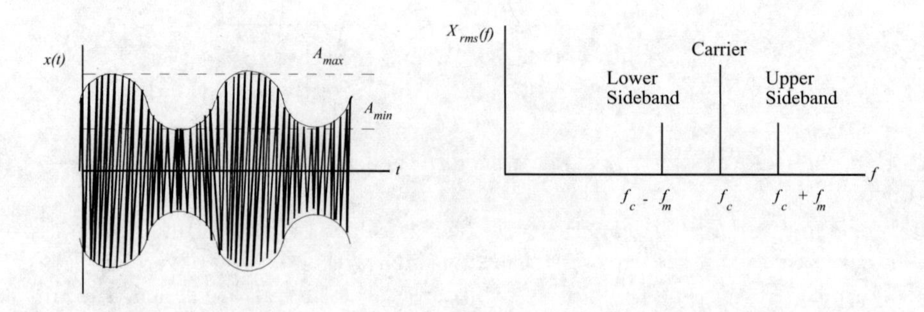

Figure 11.8: Conventional amplitude modulation (AM).

11.3.3 Conventional Amplitude Modulation (AM)

The waveform shown in Fig. 11.8

$$x(t) = A_c \left[1 + \alpha m(t)\right] \cos \omega_c t \tag{11.70}$$

is known as an amplitude modulated waveform. Such waveforms are of central importance in telecommunications. The sinusoidal $\cos \omega_c t$ is known as the carrier, $m(t)$ is known as the message or information bearing signal, and $x(t)$ is the AM modulated waveform. In commercial AM radio, the carrier frequency, $f_c = \omega_c / 2\pi$, lies between approximately 500 to $1,700$ kHz (a different frequency is assigned for each radio station) and $m(t)$ represents either speech or music and can be considered to be a signal bandlimited to $W = 5\ kHz$, i.e. $M(f) = 0 \forall |f| > W = 5\ kHz$. The message signal is normalized so that the maximum value of $m(t)$ is $+1$ and the minimum value is -1. The parameter α is a dimensionless quantity which usually lies in the range $0 \leq \alpha \leq 1$ and is known as the AM modulation index. The parameter A_c is the amplitude of the unmodulated waveform, i.e. the carrier.

For $\alpha \leq 1$ the envelope of $x(t)$ has the same shape as $m(t)$. The envelope is a continuous curve passing through the tips of the relative maxima and minima of $x(t)$. Since $m(t)$ has been normalized to have a maximum value of $+1$ and a minimum value of -1, the maximum value of the positive envelope is

$$A_{\max} = A_c \left[1 + \alpha\right] \tag{11.71}$$

and the minimum value is

$$A_{\min} = A_c \left[1 - \alpha\right] \tag{11.72}$$

which yields

$$\alpha = \frac{A_{\max} - A_{\min}}{A_{\max} + A_{\min}} \tag{11.73}$$

as the value of the AM modulation index; assuming that the modulation index lies between 0 and 1.

A conventional AM modulated wave is shown in Fig. 11.8 for sinusoidal or tone modulation for which $m(t) = \cos \omega_m t$. The expression for $x(t)$ becomes

$$x(t) = A_c \left[1 + \alpha \cos \omega_m t\right] \cos \omega_c t = A_c \cos \omega_c t + \frac{A_c \alpha}{2} \left[\cos(\omega_c + \omega_m)t + \cos(\omega_c - \omega_m)t\right] \tag{11.74}$$

by virtue of Eq. 11.60. For tone modulation, $x(t)$ consists of the sum of three sinusoidals at frequencies f_c, the carrier; $f_c - f_m$, the lower side band (LSB); and $f_c + f_m$, the upper side band (USB).

For messages signals more complicated than a single frequency, the spectra of $x(t)$ may be determined by using the modulation property of the Fourier transform, i.e.

$$\mathfrak{F}\left\{m(t) \cos \omega_c t\right\} = \frac{1}{2}\left[M(f - f_c) + M(f + f_c)\right] \tag{11.75}$$

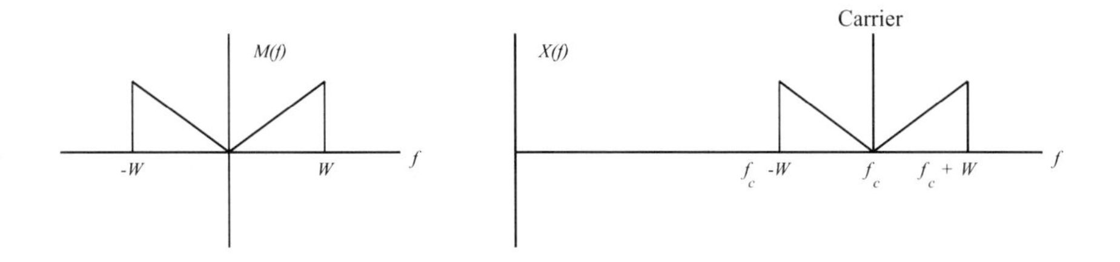

Figure 11.9: Spectra for AM.

which yields

$$X(f) = \frac{A_c}{2}\left[\delta(f - f_c) + \delta(f + f_c) + \alpha\left(M(f - f_c) + M(f + f_c)\right)\right] \tag{11.76}$$

as the spectra for AM.

If $m(t)$ represents a physical signal such as speech or music, it must be modeled probabilistically as a random process or signal. However, as long as it is bandlimited to $W\ Hz$, the spectra of $x(t)$ is as shown in Fig. 11.9. This consists of a translation or shift up the frequency axis to f_c. When this signal is demodulated f_c must be greater than W to avoid something known as foldover distortion which is similar to the aliasing phenomenon to be examined in connection with sampling. This is certainly not a problem with commercial AM radio where $W = 5\ kHz$ and the minimum value of $f_c = 600\ kHz$. The bandwidth of the AM modulated signal is $2W$ or twice the highest frequency present in the message, $m(t)$.

The spectra of conventional AM consists of a carrier and an upper and lower sideband. The carrier conveys no information; all of the information is contained in the sidebands. The sidebands are mirror images about the carrier. Therefore, the information is contained in either sideband. This means that to convey information only one side band needs to be transmitted. Transmitting both sidebands and the carrier is wasteful of both power and bandwidth. The transmitted bandwidth for AM is $BW = 2W$.

Systems which transmit both sidebands but not the carrier are known as DSB or DSB-SC (Double Side Band Suppressed Carrier). Systems which transmit only one sideband and no carrier are known as SSB or SSB-SC (Single Side Band Suppressed Carrier). Two choices are available with single side band: SSB-USB, single side band, the upper side band and SSB-LSB, single side band, the lower side band.

The advantage of conventional AM over DSB-SC or SSB-SC is it can be demodulated with a simple envelope demodulator as long as the AM modulation index is less than 1. This envelope demodulator consists of a diode and a low-pass filter (it will be considered in a later experiment). Other forms of modulation require a more complicated demodulator known as a product demodulator or synchronous detector.

11.3.4 Frequency Modulation (FM)

The waveform $x(t)$ shown in Fig. 11.10

$$x(t) = A_c \cos\left[\omega_c t + \omega_\Delta \int_{-\infty}^{t} m(u)du\right] \tag{11.77}$$

is known as a frequency modulated waveform. The parameters A_c and $f_c = \omega_c/2\pi$ and known as the amplitude and frequency of the carrier. The signal $m(t)$ is the message or information bearing signal and has been normalized so that its maximum value is $+1$ and its minimum value is -1. The parameter $f_\Delta = \omega_\Delta/2\pi$ is known as the maximum frequency deviation or simply the frequency deviation.

Eq. 11.77 can be expressed as

$$x(t) = A_c \cos\theta(t) \tag{11.78}$$

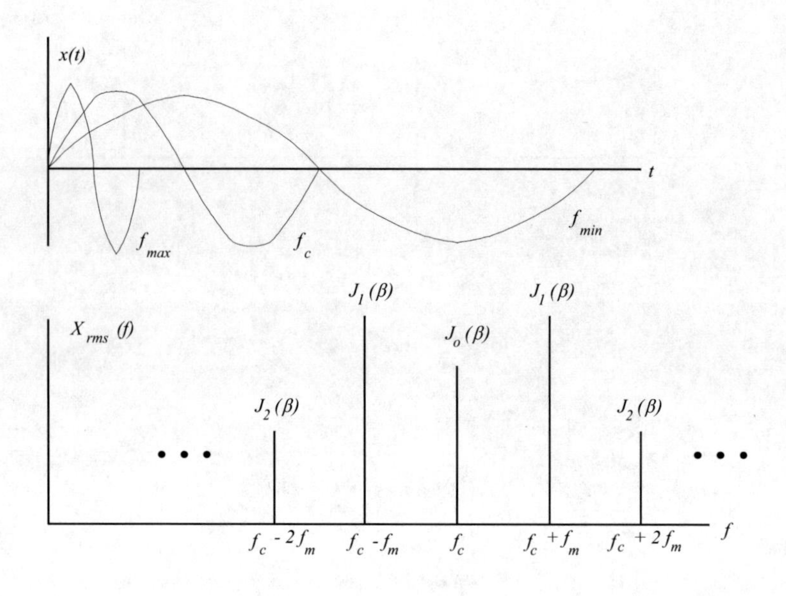

Figure 11.10: FM modulation.

where

$$\theta(t) = \omega_c t + \omega_\Delta \int_{-\infty}^{t} m(u)du \tag{11.79}$$

is known as the instantaneous phase of $x(t)$. The instantaneous frequency $f(t)$ is then given by

$$f(t) = \frac{1}{2\pi}\frac{d\theta(t)}{dt} = f_c + f_\Delta m(t) \tag{11.80}$$

which illustrates that the instantaneous frequency varies about the carrier frequency in accordance with the message. The instantaneous frequency swings above f_c by an amount f_Δ and below f_c by an amount f_Δ. Since the function $m(t)$ has been normalized to have a maximum of $+1$ and a minimum of -1, the maximum value of the instantaneous frequency is

$$f_{max} = f_c + f_\Delta \tag{11.81}$$

and the minimum value is

$$f_{min} = f_c - f_\Delta \tag{11.82}$$

which means that

$$f_\Delta = \frac{f_{max} - f_{min}}{2} \tag{11.83}$$

which is independent of the shape of $m(t)$.

The spectra of FM is considerably more complicated than that of AM. For tone modulation, $m(t) = \cos\omega_m t$, Eq. 11.77 becomes

$$x(t) = A_c \cos\left[\omega_c t + \beta \sin\omega_m t\right] \tag{11.84}$$

where

$$\beta = \frac{\omega_\Delta}{\omega_m} = \frac{f_\Delta}{f_m} \tag{11.85}$$

is known as the FM modulation index. Eq. 11.84 may now be expressed as

$$x(t) = Re\left\{e^{j[\omega_c t + \beta\sin\omega_m t]}\right\} = Re\left\{e^{j\omega_c t}e^{j\beta\sin\omega_m t}\right\} \tag{11.86}$$

n	$J_n(0.1)$	$J_n(1.0)$	$J_n(2.0)$	$J_n(5.0)$
0	0.998	0.765	0.224	−0.177
1	0.050	0.440	0.577	−0.328
2	0.001	0.115	0.353	0.047
3		0.020	0.129	0.365
4		0.002	0.034	0.391
5			0.007	0.261
6			0.001	0.131
7				0.053
8				0.018

Table 11.1: Bessel Function Sampling

where $Re\{\}$ is the real part operator. The complex exponential $e^{j\beta \sin \omega_m t}$ is a periodic function with period $T_m = 2\pi/\omega_m$ and may be expanded in a Fourier series

$$e^{j\beta \sin \omega_m t} = \sum_{n=-\infty}^{\infty} \overline{c}_n e^{jn\omega_m t} \tag{11.87}$$

and

$$\overline{c}_n = \frac{1}{T_m} \int_{-T_m/2}^{T_m/2} e^{j(\beta \sin \omega_m t - n\omega_m t)} dt \tag{11.88}$$

are the expansion coefficients. If a change of variable $u = \omega_m t$ is made in Eq. 11.88 it becomes

$$\overline{c}_n = \frac{1}{2\pi} \int_{-\pi}^{\pi} e^{j(\beta \sin u - nu)} du = J_n(\beta) \tag{11.89}$$

where $J_n(\beta)$ is a very well known and celebrated function of higher mathematics known as the Bessel function of the first kind, order n, and argument β. Therefore, the expression for $x(t)$ for tone modulation becomes

$$x(t) = A_c \cos[\omega_c t + \beta \sin \omega_m t] = A_c \sum_{n=-\infty}^{\infty} J_n(\beta) \cos[\omega_c + n\omega_m] t \tag{11.90}$$

which consists of a carrier with amplitude $A_c J_0(\beta)$ and an infinite number of sidebands separated by frequencies that are integer multiples of the modulating frequency with amplitudes of $A_c J_n(\beta)$. The theoretical transmitted bandwidth of FM in infinity. This is a sharp contrast with AM which has only 2 sidebands and a transmitted bandwidth of $2W$. Indeed, if β is selected so that it is a root of $J_0(\beta)$ there is no carrier. Fortunately, the magnitude of $J_n(\beta)$ decreases rapidly with n for fixed β.

Some value of $J_n(\beta)$ for values of n and β are given in Table 1.

where the blank slots are for values less than 10^{-3}. The value of $J_n(\beta)$ for negative values of n can be obtained from the relationship

$$J_{-n}(\beta) = (-1)^n J_n(\beta) \tag{11.91}$$

The effective transmitted bandwidth for FM is given by Carson's Rule. This states that the transmitted bandwidth, BW, is given by

$$BW = (2\beta + 1)W \tag{11.92}$$

where W is the highest frequency present in $m(t)$ and β is generalized to $\beta = f_\Delta/W$. This formula is valid for any $m(t)$ as long as it is bandlimited to W. Intuitively this is pleasing since, for large β, this is simply twice the frequency deviation or the frequency band through which $x(t)$ is swinging.

Commercial FM uses the frequency band 88 MHz to 108 MHz. The transmission bandwidth of each FM station is 200 kHz. Frequency modulation is able to obtain higher signal to noise ratios than amplitude modulation with less transmitted power. However, it requires a considerably more complicated receiver.

11.3.5 Binary Frequency Shift Key

Binary frequency shift key (FSK) is a modulation technique used to transmit binary digital data over a communications link such a telephone line with a modem. When a binary one is to be sent, this is represented as a sinusoidal at a frequency f_1 and when a binary zero is to be sent this is represented as a sinusoidal at a frequency f_o. It will be assumed that $f_1 > f_0$. (In addition to the terms "binary one" and "binary zero" the terms mark and space are sometimes used; these terms date from the early days of telegraphy.) The transmitted signal $x(t)$ is given by

$$x(t) = \begin{cases} A_c \cos \omega_1 t & 0 \leq t \leq T_b \quad \textit{binary } 1 \\ A_c \cos \omega_0 t & 0 \leq t \leq T_b \quad \textit{binary } 0 \end{cases} \tag{11.93}$$

where T_b is the duration of the binary symbol. The data rate, bit rate, or baud rate R is

$$R = \frac{1}{T_b} \tag{11.94}$$

which are all equivalent for binary digital systems. (It will be assumed for simplicity's sake that $T_1 = 2\pi/\omega_1$ and $T_0 = 2\pi/\omega_0$ are integral multiples of T_b.) Such a system is equivalent to FM where the message signal $m(t)$ is plus one for a transmitted binary one and minus one for a transmitted binary zero. Such a signal would have a complicated spectra since $m(t)$ would have to be modeled as a binary random signal if $m(t)$ represented digitized data. However, a worst case spectra can be obtained by considering $m(t)$ to be an alternating sequence of plus and minus one, i.e. a square wave with a peak-to-peak value of 2, a *DC* level of zero, and a period of $T = 2T_b$.

The spectra of $x(t)$ for square wave modulation could be obtained directly from Eq. 11.77. However, it can easily be obtained by considering $x(t)$ to be the sum of two RF pulse trains with duty cycles one half, RF frequencies f_1 and f_0, and a phase shift of 180°, i.e. one RF pulse train is delayed one half cycle compared with the other. This yields, after a fusillade of algebra,

$$x(t) = \sum_{n=-\infty}^{\infty} \lambda_n \cos 2\pi \left(f_c + \frac{nR}{2} \right) t \tag{11.95}$$

where $f_c = (f_1 + f_0)/2$ and

$$\lambda_n = \frac{A_c}{2} \left[sinc\left(\frac{\beta - n}{2}\right) + (-1)^n sinc\left(\frac{\beta + n}{2}\right) \right] \tag{11.96}$$

are the Fourier expansion coefficients for $x(t)$ where $\beta = 2f_\Delta/R$ and $f_\Delta = (f_1 - f_0)/2$. This is a line spectra centered about f_c.

The amount of constructive and destructive interference between the *sinc* functions in Eq. 11.96 is determined by the frequency separation of f_1 and f_0 and the data rate R, i.e. β. For large β distinct spectral components can be seen centered about the two RF frequencies; small β cause considerable interference and a more complex spectra. The first *sinc* function in Eq. 11.96 peaks at $n = \beta$ for which

$$f = f_c + \frac{nR}{2} = f_c + \frac{\beta R}{2} = f_c + f_\Delta = f_1 \tag{11.97}$$

and the second at $n = -\beta$

$$f = f_c + \frac{nR}{2} = f_c - \frac{\beta R}{2} = f_c - f_\Delta = f_0 \tag{11.98}$$

assuming, for simplicity, that β is an integer. The approximate bandwidth can be obtained from Carson's rule where $f_\Delta = (f_1 - f_0)/2$ and $W = R$.

11.3.6 Noise

Noise, as used in the context of electrical engineering, is an undesired signal; it can either originate in nature or systems designed by man. Random noise is noise which must be analyzed using statistical means. Although time plots of truly random noise are intractable, frequency domain plots are often reasonably well behaved functions. Indeed, a frequency domain analysis is often more illuminating than a time domain analysis.

Noise is ubiquitous in nature. Cosmic noise is produced by a variety of extraterrestrial mechanisms and impinges on all electrical and electronic circuits. Were it not for this noise it would be possible to detect modulated or information bearing signals emanating from anywhere in the universe. Despite the assertions of bug-eyed searchers for extra terrestrial intelligence, cosmic noise sets a limit on the distance at which any transmitted signal may be demodulated or detected. Other natural noise sources are found in the earth's atmosphere and crust.

Electrical and electronic circuits also generate noise. There are basically three types: thermal, shot, and flicker noise. Thermal noise or Johnson noise is produce by the random thermal motion of charge carriers, specifically electrons, in a resistor. Shot noise is produced when charge carriers cross a pn junction in a semiconductor or journey from cathode to anode in a vacuum tube. Flicker noise is produced by statistical variations in the path or resistance through which charge carriers move through a resistor.

Each of the above are examples of analog noise. Assuming a binary alphabet, digital noise is a totally random sequence of ones and zeroes; i.e. the probability of the next member of the sequence being a one or zero, conditioned each previous member of the sequence being exactly known, is still only one half which is simply maximum uncertainty or randomness. This is what would be obtained by tossing a unbiased coin with heads representing a binary one and tails a binary zero.

Some function generators have a noise setting for which the output is indicated as analog noise. Such noise is normally generated by taking digital pseudo noise (an almost random sequence of ones and zeroes) and passing it at a high rate through a low-pass filter. The pseudo noise is generated by a very long digital feedback shift register for which the degree of randomness is a function of its length. For such a properly designed pseudo random sequence generator, the cycle time is $T = 2^K T_{clock}$ where K is the number of stages in the shift register and T_{clock} is the period of the clock. Lest the tyro fret over the randomness of such as sequence, it should be pointed out if the clock had a frequency of 10 MHz and $K = 100$ this would yield a cycle time of 4×10^{15} years which is considerably older than the estimated age of the universe.

The type of noise is classified mathematically by its probability density function and rms spectra. The probability density function is that obtained at a sample instant or single value of time for the random noise signal. A random process sampled at only one instant of time is a random variable. The most common probability density function is Gaussian. Other probability density functions are the Rayleigh and Rician which describe the envelope of Gaussian noise.

Analog noise for which the plot of the rms noise voltage versus frequency is a constant is known as white noise. Thus a white noise source is one which has all frequencies present in equal amplitude. It is called white noise due to the analogy with white light where all colors are present in equal amounts. A plot of $X^2_{rms}(f)$ versus f is a horizontal line; such a random process is also said to have a flat spectra. The function $S(f) = X^2_{rms}(f)$ is known as the power spectral density. If a white noise source is connected as the input to a filter with transfer function, $T(s)$, the magnitude of the output spectrum will have the same shape as the transfer function of the filter.

A resistor with resistance R produces thermal or Johnson noise even if there is no current flowing through it. If a true rms voltmeter were connected across the terminals of such a resistor it would measure an *rms* voltage of

$$V_{rms} = \sqrt{4kTRB} \tag{11.99}$$

where $k = 1.38 \times 10^{-23}$ $Joules/Kelvin$, T is the temperature in degrees Kelvin, and B is the single-sided bandwidth over which the measurement is made. For instance, a 100 $k\Omega$ resistor would have a noise voltage of about 13 μV over a single sided bandwidth of 100 kHz at a temperature of $300°K$. Eq. 11.99 is valid for frequencies below $1,000\,GHz$ and temperatures above absolute zero. This noise cannot be eliminated and is independent of the type of resistor. It has a Gaussian probability density function and a flat power spectral density. The power spectral density for Johnson noise voltage is

$$S(f) = 4kTR \tag{11.100}$$

and has units of $watts/Hertz$. The voltage spectral density is given by

$$V_{rms}(f) = \sqrt{4kTR} \tag{11.101}$$

and is said to have units of $watts/\sqrt{Hz}$ and is read as "watts per root Hertz".

Shot noise is due to the discrete nature of the charge flow to form a current. If ears worked at the frequencies at which it occurs it would sound like hail on a tin roof. If a DC current I is flowing this discrete nature of the charge producing the current results in

$$I_{rms} = \sqrt{2qIB} \tag{11.102}$$

where $q = 1.6 \times 10^{-19}$ $coulombs$ is the charge on an electron. This formulae is valid only when the charge carriers are independent; this is the case for carrier crossing a p-n junction or traveling in a vacuum tube; it is not valid for carriers flowing inside a conductor. Shot noise is both Gaussian and flat. For instance, the shot noise in a 1 mA current over a bandwidth of 100 kHz is about 5.7 nA.

Flicker noise is highly dependent on the type of resistor through which current is flowing; it is largest in carbon composition and least in wire wound resistors. Unlike Johnson and shot noise, flicker noise has a $1/f$ power spectral density. This makes it the most prominent at lower frequencies.

11.3.7 Sampling

Digital instruments do not directly process continuous waveforms. In order for a digital instrument, such as a digital oscilloscope, aka DSO (Digital Storage Oscilloscope), or any computer or microprocessor based instrument, to analyze a continuous waveform it must first be digitized. Digitization consists of sampling the continuous signal and then quantizing the sample to a set of discrete levels. Sampling converts the continuous time variable to a discrete variable and quantization converts a continuous level of voltage or current to a discrete level. Sampling will be considered first.

The sampling process converts a continuous time waveform into a discrete time signal or sequence. If the sampling is performed at a sufficiently high rate, all of the information in the continuous time signal is preserved in the samples. If the sampling is performed at too low a rate, a distortion known as "aliasing" or foldover distortion occurs.

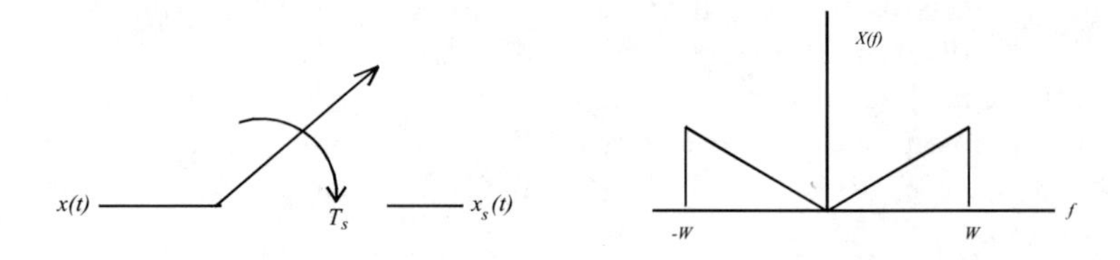

Figure 11.11: Sampling.

The signal $x(t)$ shown in Fig. 11.11 is assumed to be band limited to W Hertz. This simply means $X(f) = 0 \forall |f| > W$. Practically, this means that $x(t)$ has no spectral content above W Hertz.

This signal $x(t)$ is connected to the input of an ideal sampler which is represented as a switch that is closed every nT_s seconds where n is an integer and $T_s = 1/f_s$ where f_s is the sampling rate and T_s, the time interval between samples, is the sampling interval. The assumption of an ideal sampler means that the switch is closed for $t = nT_s$ and open for all other t. Mathematically, this means that $x_s(t)$ is given by

$$x_s(t) = \begin{cases} x(t) & t = nT_s \\ 0 & t \neq nT_s \end{cases} \tag{11.103}$$

which means $x_s(t)$ equals $x(t)$ at the sampling instants and is zero elsewhere.

Using the Dirac delta function $\delta(t)$ Eq. 11.103 can be expressed

$$x_s(t) = x(t) \sum_{n=-\infty}^{\infty} \delta(t - nT_s) \tag{11.104}$$

where the infinite summation of Dirac delta functions is a comb function in time. Taking the Fourier transform of this equation yields

$$X_s(f) = X(f) * \left[\frac{1}{T_s} \sum_{n=-\infty}^{\infty} \delta(f - nf_s) \right] \tag{11.105}$$

since the Fourier transform of the product of two functions is the convolution of their Fourier transforms and

$$\sum_{n=-\infty}^{\infty} \delta(t - nT_s) \leftrightarrow \frac{1}{T_s} \sum_{n=-\infty}^{\infty} \delta(f - nf_s) \tag{11.106}$$

form a Fourier transform pair. Since the convolution of a function of frequency with a delta function of frequency simply involves a translation of the function along the frequency axis, Eq. 11.105 yields

$$X_s(f) = \frac{1}{T_s} \sum_{n=-\infty}^{\infty} X(f - nf_s) \tag{11.107}$$

as the spectra of $x_s(t)$.

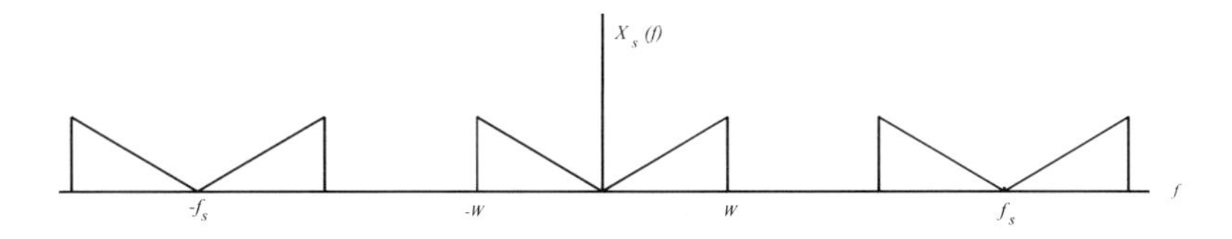

Figure 11.12: Spectra of a sampled signal.

A sketch of $X_s(f)$ versus f is shown in Fig. 11.12. As shown the terms in the infinite series 11.107 do not overlap as long as $f_s > 2W$. Therefore, $x(t)$ could be recovered or obtained from $x_s(t)$ by passing the sampled signal through a perfect low-pass filter for which the cutoff frequency is $f_s/2$ Hertz. This low-pass filtered signal is equal to $x(t)$ excepts for a scale factor. This is the Nyquist sampling theorem.

The Nyquist sampling theorem states that a signal band limited to W Hertz that is sampled at a rate greater than or equal to $2W$ Hertz can be obtained from the sampled signal by low-pass filtering with an ideal low-pass filter with cutoff frequency $f_s/2$ Hertz. The sampling rate $f_s = 2W$ is known as the Nyquist sampling rate, the Nyquist limit, or the Nyquist critical frequency. The frequency $f_s/2$ is known as the foldover or fold back frequency. If the signal is sampled at a rate less than the Nyquist rate overlap occurs in the frequency domain which produces a form of distortion known as aliasing or fold back distortion.

As a numerical example of aliasing consider the case of a 60 kHz sine wave that is sampled at a frequency of 100 kHz. If this sampled waveform is passed through an ideal low-pass filter with a cutoff frequency one-half the sampling rate, i.e. 50 kHz, a 40 kHz sine wave would be obtained. Thus the 60 kHz shows up as the alias of a 40 kHz sine wave.

If the sampling rate $f_s > 2W$, $X(f)$ can be expressed as

$$X(f) = T_s u\left[\frac{f_s}{2} - |f|\right] X_s(f) \tag{11.108}$$

where $u(\cdot)$ is the unit step function. The continuous signal can now be expressed as

$$x(t) = T_s \mathfrak{F}^{-1}\left\{u\left(\frac{f_s}{2} - |f|\right)\right\} * \mathfrak{F}^{-1}\{X_s(f)\} \tag{11.109}$$

which using Eq. 11.13 becomes

$$x(t) = T_s\left[2\frac{f_s}{2} sinc(f_s t)\right] * \left[\sum_{n=-\infty}^{\infty} x(nT_s)\delta(t - nT_s)\right] \tag{11.110}$$

yielding

$$x(t) = \sum_{n=-\infty}^{\infty} x(nT_s) sinc\left(\frac{t - nT_s}{T_s}\right) \tag{11.111}$$

as an alternative expression for $x(t)$. Eq. 11.111 illustrates that the continuous analog signal may be considered to be an infinite series of appropriately weighted $sinc(\cdot)$ functions. At the sampling instants each term in Eq. 11.111 is zero except for the term corresponding to $t = nT_s$. Between sampling instants $x(t)$ is an interpolation of the sampled values with the terms dropping off rapidly for samples not near the value of t.

Practically, an instrument, such as a digital oscilloscope, or computer must approximate Eq. 11.111 using a finite number of terms to approximate $x(t)$ at instants of time other than sample points. The accuracy of such an approximation is a function of the number of terms. To reduce the number of terms required to accurately approximate the function the sampling is usually done at a higher rate than the Nyquist rate. Many digital oscilloscopes have an effective sampling rate that is five times the Nyquist rate which permits the instrument to simply connect the dots between samples rather than use Eq. 11.111 to produce $x(t)$ at values of t that are not sample points. Other oscilloscopes use the eight nearest terms of Eq. 11.111 for values of time other than the sample points which permits sampling at a lower frequency.

The purpose of the trigger in a digital oscilloscope is to set the time origin for the samples. The time origin on the displayed waveforms is determined by a trigger event. A $TRIGGER\ LEVEL$ and $TRIGGER\ SLOPE$ controls allow the user to set the voltage level and slope for which a trigger event, aka trigger pulse, aka trigger, occurs. The samples are displayed relative to this trigger event. In general, the time origin may be located at the left, right, or center of the screen.

Three methods are used to acquire samples in digital oscilloscopes: real-time sampling, sequential repetitive sampling, and random repetitive sampling. Real-time sampling can be used with any type of band limited signal whereas repetitive sampling can only be used with periodic or repetitive signals. Repetitive sampling permits sampling at a rate lower than the Nyquist rate by exploiting the periodic nature of the signals.

Real-time sampling is the direct application of Nyquist sampling theory. The sampling must be done at rate which is greater than at least twice the highest frequency present in the signal. Samples are continuously written to the acquisition memory which a serially addressed device similar to a shift register. Once the trigger signal is acquired sampling is halted and the results displayed. For signals with significant high frequency content this requires fast analog to digital converters and memories which means an expensive instrument.

Sequential repetitive sampling samples at the trigger point on the first cycle. On the second cycle the sample point is advanced forward in time from the trigger point. This process is repeated with the sample point being advanced linearly in time with respect to the trigger point. Eventually enough samples are taken at which time the sampling process stops and the results displayed. Sequential repetitive sampling can only obtain samples after the trigger point which is a severe disadvantage.

Random repetitive sampling continuously samples the signal. In addition to the samples, the distance in time from the trigger point must be recorded. This permits the reconstruction of the signal once enough samples are taken. This type of sampling results in samples before and after the trigger point.

Neither of the repetitive sampling techniques violates Nyquist theory since the samples are obtained on different cycles of a time periodic function. Instruments that use this sampling technique display an "effective sampling rate" which is simply the reciprocal of the time spacing between sample points. Repetitive sampling permits equivalent sampling rates that are considerably higher than the actual sample rate of the analog to digital converter system.

11.3.8 Quantization

Although sampling digitizes the time variable, the signal level is still an analog quantity. To completely digitize the signal the amplitude must be quantized to one of a finite number of levels. If a K bit quantizer is used, this converts the continuous analog signal level into one of 2^K levels. The value of K used for digital oscilloscopes varies from 6 to 12 with 8 being the most common value.

The resolution is the smallest change in the voltage level that can be measured and/or displayed. For an oscilloscope using 8 bit quantization and set to the 5 $Volts/div$ this results in a resolution of

$$\Delta V = \frac{8 \times 5}{2^8} = 0.15625\ Volt/div \tag{11.112}$$

since there are 8 major vertical divisions. If the analog voltages are rounded off to the nearest of the 2^K levels the resolution is one half of this value.

11.3.9 Discrete Fourier Transform

Instruments which sample waveforms and store the sampled values in memory can perform mathematical analyses on the stored data. One of the most useful is the discrete Fourier transform. Since the instrument is digital both the time function and its Fourier transform must be discrete. This discrete Fourier transform is useful only if it is sufficiently close to the continuous Fourier transform that would be obtained by transforming the continuous time function.

A finite time record, T, must be used to implement the discrete Fourier transform. If N equally spaced points are taken in the time interval T, then the spacing between time points is

$$\Delta t = \frac{T}{N} = T_s \tag{11.113}$$

assuming that each sample is used in the algorithm to implement the discrete Fourier transform. The discrete values of time that will be used are $t = i\Delta t$ where $i = 1 \cdots N$.

If sampling of the bandlimited signal with bandwidth W is done at the Nyquist rate, then it is reasonable to pick the separation between the discrete values of frequency used to implement the discrete Fourier transform as

$$\Delta f = \frac{2W}{N} = \frac{f_s}{N} = \frac{1}{NT_s} = \frac{1}{N\Delta t} = \frac{1}{T} \tag{11.114}$$

since this will result in an $N \times N$ matrix for the Fourier transform pair. The discrete values of frequency that will be used are $f = k\Delta f$ where $k = 1 \cdots N$.

The discrete Fourier transform may be now be obtained from Eqs. 11.1 and 11.2 as

$$X(k) = \Delta t \sum_{i=1}^{N} x(i) e^{-j2\pi ik/N} \tag{11.115}$$

and

$$x(i) = \Delta f \sum_{k=1}^{N} X(k) e^{j2\pi ik/N} \tag{11.116}$$

where $X(k)$ is a short hand notation for $X(k\Delta f)$, $x(i)$ is a short hand notation for $x(i\Delta t)$, and the relation $\Delta t \Delta f = 1/N$ has been used. These equations illustrate that then N values for $X(k)$ can be obtained from the N values of $x(i)$ and vice versa.

The discrete Fourier transform pair are periodic in both time and frequency with period N. Moreover, the positive frequency spectra in contained in the first $N/2$ values of $X(k)$ and the negative frequency values are in the latter $N/2$ values (assuming that N is an even number). For this reason, normally only the first $N/2$ values of $X(k)$ are displayed.

Examination of Eqs. 11.115 and 11.116 shows that the matrix required to obtain the N $X(k)$ from the N $x(i)$ contains numerous duplicate evaluations of complex exponentials. A numerically efficient algorithm that utilizes these duplications known as the Fast Fourier Transform (FFT) is used to implement the discrete Fourier transform since it greatly decreases the computational effort. Indeed, the term FFT is often used instead of DFT (Discrete Fourier Transform).

If the signal $x(t)$ does not fall to zero at the end and beginning of the time record, spectral smearing may occur. Spectral smearing or leakage means that the spectra of the discrete Fourier transform has spread out or leaked into portions of the frequency spectra for which the continuous Fourier transform predicts no spectral content. For this reason, the original signal is usually multiplied by a data window function prior to implementing the transform. This windowed signal is given by

$$x_T(t) = w(t)x(t) \tag{11.117}$$

where the subscript T means that the original signal has been truncated to a length of T seconds and weighted by the data window $w(t)$.

One common window is the cosine bell window defined as

$$w(t) = \begin{cases} \frac{1}{2} - \frac{1}{2}\cos 2\pi \frac{t}{\alpha T} & 0 < t < \frac{\alpha T}{2} \\ 1 & \frac{\alpha T}{2} < t < T - \frac{\alpha T}{2} \\ \frac{1}{2} - \frac{1}{2}\cos 2\pi \frac{T-t}{\alpha T} & T - \frac{\alpha T}{2} < t < T \end{cases} \tag{11.118}$$

where $\alpha = 0.1$ and 0.2 are common values.

Another common window is the generalized Hamming window

$$w(t) = \begin{cases} \alpha - (1-\alpha)\cos \frac{2\pi t}{T} & 0 < t < T \\ 0 & \text{elsewhere} \end{cases} \tag{11.119}$$

where when $\alpha = 0.5$ it is known as the Hanning window (one should not become confused between the Hamming and Hanning windows). When $\alpha = 0.54$ it is known as the Hamming window (the term generalized is removed). The Hanning window is the simplest to implement while the Hamming window produces the least spectral smearing.

Another representation of the cosine bell window function is

$$w(t) = \sum_{n=0}^{\infty} a_n \cos\left(\frac{2\pi n\,[t - T/2]}{T}\right) \tag{11.120}$$

where the coefficients a_n are given by

n	0	1	2	3	4
Uniform or Rectangular	1	0	0	0	0
Flattop	1	1.9411	1.3084	0.4040	0.03511
Hanning	1	1	0	0	0
Gausstop	1	1.4588	0.5431	0.0882	0.0039

where the coefficients that are not shown may be taken as 0.

Another window function of interest is the exponential which is used with transient type signals. It is defined as

$$w(t) = \frac{e^{-\beta t} - e^{-\beta T}}{1 - e^{-\beta T}} \tag{11.121}$$

where β is a parameter set by the rate of decay of the transient signal.

The spectra of $x_T(t)$ may be obtained by convolving the spectra of $x(t)$ and $w(t)$. This results in some smearing of the spectrum of $x(t)$. The rectangular window has significant smearing and is only used with signals that are self windowing such as RF pulse trains. Exponential windows are used only with functions such as damped sinusoidals to insure that the windowed function drops to zero at the end of the time record. The most commonly used windows are the Hanning and flattop. The flattop has the best amplitude spectra accuracy while the Hanning has the best frequency resolution (the ability to distinguish between two spectra components that are close to each other in frequency).

When viewing the spectra of a signal with an instrument that performs the FFT the sampling rate must be carefully selected. The sampling rate must be high enough to prevent aliasing but low enough for good spectra resolution. The smallest frequency difference that such an instrument can display is

$$\Delta f = \frac{f_s}{N} \tag{11.122}$$

Thus, if a small resolution is required either the sampling rate must be small or the number of points used large.

11.4 SPICE

SPICE can be used to obtain the Fourier series expansion coefficients of a periodic function of time. The Fourier analysis must be performed in conjunction with a transient analysis. Since SPICE can only analyze circuits, a circuit of some kind must be used with the periodic function as the excitation. Below is the SPICE source code to perform the Fourier analysis of a symmetric square wave with a frequency of 1 Hz, a peak-to-peak value of 2 V, and a DC level of 0 V. The first nine Fourier series expansion coefficients will be printed to the text file with the extension `.OUT`. The square wave is in series with a 1 Ω resistor. The control statement .FOUR is used to request a Fourier analysis of the voltage at node 1 which has a fundamental frequency of $1.Hz$.

```
TITLE LINE
VI 1 0 PWL(0 -1 1F 1 0.5 1 0.50001 -1 1 -1)
*VI 1 0 PULSE(-1 1 0 1F 1F 0.5 1)
R 1 0 1
.TRAN 1M 1
.FOUR 1 V(1)
.END
```

Two ways of specifying the source are shown. One uses the SPICE function `PWL` (Piece Wise Linear) and the other `PULSE`. Either may be used. The asterisk at the beginning of a line converts it into a comment line.

The Fourier SPICE analysis will also yield the THD (Total Harmonic Distortion). Mathematically this is given by

$$THD = 100\sqrt{\sum_{n=2}^{\infty} |\bar{c}_n/\bar{c}_1|^2} \tag{11.123}$$

where the 100 is used to express it in terms of percent. This is simply a measure of how much the waveform deviates from a pure sine wave. Namely, a pure sine wave would have $c_n = 0 \forall n \geq 2$ and therefore a THD of zero.

11.5 Procedure

11.5.1 Function Generator Adjustment

The **Tektronix 3022B** function generator displays the amplitude of the voltage available at its output terminals. In order for the **Tektronix 3022B** function generator to indicate the open circuit or Thévenin voltage in the display when amplitude is selected the output termination must be set to *HIGH Z*. The default or power-on setting is 50 Ω which would then result in a display one half of the Thévenin source voltage.

Terminating Impedance Adjustment for Function Generator

Turn on the **Tektronix 3022B** function generator and wait until it finishes its self-test on power-on; this should only take a few seconds. Enable the output and configure it for High Z. The function generator is now configured so that it will display the Thévenin source voltage when the amplitude display is selected.

Set the amplitude of the waveform produced by the **Tektronix 3022B** function generator to 2 V peak-to-peak. Set the frequency of the function generator to 10 kHz.

11.5.2 Oscilloscope Adjustment for Spectrum Analyzer

Turn the **Tektronix 3012B** oscilloscope on and wait until it boots. Connect the CH1 output of the **Tektronix 3022B** function generator directly to the *Ch*1 input of the oscilloscope with a *BNC* to *BNC* connector. Nothing should be connected to *CH*2 of the oscilloscope and it should be turned off. Press *AUTOSET*. Use the vertical position and scale controls to position the waveform to the upper portion of the screen. Do not position the waveform off the screen; this will clip it and produce a severely distorted spectrum. Manually set the time per division to 40 μs.

The oscilloscope will now be configured to display the frequency spectrum of the waveform connected to the *CH*1 input. Press the *MATH* button followed by FFT. Use the vertical position control to place the frequency display in the lower half of the screen. Use the horizontal scale and position controls to position the bracket that appears at the top of the screen to the far left. The setting of this control can be speeded up by pressing the *COARSE* button. The bracket indicates what portion of the display is being viewed. It is color coded to indicate which trace is being examined. Because this is a sine wave the desired portion is near the origin of the frequency axis. A spike one horizontal division to the right of the left most graticule line should be visible.

Turn the cursors on. Select vertical bars. Press Bring Both Cursors On Screen. Vary the position of one until it passes through the spectral peak. The cursor should indicate that it is at approximately 10 kHz.

11.5.3 Windows

The effect of the choice of the data window will now be examined. With the **Tektronx 3012B** oscilloscope used as a spectrum analyzer, the Hanning, rectangular, exponential, and flattop data windows are available. They are selected by toggling the *Window* softkey found under the *FFT* menu.

Under the *Window* softkey switch the window to Rectangular, Hamming, Blackman-Harris, and Hanning and note the effect that it has on the display and describe it in the laboratory report.

Set the window to rectangular and leave it there for the remainder of the experiment.

11.5.4 Fourier Coefficients

Sine

Set the frequency of the **Tektronix 3022B** function generator to 10 kHz and the amplitude to 1 V rms. Do this by pressing the CH1 menu and then select units and change to Vrms. Then set the amplitude to 1 Vrms. Press the *CURSORS* button. Select Vertical Bars. The cursors can be switched from *CH*1, *CH*2, and *MATH*.The color of the cursor indicates what it is measuring. Vary the position of the cursor until the frequency is indicated which should be close to 10 kHz. Switch the cursors to horizontal bars and measure the amplitude of the spectral component (put the cursor at the top of the spike). Record the indicated frequency and amplitude (dBV rms) in the laboratory report. Press *MATH*, FFT and note the effect of switching the FFT Vertical Scale from dBV rms to Linear rms. Switch the vertical scale back to dBV.

Use the **Agilent 34401A DMM** to measure and record (in the laboratory report) the *rms* value of the output of the function generator. (Remember to press *AC V*.) Also measure and record the dc level.

Square

Switch the function on the **Tektronix 3022B** function generator from sine to square. Use the cursors to measure and record (in the tables in the laboratory report) the amplitude of the spectral components.

Use the **Agilent 34401A DMM** to measure and record (in the laboratory report) the *rms* value of the output of the function generator. (Remember to press *AC V*.) Also measure and record the dc level.

Triangular

Switch the function on the **Tektronix 3022B** function generator to triangular (ramp with 50% symmetry) and use the cursors to measure and record (in the laboratory report) the amplitude of the spectral components.

Use the **Agilent 34401A DMM** to measure and record (in the laboratory report) the rms value of the output of the function generator. (Remember to press *AC V*.) Also measure and record the dc level.

Ramp

Switch the function on the **Tektronix 3022B** function generator to ramp (0% symmetry) and use the cursors to measure and record (in the laboratory report) the amplitude of the spectral components.

Use the **Agilent 34401A DMM** to measure and record (in the laboratory report) the rms value of the output of the function generator. (Remember to press *AC V*.) Also measure and record the dc level.

AM

Connect the output of the **Tektronix 3022B** function generator to *CH*1 on the **Tektronix 3012B** oscilloscope. Set the function on the **Tektronix 3022B** function generator to sine, the amplitude to 2 V peak-to-peak, and the frequency to 60 kHz. The spike should now be near the center of the screen.

On the **Tektronix 3022B** function generator press the *Mod* button. Select type AM, AM depth 50%, and AM frequency 100 Hz.

Press *AUTOSET* on the oscilloscope and rotate the Horizontal Scale knob until the TIME/DIV indicated for *CH*1 is 4 ms. Connect the Sync Output (TTL) of the function generator to the EXT TRIG on the oscilloscope and select EXT/10 as the trigger source and use TTL as the level. Position *CH*1 so that it is in the upper half of the display. Press *MATH* and then position the spectra so that it is in the bottom half of the display. Press the icon for the magnifying glass and change the horizontal scale and position until the bracket on the display on the top just encloses the spike. The lowest display should consists of the carrier and upper and lower sidebands. Use the horizontal bars cursors to measure the amplitude of the spectral components. Print the display.

Turn the AM modulation off by pressing Continuous on the Run Mode. Turn the cursors off.

FSK (Frequency Shift Key)

Set the frequency of the function generator to 60 kHz. Press *Mod*, select type as FSK, set the Hop Frequency to 40 kHz, and the FSK rate to 2 kHz.

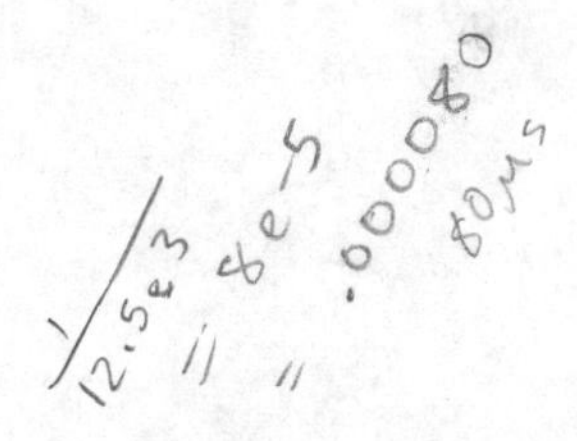

Vary the horizontal scale and position until the spectrum indicates that each major horizontal division represents 12.5 kHz and the display is centered. Print the display. Turn off the modulation by pressing Continuous on the Run Mode..

RF Pulse Train

Set the frequency of the function generator to 60 kHz. There should now be a spike near the center of the screen. If not, press MATH and then the Horizontal Position Control until the spike is centered.

Press the *BURST* button and set the Trigger Interval to 1 ms. Set the number of cycles to 5.

Print the display

Turn the burst off by pressing Continuous on Run Mode. Turn the Tektronix function generator off.

ELVIS

Connect the FG output of **ELVIS** to the AI7+ input on **ELVIS**. Connect the AI7- to the **ELVIS** ground. Launch **ELVIS** on the pc. Start the **ELVIS** FG. Set the **ELVIS** FG to produce a sine wave with a peak-to-peak value of 1 V and a frequency of 5 kHz. Start the **ELVIS** Dynamic Signal Analyzer (DSA) and set its input to AI7. This displays the frequency spectra of a signal. Press Run on both the FG and DSA. The spectra should be a spike at a frequency of 5 kHz.

Switch the function on the function generator to square. Do a Print Screen on the pc and use paint to paste the display into the laboratory report.

Switch the function to triangular and repeat.

11.5.5 Sound

Set the **Tektronix 3022B** function generator to produce a sine wave with a dc level of 0 (its default mode) and a frequency of 1 kHz. Connect the output of the **Agilent 33220A** function generator directly to the speaker provided. Adjust the amplitude of the waveform produced by the function generator to a level that is loud enough to hear but not loud enough to annoy. Listen to the sound emanating from the speaker while observing the waveform on the oscilloscope.

Change the function to square and repeat.

Change the function to triangular (ramp symmetry 50%) and repeat.

Change the function to ramp (symmetry 0%) and repeat.

Record which of the three waveforms (square, triangular, or ramp) "sounds" closest to the sine wave in the laboratory report?

Change the frequency of the function generator to 3 kHz and listen to the sound produced by all four functions.

Turn on the Sweep on the function generator (button labeled sweep). Pick a start frequency of 1 kHz and a stop frequency 10 kHz and a sweep time of 1 s. Reduce the sweep time and note the effect. Change the sweep mode from linear to log and describe the difference in sound.

11.6 Laboratory Report

The laboratory report should contain the following:

11.6.1 Windows

Describe the appearance of the spectrum for the following data windows:

Hamming

____

____

Hanning

Rectangular

Blackman-Harris

11.6.2 Signal Levels for Periodic Waveforms

The measured value of the dc Level and rms values of the periodic waveforms.

Parameter	Sine	Square	Triangular	Ramp
DC Level				
RMS Value				
Peak Value				

Table 11.2: Voltage Levels

11.6.3 Spectra Measurement

A comparison of the experimentally measured frequency spectra and the corresponding theoretical values. All spectral components should be expressed in terms of rms voltages and then converted to decibels. Recall that:

Square Wave

$$A_n = 2\left|\overline{c}_n\right| = \begin{cases} \dfrac{4A}{n\pi} & n \; odd \\ 0 & n \; even \end{cases}$$

Triangular Wave

$$A_n = 2\left|\overline{c}_n\right| = \begin{cases} \dfrac{8A}{n^2\pi^2} & n \; odd \\ 0 & n \; even \end{cases}$$

Ramp Wave

$$A_n = 2\left|\overline{c}_n\right| = \begin{cases} \dfrac{2A}{n\pi} \; \forall n \end{cases}$$

Harmonic	$A_{n,dB} = 20\log_{10}(A_n/\sqrt{2})$ **(Theory)**	$A_{n,dB}$ **(Exp)**	A_n/A_1**(Exp)**	**% Error**
1				
2				
3				
4				
5				
6				
7				

Table 11.3: Square Wave

Harmonic	$A_{n,dB} = 20\log_{10}(A_n/\sqrt{2})$ **(Theory)**	$\mathbf{A}_{n,dB}$ **(Exp)**	A_n/A_1**(Exp)**	**% Error**
1				
2				
3				
4				
5				
6				
7				

Table 11.4: Triangular Wave

11.6.4 ELVIS

Compare the results obtained for the spectra with **ELVIS** with that measured with the **Tektronix 3012B** oscilloscope.

11.6.5 AM

Sketch of spectrum of AM modulated waveform.

60.0 kHz
-9.6 dB
59.9 kHz
-21.6 dB
60.1 kHz
-25.2 dB

11.6.6 FSK

Sketch of spectrum of FSK modulated waveform.

11.6.7 RF Pulse Train

Sketch of spectrum of RF pulse train.

11.6.8 Aural Analysis

Unmodulated Waveforms

Discuss which of the three waveforms (square, triangular, and ramp) sounded the most like a sine wave and why.

Harmonic	$A_{n,dB} = 20\log_{10}(A_n/\sqrt{2})$ (Theory)	$A_{n,dB}$ (Exp)	A_n/A_1(Exp)	% Error
1				
2				
3				
4				
5				
6				
7				

Table 11.5: Ramp Wave

Sine Wave Modulation

When the function generator was frequency swept, was the sinusoidal nature discernible or did the gliding tone transmogrify into a warble?

____

____

____

If the laboratory experiment and/or report cannot be completed by the end of the three hour lab session, turn in what has been completed to the laboratory instructor.

11.7 References

1. Arfken, G., *Mathematical Methods for Physicists*, Academic Press, 1966.
2. Banzhaf, W., *Computer-Aided Circuit Analysis Using PSpice*, 2nd edition, Prentice-Hall, 1992.
3. Bracewell, R., *The Fourier Transform and Its Applications*, McGraw-Hill, 1965.
4. Carlson, A. B., *Circuits*, Wiley, 1996.
5. Carlson, A. B., *Communication Systems*, 2nd ed.,McGraw-Hill, 1975.
6. Coombs, C. F., Jr., *Electronic Instrument Handbook*, 2nd ed., McGraw-Hill, 1995.
7. Couch, L. W., II, *Digital and Analog Communication Systems*, 4th ed., Macmillan, 1993.
8. Churchill, R. V., *Fourier Series and Boundary Value Problems*, McGraw-Hill, 1941.
9. Dorf, R. C., *Introduction to Electric Circuits*, 2nd edition, Wiley, 1993.
10. Jayant, N. S. and Noll, P., *Digital Coding of Waveforms*, Prentice-Hall, 1984.
11. Keown, J., *PSpice and Circuit Analysis*, Merrill, 1991.
12. Kerchner, R. M., and Corcoran, G. F., *Alternating Current Circuits*, Wiley, 1962.
13. Krenz, J. H., *Introduction to Electrical Circuits and Electronic Devices: A Laboratory Approach*, Prentice-Hall, 1987.
14. Lathi, B. P., *Modern Digital and Analog Communications Systems*, 2nd ed., Holt, Rhinehart, and Winston, 1989.
15. Lago, G. V., and Waidelich, D. L., *Transients in Electrical Circuits*, Ronald, 1958.
16. LePage, W. R., *Complex Variables and the Laplace Transform for Engineers*, Dover, 1961.
17. McGillem, C. D., and Cooper, G. R., *Continuous and Discrete Signal and System Analysis*, Holt Rinehart and Winston, 1974.
18. MicroSim Corporation, *PSpice Circuit Analysis*, MicroSim, 1990.
19. Papoulis, A., *The Fourier Integral and its Applications*, McGraw-Hill, 1962.
20. Proakis, J. G., *Digital Communications*, McGraw-Hill, 1983.
21. Rashid, *SPICE for Circuits and Electronics Using PSpice*, Prentice-Hall, 1990.
22. Reddick, H. W., and Miller, F. H., *Advanced Mathematics for Engineers*, Wiley, 1938.

23. Riaz, M., *Electrical Engineering Laboratory Manual*, McGraw-Hill, 1965.
24. Sklar, B., *Digital Communications*, Prentice-Hall, 1988.
25. Start, H., Tuteur, F. B., and Anderson, J. B., *Modern Electrical Communications*, Prentice-Hall, 1988.
26. Su, K. L., *Fundamentals of Circuit Analysis*, Waveland Press, 1993.
27. Tuinenga, P. W., *SPICE: A Guide to Circuit Simulation & Analysis Using PSpice*, 3rd edition, Prentice-Hall, 1995.
28. Witte, R. A., *Electronic Test Instruments*, Prentice-Hall, 1993.

Chapter 12

Diodes

12.1 Objective

The objective of this experiment is to obtain the terminal $I - V$ characteristic curves for some common semiconductor diodes and to examine applications of diodes in rectifier, voltage doubler, waveshaper, and envelope detector circuits.

12.2 Theory

12.2.1 Ideal Diode

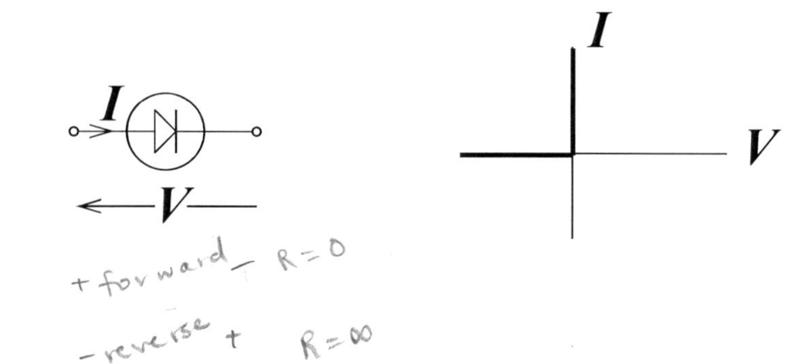

Figure 12.1: Ideal diode; (a) circuit symbol, (b) $I - V$ characteristic.

Shown in Fig. 12.1a is the circuit symbol and $I - V$ terminal characteristics for the ideal diode. The circuit symbol consists of a triangle touching a straight line surrounded by a circle. The side with the triangle is known as the anode, p side, high side, or plus side. The other side with the straight line is know as the cathode, n side, low side, or minus side. The reasons for the terms for the two sides of the ideal diode will become apparent.

The ideal diode is a two terminal nonlinear, nonbilateral circuit element. When current flows from left to right or from the anode to the cathode the ideal diode is said to be forward biased. When it is forward biased the voltage drop across the ideal diode is exactly zero volts and it is equivalent to a short circuit. When the cathode is positive with respect to the anode, the ideal diode is said to be reverse biased and no current flows no matter how large this reverse bias voltage is. Thus, a reverse biased ideal diode is equivalent to an open circuit.

The $I - V$ characteristic for the ideal diode is shown in Fig. 12.1b. It consists of the positive current axis and the negative voltage axis with the values of the ideal diode current and voltage determined by the external electrical circuit. When the ideal diode is forward biased the operating point lies on the positive current axis and when it is reverse biased it lies on the negative voltage axis. The origin of this characteristic corresponds to the point at which the ideal diode switches from an open circuit to a short circuit when both the voltage drop across it and the current flowing through it are exactly zero.

Ideal diodes are obviously an idealization. Physical diodes will not have characteristics that exactly match those of ideal diodes but ideal diodes can be used to construct simple but useful models of physical diodes. Physical diodes are implemented with vacuum tubes and semiconductors. Anent vacuum tubes, these are used primarily in very high power or high frequency applications and esoteric audio equipment. Almost all modern diodes are constructed using semiconductors.

12.2.2 Semiconductor Diodes

Two types of semiconductor diodes are widely used: the *pn* junction diode and Schottky-barrier diodes. Schottky-barrier diodes are formed from the metallurgical junction of a doped semiconductor and a metal and are used primarily in integrated circuits and high speed switching applications. Diodes formed from the metallurgical junction of a p type and n type semiconductors are know as *pn* junction diodes and are the most commonly used diodes. Only *pn* junction diodes will be considered.

Semiconductors are crystalline materials with electrical conductivities that lie between the conductivity of insulators and conductors. Semiconductors important to electronics are silicon, germanium, and gallium arsenide. Only silicon devices will be considered.

Silicon is a tetravalent element that has a partial filled valence band containing four electrons at absolute zero. Crystals of silicon have their atoms arranged in a three dimensional periodic array known as a lattice in which adjacent atoms share electrons via covalent bonding with their four nearest neighbors. At room temperatures, some of these electrons acquire sufficient thermal energy to break these covalent bonds and they enter the conduction band and leave a vacancy in the crystalline structure known as a hole. Both the free electrons in the conduction band and holes in the lattice contribute to the conduction process because they are free to wander about the crystal. Semiconductors which have equal numbers of free electrons and holes per unit volume are known as intrinsic. As temperature increases the conductivity of the semiconductor increases.

The conductivity of semiconductors may be altered by adding chemical impurities known as dopants; it is this property of semiconductors that makes possible sold-state electronic devices such as diodes, transistors, and integrated circuits. Even small amounts of chemical impurities can greatly alter the electrical characteristics of the semiconductor. Dopants with three electrons in the valence band, such as such as the trivalent element boron, create additional holes in the crystal lattice and semiconductors with this type of impurity are known as p type because the number of holes per unit volume exceeds the number of free electrons per unit volume. Dopants with five electrons in the valence band, such as the pentavalent element phosphorus, add additional electrons to the conduction band and semiconductors with this type of impurity are known as n type because the number of free electrons per unit volume exceeds the number of holes per unit volume.

Figure 12.2: *pn* junction diode; (a) physical structure, (b) circuit symbol.

A metallurgical bond of p type and n type semiconductor can be formed by adding acceptor dopants to one side of a semiconductor crystal and donor dopants to the other as shown in Fig. 12.2a. This *pn* junction acts as a diode for certain ranges of applied voltages in that it is easy for current to flow from the p to the

n side and difficult for current to flow from the n to the p side. The circuit symbol for a pn junction diode is shown in Fig. 12.2b.

When the junction of p and n type semiconductor is formed holes diffuse from the p to the n side and free electrons diffuse from the n to the p side leaving behind immobile ion cores from which they were ionized. This diffusion process is analogous to opening a door between two rooms which have different air pressures. The movement of the charge carriers constitutes an electrical current. The area immediately around the junction becomes depleted of mobile charge carriers leaving the immobile ion cores that are present due to the addition of the dopants. This sets up an electric field from the n side to the p side of the pn junction and a voltage across the junction making the n side positive with respect to the p side. This sets up another current flow mechanism known as drift current for which thermally generated electrons and holes are swept through the depletion region due to the presence of the electric field from the n to the p side of the junction.

With no applied voltage to the diode the current must be zero which means that the diffusion and drift current have equal magnitudes and opposite directions. The electric field from the n side to the p side prevents the diffusion current from exceeding the drift current. The voltage that exists across the junction is known as the barrier potential or built in potential of the diode and has a value of about 0.5 V. This voltage cannot be measured with a voltmeter since the metal contacts used to connect the p and n side of the diode to an external circuit would also have a metal-semiconductor voltage equal to the barrier potential and in the opposite direction.

When the diode is forward biased the p terminal is higher in potential than the n side and this external voltage acts in series with the barrier potential and has the opposite direction which has the net effect of reducing the voltage from the n side to the p side of the junction. This reduces the electric field in the depletion region and permits the diffusion current from the p to the n side to exceed the drift current from the n to the p side. The net effect is that a large current flow can be obtained from the p side to the n side with a relatively small applied voltage.

A reversed biased diode has the n side higher in potential than the p side. This external applied voltage adds in series with the junction potential and increases the electric field from the n side to the p side. This reduces the diffusion current to essentially zero and the only current that flows is the drift current from the n side to the p side. This current is rather small and is known as the reverse saturation current. The reverse saturation current is essentially independent of the applied reverse voltage. Typical reverse saturation currents lie in the range 10^{-8} to $10^{-16}A$. If the reverse voltage is made sufficiently large, the diode breaks down and large reverse currents can be obtained via the Zener or avalanche mechanisms.

A solid state physics analysis of a pn junction discloses that for sufficiently small currents and voltages the terminal characteristics are well modeled by

$$I = I_s[e^{V/\eta V_T} - 1] \tag{12.1}$$

where I is the diode current, V the diode voltage, I_s the reverse saturation current, η is the emission coefficient, and V_T is the thermal voltage. The emission coefficient is a dimensionless constant which ranges from 1 to 2 with 2 being typical for discrete diodes and 1 being typical for integrated circuit diodes. The thermal voltage V_T is given by

$$V_T = kT/q \tag{12.2}$$

where k is the Boltzmann constant $k = 1.38 \times 10^{-23} J/K$, T is the absolute temperature in degrees Kelvin, and q is the electronic charge $q = 1.6 \times 10^{-19}C$. At room temperature of $300^o K$, $V_T = 25.9\ mV$. When V is negative and larger in magnitude than 10 times V_T, the current is given by

$$I \simeq -I_s \tag{12.3}$$

and when V is positive and larger in magnitude than 10 V_T the current is given by

$$I \simeq I_s e^{V/\eta V_T} \tag{12.4}$$

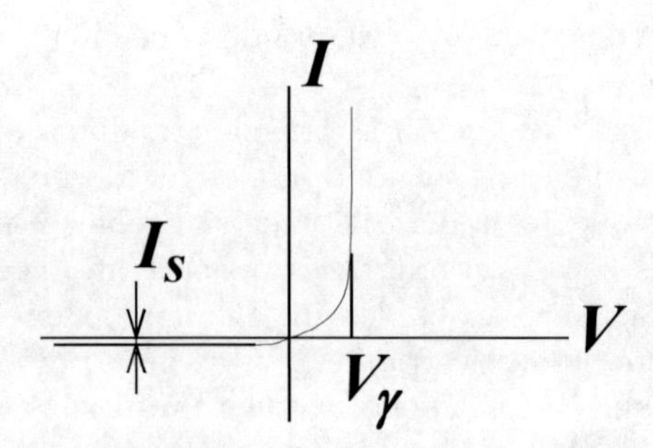

Figure 12.3: $I - V$ characteristic for a semiconductor diode.

If the diode were forward biased and a plot is made of $\ln(I)$ versus V, the parameters η and I_s can be obtained.

A typical plot of Eq. 12.1 is given in Fig. 12.3. Different current scales are used for positive and negative currents. When this characteristic is compared to the $I - V$ characteristic of the ideal diode in Fig. 12.1b, it is seen that semiconductor diode approximates the ideal diode. When the semiconductor diode is reversed biased the current is not zero but nevertheless very small. When the diode is forward biased, the voltage drop across the diode is not zero but has a value that depends on the current. For the currents in the milliampere range this voltage drop is almost constant and is given by V_γ which is known as the "on" or "turn-on" voltage for the diode. For silicon diodes, this "on" voltage is in the range 0.6 V to 0.7 V.

12.2.3 Model for Diodes

To analyze circuits containing diodes a model for the diode must be used. Eq. 12.1 provides an excellent model for the $I - V$ characteristic of a diode. However, since this is a nonlinear transcendental equation, it will be difficult to solve problems using this model unless a computer is employed. A simpler model is desirable for approximate solutions of circuits contains diodes and to obtain insight into constraints that diodes introduce into circuit analysis.

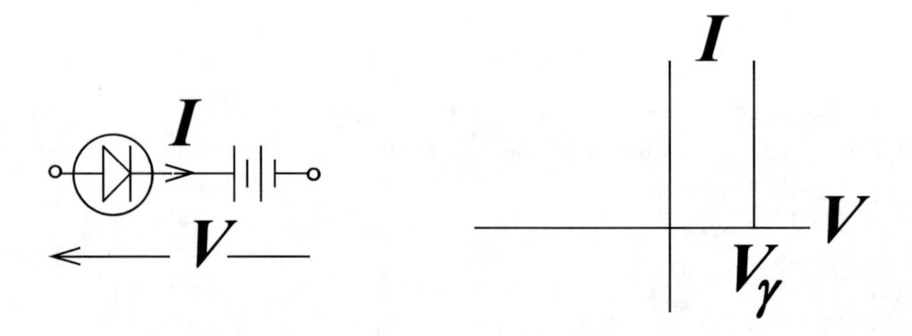

Figure 12.4: Diode model; (a) circuit, (b) $I - V$ characteristic.

The simplest model of the diode is the ideal diode. This is a good model for circuits in which the voltages sources are quite large compared to the "on" voltage of the diode, V_γ. If the voltage sources are comparable in magnitude to V_γ, then a better model is an ideal diode in series with a battery V_γ. This model is shown in Fig. 12.4 along with the $I - V$ curve for this model. No current flows through this model until the forward voltage drop across the model is V_γ and when the voltage drop is V_γ the current is set by the external circuit.

When a semiconductor diode is reverse biased and the reverse voltage is sufficiently large the reverse current no longer is equal to the saturation current because the diode begins to break down. There are two breakdown mechanisms: Zener breakdown and avalanche breakdown. Zener breakdown occurs when the electric field in the depletion region becomes so large that electrons are ripped out of their covalent bonds. Avalanche breakdown occurs when electrons are knocked out of their covalent bonds by other electrons in much the same fashion of a bowling ball hitting the pins. If the external circuit limits the current to safe levels, neither of these breakdown mechanisms damages the diode.

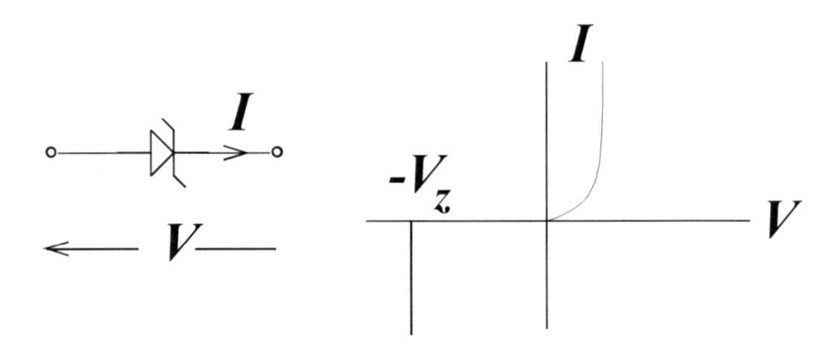

Figure 12.5: Zener diode; (a) circuit symbol, (b) $I - V$ characteristic.

Some diodes are designed to breakdown at a specific voltage and are used in their breakdown region. These diodes are termed breakdown diodes, avalanche diodes, or Zener diodes. If the breakdown voltage is less than 5 V, the mechanism is usually Zener breakdown and if the breakdown voltage is greater than 7 V the mechanism is avalanche breakdown. For diodes that breakdown between 5 and 7 V, a combination of Zener and avalanche breakdown is involved. The term Zener diode is commonly used to describe breakdown diodes no matter what the breakdown mechanism is. The circuit symbol and current voltage characteristic of a Zener diode are shown in Fig. 12.5 where the breakdown voltage is $V = -V_z$. Zener diodes are used as voltage references, voltage regulators, and circuit protection devices.

12.2.4 Rectifiers

Rectifiers are one of the most important applications of diodes. A rectifier is a circuit that converts a signal that has a DC level or zero average component into a signal with a DC level or nonzero average component. This is required when a DC voltage must be obtained from an ac source. Power supplies for electronic devices produce dc voltages for the operation of solid-state devices by rectifying the ac line voltage.

Shown in Fig. 12.6 is a half-wave rectifier. The voltage source $e(t)$ will be assumed to be $e(t) = E_p \sin(\omega t)$. If the diode in Fig. 12.6 is assumed to be an ideal diode, then current will flow in the clockwise direction in the circuit during the positive half cycles of $e(t)$ and no current will flow during the negative half cycles. Thus $i(t)$ has the shape shown in Fig. 12.6b and the resistor voltage would have the same shape. This circuit is known as a rectifier because it has "rectified" $e(t)$. The current, $i(t)$, and the resistor voltage, $v_R(t)$, are periodic signals with the same period as $e(t)$.

The voltage source $e(t)$ has a dc level of zero but $v_R(t)$ and $i(t)$ are not symmetric about the time axis and therefore do not have an average value of zero. The dc level of the resistor voltage is given by

$$V_{DC} = \frac{1}{T}\int_0^T R\, i(t)dt = \frac{E_p}{\pi} \tag{12.5}$$

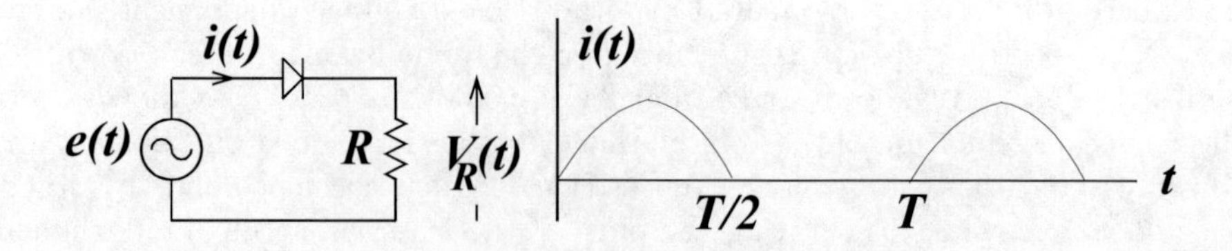

Figure 12.6: Half-wave rectifier; (a) circuit, (b) current versus time.

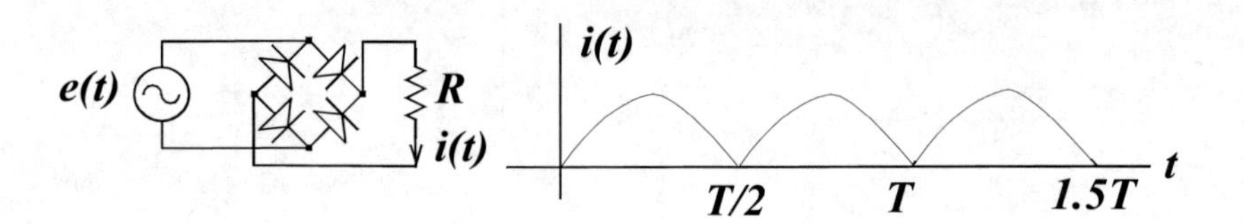

Figure 12.7: Full-wave rectifier; (a) circuit, (b) current versus time.

where E_p is the peak value of $e(t)$.

The circuit shown in Fig. 12.7 is known as a full-wave rectifier. The arrangement of the four diodes is known as a bridge circuit. Two of the diodes in this circuit are always forward biased and two are reverse biased so that current is always flowing through the resistor in the direction shown. The resistor current and voltage are periodic signals with periods that are one-half the period of $e(t)$. The dc level of the voltage drop across the resistor R is twice that of the half-wave rectifier.

The voltage across the resistor for either the half-wave or full-wave rectifier consists of a dc component and an ac component. For use in a dc power supply circuit the ac component should be minimized. This can be accomplished by placing a large capacitor across the resistor. If C is chosen so that $\tau = RC >> T$, then the capacitor will charge up to the almost the peak value of $e(t)$ and the resistor voltage will almost remain constant at E_p for the entire cycle of $e(t)$. There will be a small ac voltage superimposed upon the peak value, E_p, known as the ripple voltage. This ripple voltage has an rms value of $E_pT/(4\ \tau\sqrt{3})$ for the full-wave rectifier and twice that value for the half-wave rectifier.

If the semiconductor diode were modeled as an ideal diode in series with a battery V_γ in the rectifier circuits, the major effect would be to inhibit conduction until the diodes in the current path were turned on. Thus for the half-wave rectifier, no current would flow until $e(t)$ exceeded V_γ and the peak value of the voltage across R would be decreased by V_γ. For the full-wave rectifier, no current would flow until $e(t)$ exceeded $2V_\gamma$ and the peak value of the voltage across R in Fig. 12.7 would be decreased by $2V_\gamma$. The values for the average value of the current and resistor voltage are decreased when the effect of V_γ is included.

12.2.5 Clamping Circuit

The circuit shown in Fig. 12.8 is known as a clamping circuit or dc level restorer. Current can flow only in the counterclockwise direction. When the diode is forward biased the capacitor voltage is equal to $e(t)$ [assuming that the diode is ideal] and the diode prevents the capacitor from discharging. The capacitor is then charged to the peak value of $e(t)$ which is the voltage E_p. The diode voltage is then given by $E_p[1+\sin(\omega t)]$ as shown in Fig. 12.8b.

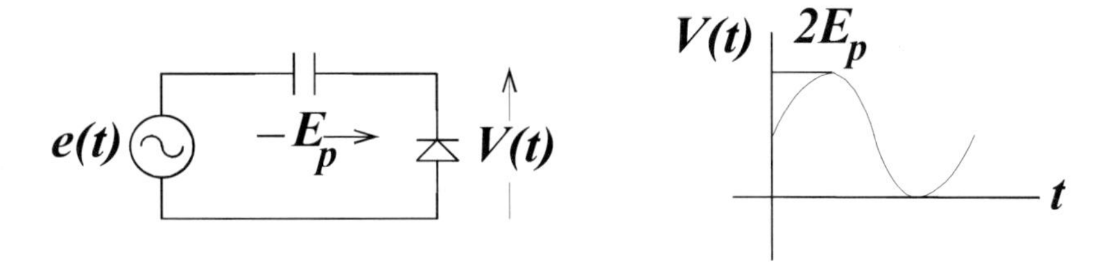

Figure 12.8: Clamper; (a) circuit, (b) output voltage versus time.

This circuit is known as a clamper because it "clamps" the negative peak of $e(t)$ to 0 V. It is known as a dc restorer because it reinserts a dc level to $e(t)$; it is an integral component in every TV set where it is used to set the voltage level in the baseband video signal corresponding to the color black.

12.2.6 Voltage Doubler

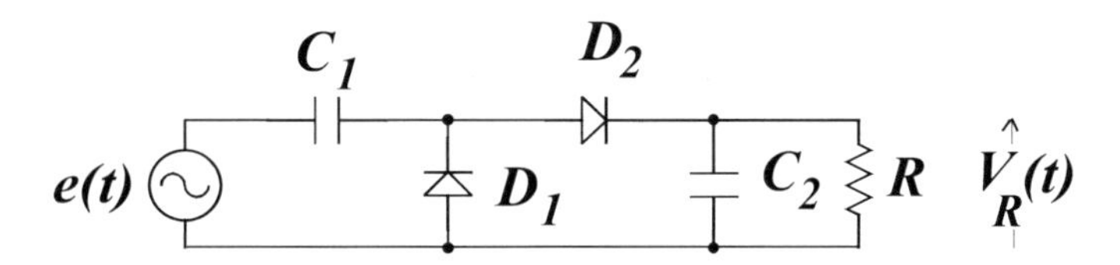

Figure 12.9: Voltage doubler.

The circuit shown in Fig. 12.9 is known as a voltage doubler. It is formed by cascading the output of a clamper with a half-wave rectifier that has a capacitor in parallel with the resistor. The diode D_1 and capacitor C_1 are the clamping circuit. The diode D_2, resistor R, and capacitor, C_2 are the half-wave rectifier. The capacitor C_2 and resistor R are chosen so that the time constant $RC_2 >> T$ where T is the period of $e(t)$. Voltage doublers and modifications known as triplers and quadruplers are used in high voltage dc power supplies such as those found in television sets.

The voltage across R is essentially a dc voltage $2E_p$ plus a small ripple component. This circuit is called a voltage doubler because the dc voltage is double the peak value of $e(t)$.

12.2.7 Waveshapers

Waveshapers are circuits that use the nonlinear, nonbilateral properties of diodes to alter the shape of a waveform. The shaping that such a circuit produces can be determined by obtaining the dc transfer characteristic or transfer function. This consists of determining the dc output voltage as a function of the input dc voltage. This transfer characteristic can be used to determine the wave shaping that such a circuit will produce. There are an infinite number of waveshapers; four will be considered here.

Shown in Fig. 12.10 is the first waveshaper to be considered. The input is the variable dc source V_i. It can be set to any positive or negative voltage. The output for this circuit will be taken as the voltage across resistor R_1. The first step in solving these types of circuits is to decide what model will be used for the semiconductor diode. The model that will be used is an ideal diode in series with a battery of V_γ volts. Once the semiconductor diode is replaced with this model, it is clear that no current will flow in this circuit

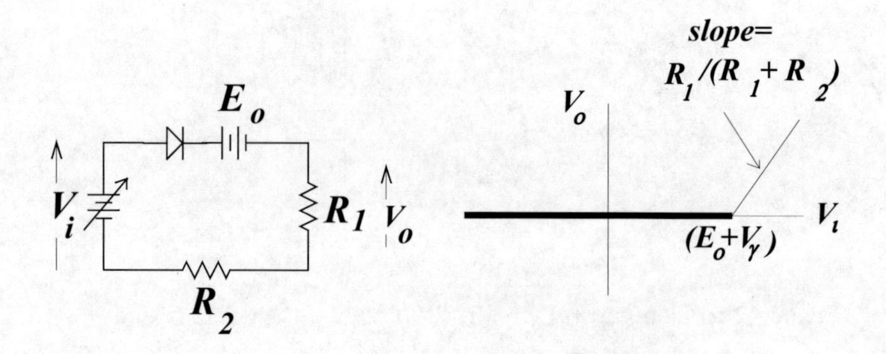

Figure 12.10: Waveshaper No. 1; (a) circuit, (b) transfer characteristic.

for values of $V_i < (E_o + V_\gamma)$. If no current is flowing through the circuit, then the voltage drop across the resistor R_1 is zero. Therefore, for $V_i < (E_o + V_\gamma)$, the output V_o is zero. For values of $V_i > (E_o + V_\gamma)$, the diode is forward biased and the output voltage is given by

$$V_o = [V_i - (E_o + V_\gamma)]\frac{R_1}{R_1 + R_2} \tag{12.6}$$

which is simply a straight line with a slope of $R_1/(R_1 + R_2)$. This plot of V_o versus V_i is a continuous plot that consists of two straight line segments; the point at which the line changes slope is the point at which the ideal diode switches from an open circuit to a short circuit. If an arbitrary voltage, $e(t)$, were connected in place of V_i, the output could be determined from this characteristic.

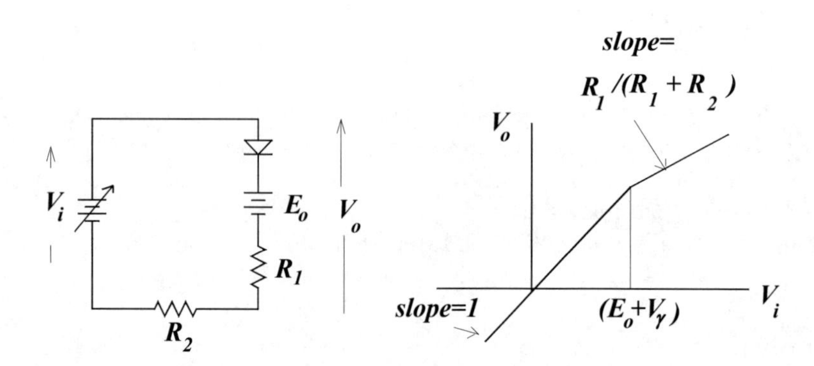

Figure 12.11: Waveshaper No. 2; (a) circuit, (b) transfer characteristic.

Shown in Fig. 12.11 is the second waveshaper circuit to be considered. This is the same as the previous circuit except that the output will now be taken as the voltage across the series combination of the diode, battery E_o, and resistor R_1. Again, no current flows until $V_i > E_o + V_\gamma$ which means that the voltage drop across the resistor R_2 is zero and that the input and output voltages are the same for $V_i < E_o + V_\gamma$. For values of $V_i > E_o + V_\gamma$, the output voltage is given by

$$V_o = (E_o + V_\gamma) + [V_i - (E_o + V_\gamma)]\frac{R_1}{R_1 + R_2} \tag{12.7}$$

which is a straight line with a slope of $R_1/(R_1 + R_2)$.

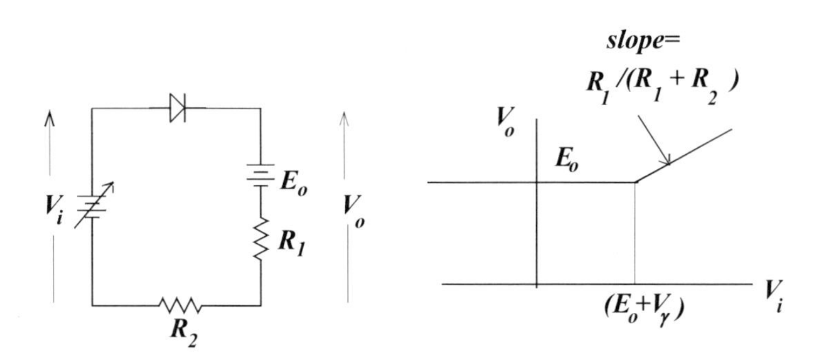

Figure 12.12: Waveshaper No. 3; (a) circuit, (b) transfer characteristic.

Shown in Fig. 12.12 is the third waveshaper circuit to be considered. Again, it is the same as the three previous circuits except that the output is taken as the voltage across the series combination of the battery E_o and the resistor R_1. As before, no current flows for $V_i < (E_o + V_\gamma)$ which means that the voltage drop across the resistor R_1 is zero and the output is E_o. For values of $V_i > (E_o + V_\gamma)$, the output is given by

$$V_o = E_o + [V_i - (E_o + V_\gamma)]\frac{R_1}{R_1 + R_2} \tag{12.8}$$

which is a straight line with a slope of $R_1/(R_1 + R_2)$.

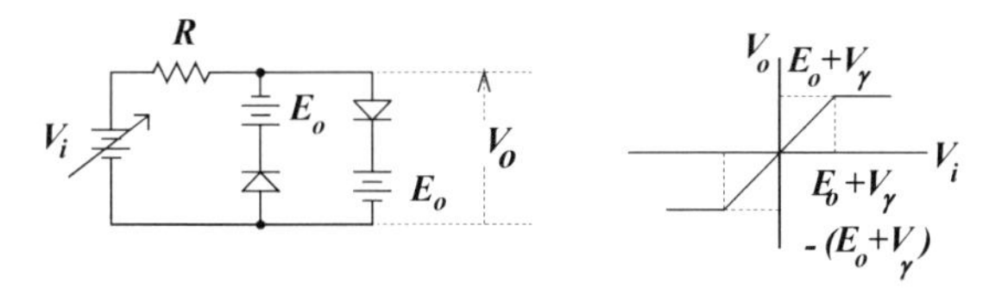

Figure 12.13: Waveshaper No. 4; (a) circuit, (b) transfer characteristic.

Shown in Fig. 12.13 is the fourth waveshaper to be considered. When $|V_i| < (E_o + V_\gamma)$ no current can flow in any branch of this circuit which makes the voltage drop across the resistor R_1 zero and $V_o = V_i$. When $V_i > (E_o + V_\gamma)$ the output is clamped to $(E_o + V_\gamma)$ and due to symmetry the same clamping occurs for negative V_i. This type of circuit is known as a limiter since it limits the magnitude of the output voltage.

12.2.8 Envelope Detectors

The heart of any AM radio is the envelope detector that demodulates an amplitude modulated waveform. An envelope detector is simply a rectifier cascaded with a low-pass filter. Such a circuit is shown in Fig.

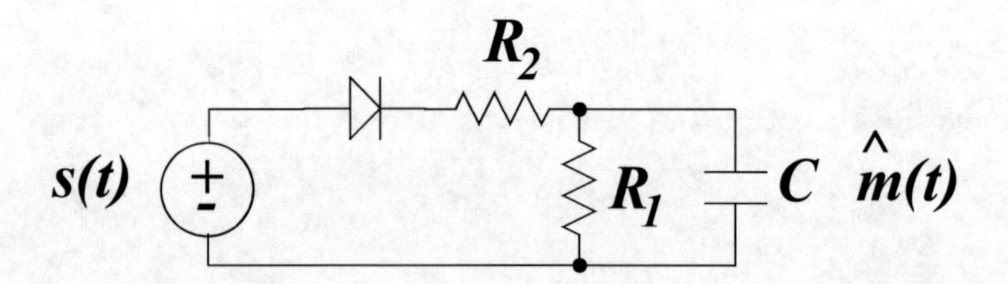

Figure 12.14: Waveshaper No. 4; (a) circuit, (b) transfer characteristic.

12.14. The resistor R_1 and capacitor C are chosen so that the time constant $\tau = R_1 C$ is small compared to the period of the high frequency carrier and large compared to the message or information.

An AM modulated waveform can be represented as

$$s(t) = A_c[1 + k\ m(t)]\cos(\omega_c t) \tag{12.9}$$

where the carrier is given by

$$c(t) = A_c \cos(\omega_c t) \tag{12.10}$$

k is a constant which sets the AM modulation index and $m(t)$ is the message or information bearing waveform. The diode rectifies Eq. 12.9 and the low-pass filter removes the high-frequency carrier so that the output is given by

$$\hat{m}(t) = A_c[1 + k\ m(t)] \tag{12.11}$$

which is the desired output $m(t)$ plus a dc level which can be removed via a blocking capacitor.

12.3 SPICE

SPICE can be used to analyze circuits with diodes. This two terminal circuit element is given a name beginning with the letter "`D`", the node numbers across which it is connected are listed, and, instead of a parameter value, a model statement is used to specify the parameters of the diode. Since the diode is not a bilateral circuit element, specification of the node numbers is crucial. The anode is listed first and then the cathode.

The source code to produce a plot of V_o versus V_i for the circuit shown in Fig. 12.13, Waveshaper No. 4 is:

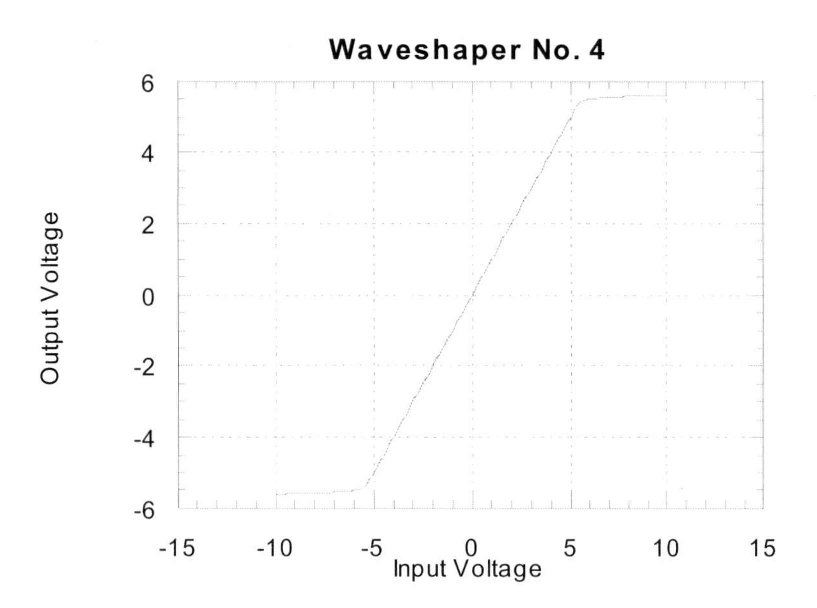

Figure 12.15: SPICE analysis of Waveshaper No. 4.

```
TITLE LINE
VI 1 0 DC 0
R 1 2 10K
D1 0 3 SIMPLE
V1 3 2 DC 5
D2 2 4 SIMPLE
V2 4 0 DC 5
.DC VI -10 10 0.1
.MODEL SIMPLE D(IS=1.7N N=1.89 BV=100)
.PROBE
.END
```

The voltage source `VI` is initially specified to have a value of 0 V since a DC sweep will be performed and its value is immaterial. The control line `.DC` specifies that the input voltage is to be swept from -10 V to $+10$ V in steps of 0.1 V. Instead of a parameter value the diodes are specified with the name `SIMPLE` and the parameters are passed through the `.MODEL` statement. To obtain the plot of V_o versus V_i, `PROBE` is run and the trace `V(2)` is plotted. The plot is shown in Fig. 12.15.

SPICE uses stored values known as the default values for all of the parameters for each semiconductor device that it can analyze. If values other than the default values are desired, they must be specified inside the parenthesis inside the `MODEL` statement. For the above example the diode parameters $I_s = 1.7$ nA, $\eta = 1.89$, and a breakdown voltage of 100 V is used. The diode turn on voltage V_γ is not a SPICE diode parameter.

The source code to analyze the half-wave rectifier shown in Figure 12.6 is:

```
TITLE LINE
VI 1 0 SIN(0 5 1K)
D 1 2 SIMPLE
R 2 0 10K
.MODEL SIMPLE D(IS=1.7N N=1.89 BV=100)
.TRAN 1U 2M 0 1U
.PROBE
.END
```

The input is a sine wave with a DC offset of 0 V, a peak value of 5 V, and a frequency of 1 kHz. PROBE would be used to plot the current `I(R)` to obtain the series current or the voltage `V(1,2)` to obtain the diode voltage.

The SPICE plot for the current in the half-wave rectifier is shown in Fig. 12.16.

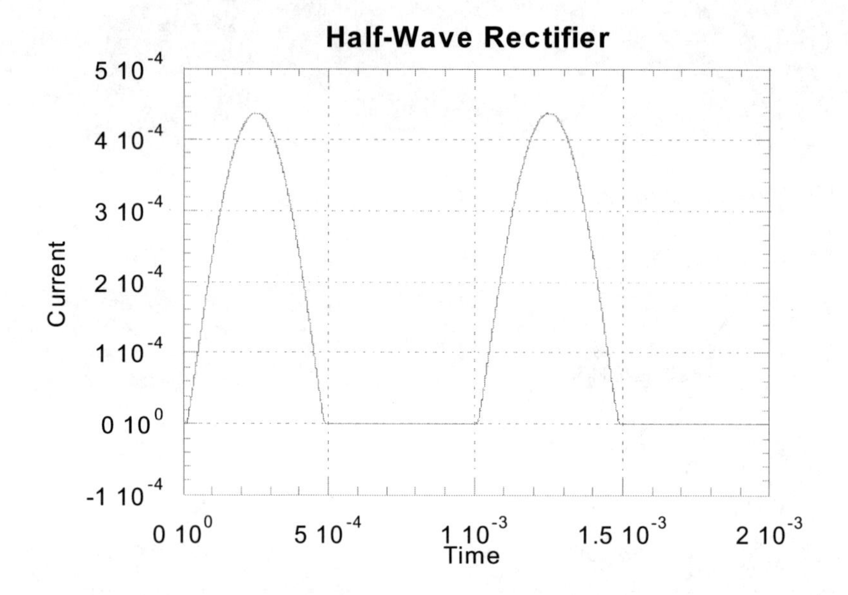

Figure 12.16: SPICE plot for current in half-wave rectifier.

12.4 Procedure

12.4.1 Resistance Measurement

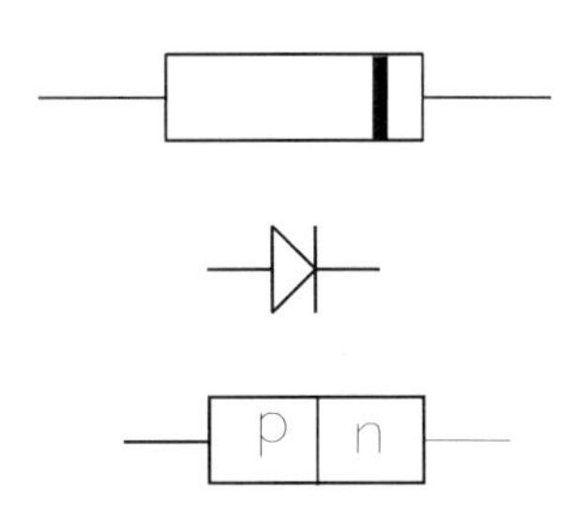

Figure 12.17: Semiconductor diode.

Measure the "resistance" of a silicon low-power signal diode shown in Fig. 12.17 with the **Fluke Model 73 Handheld Digital Multimeter**. Record the resistance obtained when the diode is forward biased and reverse biased and the **Fluke 73** is set to its Ohms "Ω" function (second column) and diode "→" function (third column). If the indication is " *O.L*" record the resistance as "over range". Be certain that only one body appendage touches the diode so that the resistance values obtained are not the parallel combination of the diode "resistance" and the body resistance.

Measure the "resistance" of the signal diode with the **Agilent 34401A Digital Multimeter**. The terminals used for resistance measurement are labeled *HI Input* (red) and *LO* (black). Press the button labeled "Ω 2*W*" to measure the resistance. Measure and record the resistance for both orientations of the diode. Press the blue shift button and then the button with the picture of a diode above it. Measure the "resistance" for both orientations of the diode.

Measure the resistance of the signal diode with the **Simpson Model 260-7**. First set the rotary knob to $R \times 10,000$ and short the two leads connected to the + and *COMMON* inputs together. Then vary the *ZERO OHMS* knob until a resistance of 0 (full scale deflection) is obtained. Measure the resistance of the diode (the resistance scale is the top scale). Repeat for the $R \times 100$ and $R \times 1$ scales (the zero ohms has to be readjusted each time the range scale is changed).

Measure the resistance of the signal diode with **ELVIS**. Launch **ELVIS** and select the DMM as the instrument. Set it to measure diodes, ⟶. Record the reading for the two orientations of the diode.

Orientation	**Fluke 73 Ω**	⟶	**Agilent 34401A Ω**	⟶	**Simpson (1)**	**(100)**	**(10,000)**	**ELVIS**
Forward Biased								
Reverse Biased								

Table 12.1: Diode Resistance Measurement

12.4.2 Measurement of Diode Parameters

Assemble the circuit shown in Fig. 12.18. Use the **HP3630A** triple output dc power supply for E, a 1N4148 diode for the diode, and a 100 kΩ resistor for R. The +20 V power supply will be used. Press the meter button to monitor the +20 V power supply. Use the knob labeled ±20 V to adjust the power supply to 10 V. The binding post labeled "+20 V" is the positive terminal and the binding post labeled "*COM*" is the minus

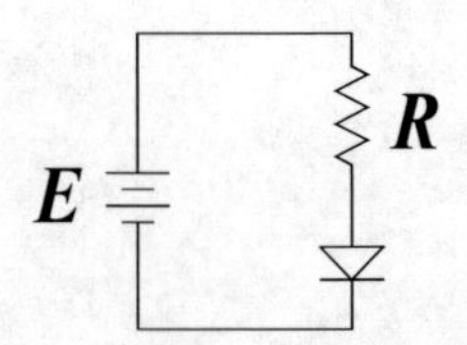

Figure 12.18: Circuit to Determine Precise $I - V$ Characteristics of a Diode.

terminal. The ground binding post on the far right it not used; do not connect anything to it. Measure and record the dc voltage drops across the resistor R (V_R) and then the diode (V_D) with the **Agilent 34401A** DMM set to measure a dc voltage. Set the power supply voltage to 7 V and repeat. Set the power supply voltage to 3 V and repeat. Set the power supply voltage to 1 V and repeat.

Change the resistor R to 10 kΩ and repeat the above measurements. Change the resistor R to 1 kΩ and repeat the above measurements. Turn off the dc power supply and disconnect all the leads connected to it; it will not be used for the remainder of the experiment. Measure and record the actual values of the nominal 100 kΩ, 10 kΩ, and 1 kΩ resistors with the **Agilent 34401A DMM** set to "Ω $2W$" resistance measurement. Compute the current I using Ohm's Law $I = V_R/R$ (the voltage across R divided by R).

E (Volts)	**R (kΩ)**	**V$_D$ (Volts)**	**I (Amps)**	V_R $(Volts)$
10	100 kΩ			
7	100 kΩ			
3	100 kΩ			
1	100 kΩ			
10	10 kΩ			
7	10 kΩ			
3	10 kΩ			
1	10 kΩ			
10	1 kΩ			
7	1 kΩ			
3	1 kΩ			
1	1 kΩ			

Table 12.2: Diode Voltage and Current

12.4.3 Function Generator Adjustment

Turn on the **Tektronix 3022B** function generator and wait until it finishes its self-test on power-on; this should only take a few seconds. Enable the CH1 output and set it for High Z.

12.4.4 $I - V$ Curve for Diode

Assemble the circuit shown in Fig. 12.19 using the **Tektronix 3022B** function generator as the signal source. Set the function generator to produce a sine wave with a peak-to-peak value of 2 V and a frequency of 100 Hz. Switch the Display to Triggered XY. Turn on the cursors. Position the cursors at two points on the display corresponding to the largest voltage and current and to where the diode is just turning on. Print the display.

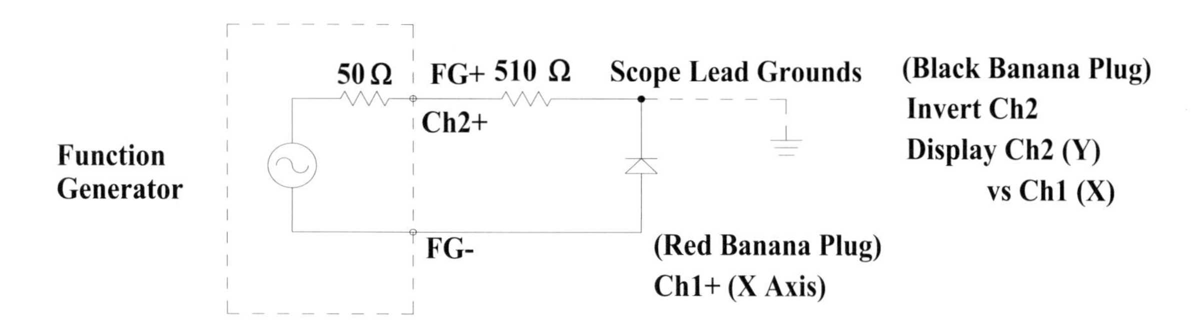

Figure 12.19: Circuit to Determine IV Characteristic for a Diode

Repeat this step for the 1N5226 3.3 V Zener diode. For this diode set the function generator to produce a voltage of 10 V peak-to-peak and set the $Volts/div$ for $CH1$ to 0.5 V and $CH2$ to 2 $Volts/div$.

12.4.5 Half Wave Rectifier

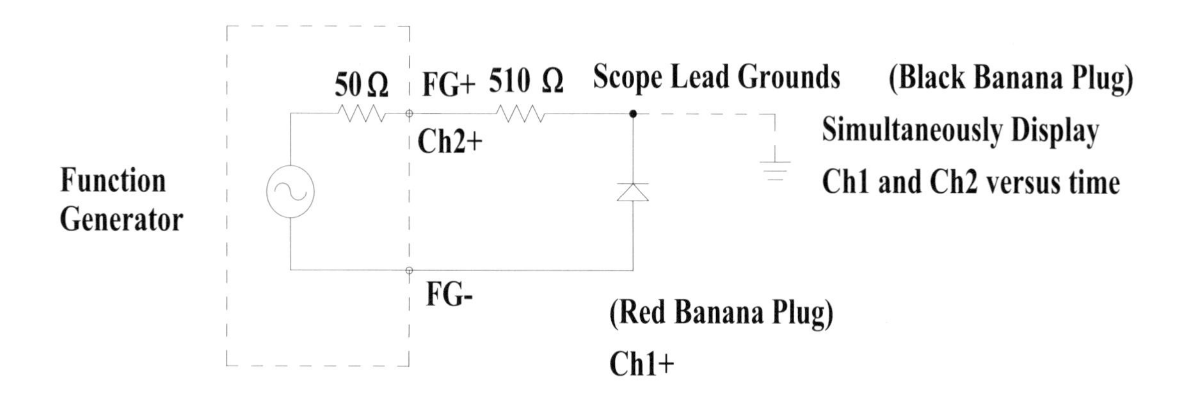

Figure 12.20: Circuit to Determine Diode Current and Diode Voltage versus Time

Assemble the half wave rectifier circuit shown in Fig. 12.20. Set the function generator to produce a sine wave with a frequency of 100 Hz and a peak-to-peak value of 10 V. Press *AUTOSET*on the oscilloscope. Press the $CH2$ button, MENU, and switch the *Invert* to *On*.

Turn on the cursors. Position the time cursors to correspond to the interval of time for which current flows. Print the display.

12.4.6 Full Wave Rectifier

Assemble the full wave rectifier circuit shown in Fig. 12.21. Do not connect the capacitor shown with the dotted line at this time. Set the function generator to produce a sine wave with an amplitude of 10 V and a frequency of 100 Hz. Press *AUTOSET* on the **Tektronix 3012B** oscilloscope. Turn on the cursors. Position the two time cursors to correspond to an interval of time for which current flows through the resistor. Print the display.

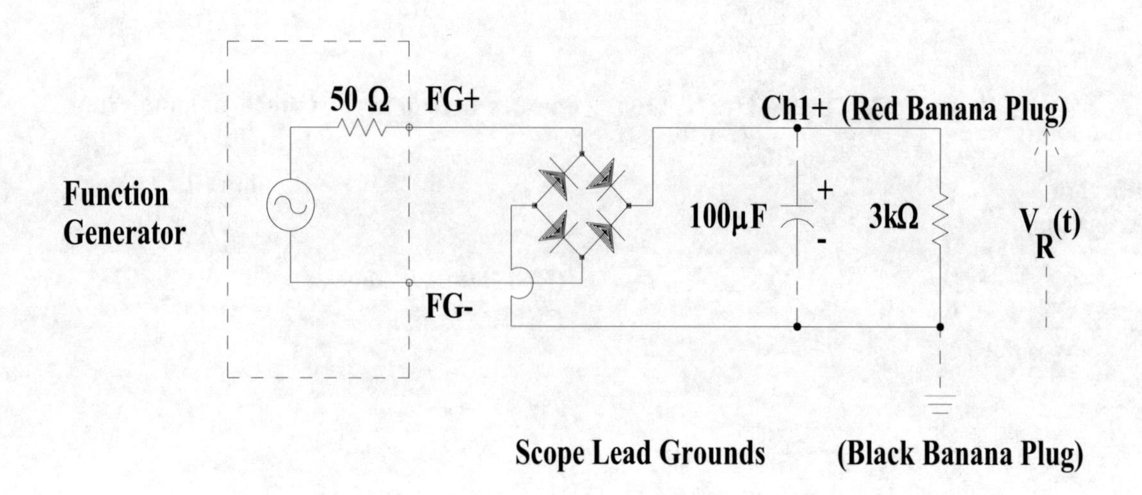

Figure 12.21: Full Wave Rectifier

Measure the rms ac value of and dc level of $v_R(t)$ with the **Agilent 34401A Digital Multimeter**. To measure a voltage use the terminals labels *HI Input* and *LO*. To measure a dc voltage press the button labeled "*DC V*" and to measure an ac voltage press the button labeled "*AC V*".

Place the 100 μF capacitor across the resistor as shown in Fig. 12.21. This type of capacitor is known as an electrolytic capacitor and has a polarity the must be observed, i.e. the terminal marked as minus should be placed on the side of the resistor that will have a lower dc voltage. If this type of capacitor were placed with an incorrect polarity in a circuit which had an energy source capable of producing a significant amount of current, it would physically explode.

Observe $v_R(t)$ as a function of time. Note that it is almost a horizontal line above the ground reference by almost the peak value of the sine wave. (The ground reference is shown in the left hand portion of the display just to the left of the grid.) Measure the dc level of $v_R(t)$ with the **Agilent 34401A** set to measure dc voltages. Also measure the rms ac component of $v_R(t)$ with the **Agilent 34401A**.

Switch the oscilloscope input coupling from DC to AC and observe how far vertically the display shifts. (This is done by pressing the $CH1$ button and then MENU). Also turn the bandwidth limit on for $CH1$). The vertical shift when the coupling is switched from DC to AC is the dc level of the waveform. Compare this with dc voltage measured with the **Agilent 34401A DMM**.

With the scope input AC coupled, change the $Volts/div$ for $CH1$ until the time varying component is adequately displayed (it should look like a ramp wave). Press *Stop*. Turn the cursors on and position them so that the period and peak-to-peak value of this waveform can be displayed.

Print the display.

12.4.7 Diode Limiter

Assemble the diode limiter circuit shown in Fig. 12.22. Use the **Tektronix 3022B** function generator as the signal source. Set the function generator to produce a sinusoidal voltage with a peak-to-peak value of 10 V and a frequency of 100 Hz.

Press *AUTOSET*. Switch the display to XY. Adjust the $VOLTS/DIV$ until the display occupies most of the screen. Turn the cursors on and position them to correspond to the break points on the display.

Print the display.

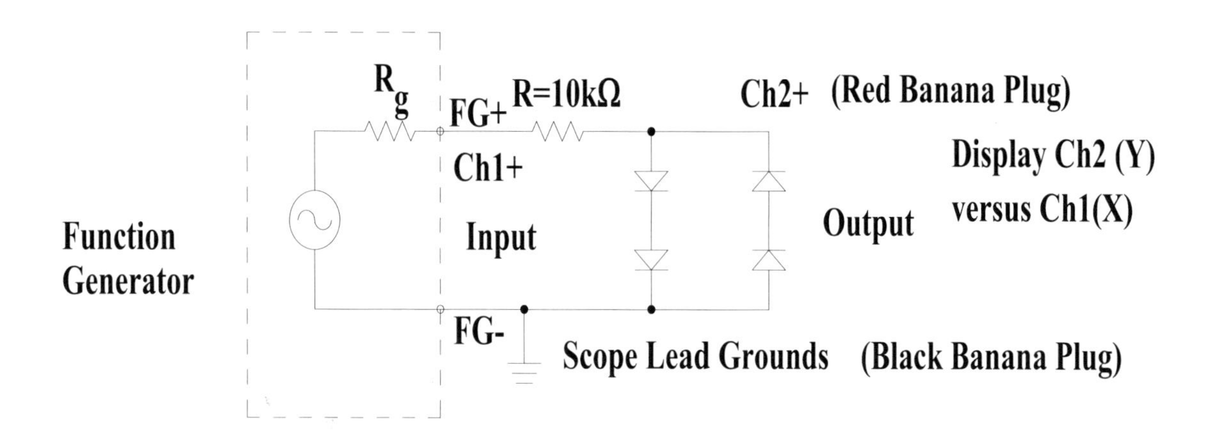

Figure 12.22: Diode Limiter

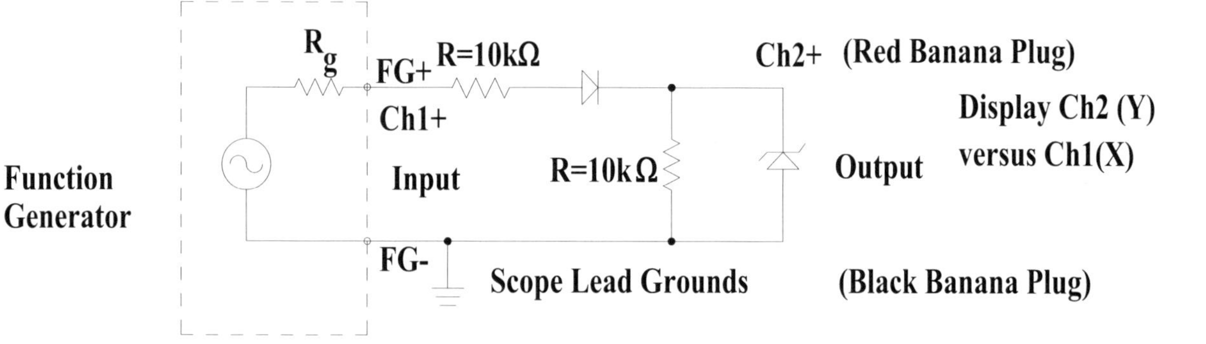

Figure 12.23: Rectifier Limiter

12.4.8 Rectifier Limiter

Assemble the rectified limiter circuit shown in Fig. 12.23. Use the **Tektronix 3022B** function generator as the signal source. Set the function generator to produce a sinusoidal voltage with a peak-to-peak value of 10 V and a frequency of 100 Hz.

Press *AUTOSET*. Switch the display to XY. Adjust the *VOLTS/DIV* until the display occupies most of the screen. Turn the cursors on and position them to correspond to the break points on the display.

Print the display.

(7)

12.4.9 Voltage Doubler

Assemble the voltage doubler circuit shown in Fig. 12.24. Press *AUTOSET* on the **Tektronix 3012B** oscilloscope. Press $CH1$ and then the $CH2$ buttons and make sure each channel is DC coupled. Manually set the $Volts/div$ for both channels to 5 V. Use the vertical position knob for $CH2$ to set the center horizontal line on the graticule to 0 $Volts$ for $CH2$. Use the vertical position knob for $CH1$ to position the sine wave so that the negative peak just touches the center horizontal line on the graticule. Turn on the cursors.

Print the display.

(8)

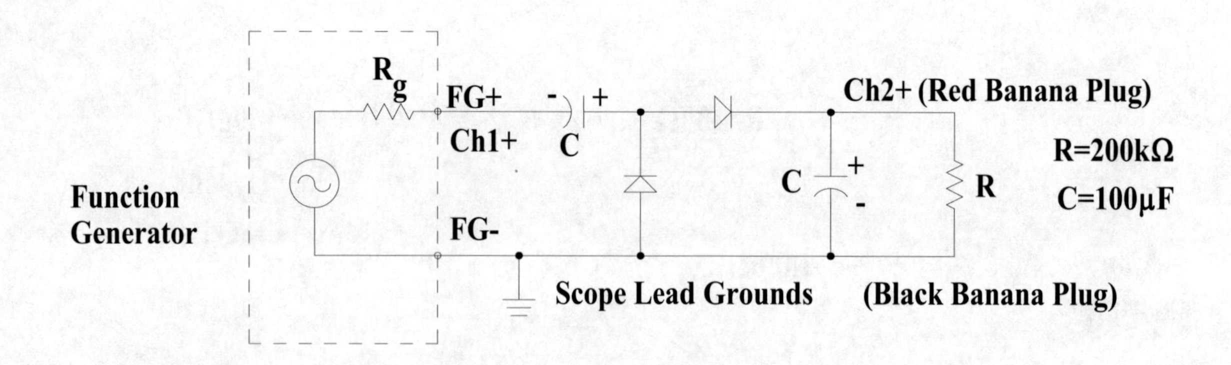

Figure 12.24: Voltage Doubler

Measure and record the dc and ac levels of the output voltage with the **Agilent 34401A DMM**. Compare the measured dc level with that indicated on the oscilloscope.

12.4.10 Envelope Detector

Set the frequency of the **Tektronix 3022B** function generator to 100 kHz; this is the carrier frequency for the AM signal. Press the blue *Mod* button followed by type *AM* button which produces an AM modulated waveform. Set the modulation index to 50%, modulating or AM frequency to 100 Hz; this is to be the modulating, information or message signal. Both the carrier and modulating signals are sine waves.

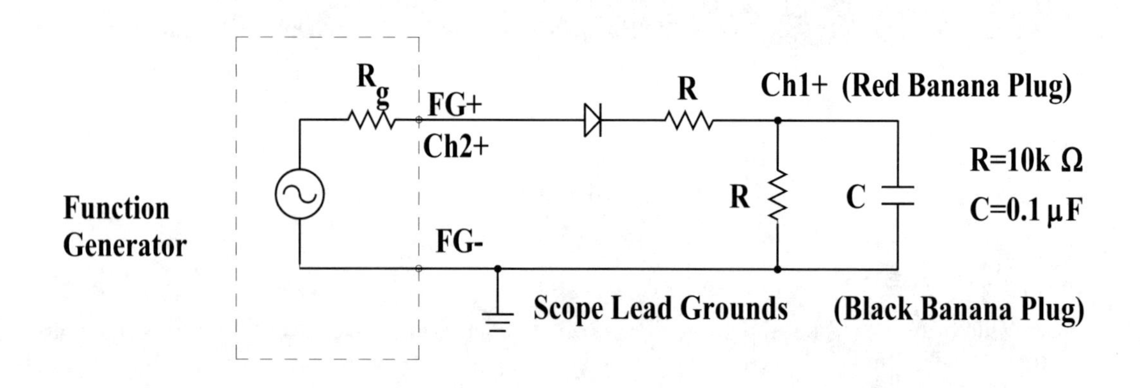

Figure 12.25: Envelope Demodulator

Assemble the circuit shown in Fig. 12.25. Press *AUTOSET* on the oscilloscope. Manually change the *TIME/DIV* to 2 ms. Press the *CH*2 button and turn the bandwidth limit on. Select external triggering by switching the trigger source to EXT/10 and the level to TTL.

Print the display.

The circuit shown in Fig. 12.25 is known as an envelope demodulator or detector. It is the fundamental component in all inexpensive AM radios. It removes the modulating waveform from the envelope and, therefore, retrieves the information.

12.5 Laboratory Report

The laboratory report should include the following:

1) Use the data obtained in step 12.4.2 to plot the natural log of the diode current $\ell n(I)$ versus the diode voltage V on linear graph paper. The diode current is the voltage across the resistor R divided by the actual value of R measured with the ohmmeter. From the slope of this line determine the emission coefficient η and from the y intercept the reverse saturation current I_s. Use a spreadsheet (or MathCad) to plot the graph. Use the linear regression feature of the spreadsheet (or MathCad) to determine the slope and y intercept of the plot. Assume that the thermal voltage V_T is 26 mV.

2) Plots of the $I - V$ characteristics of the signal diodes and Zener diode. From these plots the value of V_γ should be determined for all the diodes measured. For the Zener diode the value of the Zener breakdown voltage should also be determined.

3) Plots of diode current and voltage versus time for steps 12.4.5 and 12.4.6. From these plots, for what percentage of the period of the source voltage does it appear that the diode current is nonzero.

4) For the full wave rectifier, a comparison of the dc voltages measured across the resistor with and without the capacitor in parallel with the resistor. What should be the theoretical value of the dc voltages (assuming that the diodes can be modeled as ideal diodes in series with a battery of V_γ Volts)?

5) For the diode limiter and rectified limiter circuits, a comparison of the transfer characteristics obtained and what would have been theoretically expected using the model of the diode as an ideal diode in series with a battery of V_γ Volts.

6) For the voltage doubler, a discussion of the dc voltage measured on $Ch2$ and the waveform being produced by the function generator.

7) For the envelope demodulator, a discussion of the relationship between the waveform on Ch1 and the waveforms being produced by the function generators.

12.6 References

1. Antognetti, P., and Massobrio, G., *Semiconductor Device Modeling With Spice*, McGraw-Hill, 1988.
2. Banzhaf, W., *Computer-Aided Circuit Analysis Using PSpice*, 2nd edition, Prentice-Hall, 1992.
3. Burns, S. G. and Bond, P. R., *Principles of Electronic Circuits*, 2nd ed., PWS, 1997.
4. Fjeldly, T. A., Ytterdal, T., and Shur, M., *Introduction to Device Modeling and Circuit Simulation*, Wiley, 1998.
5. Gray, P. R., and Meyer, R. G., *Analysis and Design of Analog Integrated Circuits*, Wiley, 1993.
6. Hodges, D. A., and Jackson, H. G., *Analysis and Design of Digital Integrated Circuits*, 2nd edition, McGraw-Hill, 1988.
7. Horenstein, M. N., *Microelectronic Circuits and Devices*, Prentice-Hall, 1990.
8. Horowitz, P., and Hill, W., *The Art of Electronics*, 2nd ed, Cambridge, 1989.
9. Horowitz, P., and Robinson, I., *Laboratory Manual for The Art of Electronics*, Cambridge, 1981.
10. Howe, R. T. and Sodini, C. G., *Microelectronics*, Prentice-Hall, 1997.
11. Jaeger, R. C., *Microelectronic Circuit Design*, McGraw-Hill, 1997.
12. Keown, J., *PSpice and Circuit Analysis*, Merrill, 1991.
13. Krenz, J. H., *Introduction to Electrical Circuits and Electronic Devices: A Laboratory Approach*, Prentice-Hall, 1987.
14. MicroSim Corporation, *PSpice Circuit Analysis*, MicroSim, 1990.
15. Rashid, *SPICE for Circuits and Electronics Using PSpice*, Prentice-Hall, 1990.
16. Riaz, M., *Electrical Engineering Laboratory Manual*, McGraw-Hill, 1965.
17. Soclof, S., *Design and Applications of Analog Integrated Circuits*, Prentice-Hall, 1991.
18. Sedra, A. S., and Smith, K. C., *Microelectronic Circuits*, 4th edition, Oxford, 1998.

19. Sze, S. M., *Physics of Semiconductor Devices*, Wiley, 1981.
20. Thorpe, T. W., *Computerized Circuit Analysis with SPICE*, Wiley, 1992.
22. Tuinenga, P. W., *SPICE: A Guide to Circuit Simulation & Analysis Using PSpice*, Prentice-Hall, 1992.

Appendix A

Measurements with Grounded Instruments

Many electronic instruments used to measure physical parameters have one or more terminals of their input or output connectors that are internally connected to the earth ground. Such instruments are said to be "grounded" and require special procedures to make proper measurements. This requires an elementary understanding of the grounding system used in ac power systems.

A.1 AC Power Systems

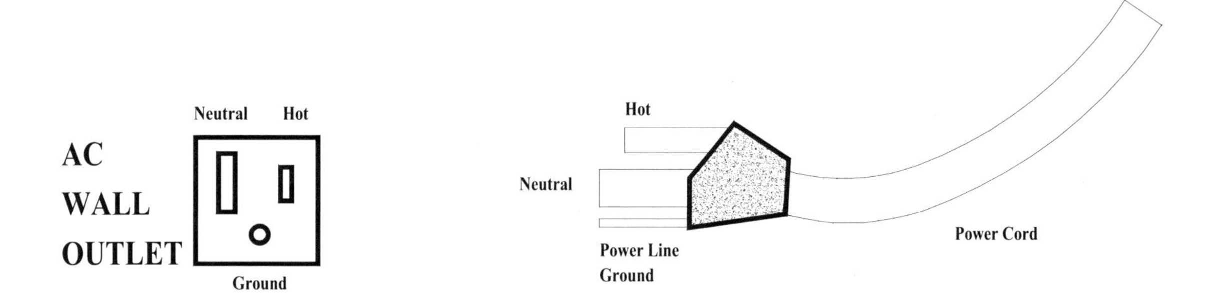

Figure A.1: 120 V ac Outlet.

Shown in Fig. A.1 is an 120 V ac receptacle or ac wall outlet found in commercial and residential facilities. It has three slots for a power cord with up to three prongs. The upper left slot is for the neutral wire in the power cord; it is rectangular and slightly larger vertically than the slot to its right. The slot on the upper right is for the hot wire which is also rectangular in shape. The lower slot is cylindrical in shape and is for the ground wire. Three types of power cords can be plugged into this 120 V ac receptacle: three wire, two wire polarized, and two wire unpolarized.

Three wire power cords have prongs for all three wires and, due to the cylindrical shape of the ground prong and slot, can only be plugged into the receptacle with one orientation. A polarized two wire power cord has the neutral prong slightly larger vertically than the hot prong and can, therefore, also only be plugged into the receptacle with only one orientation. An unpolarized two wire power cord has two identical prongs and can be plugged into the ac receptacle with either of two possible orientations.

A.2 120 V AC Power System

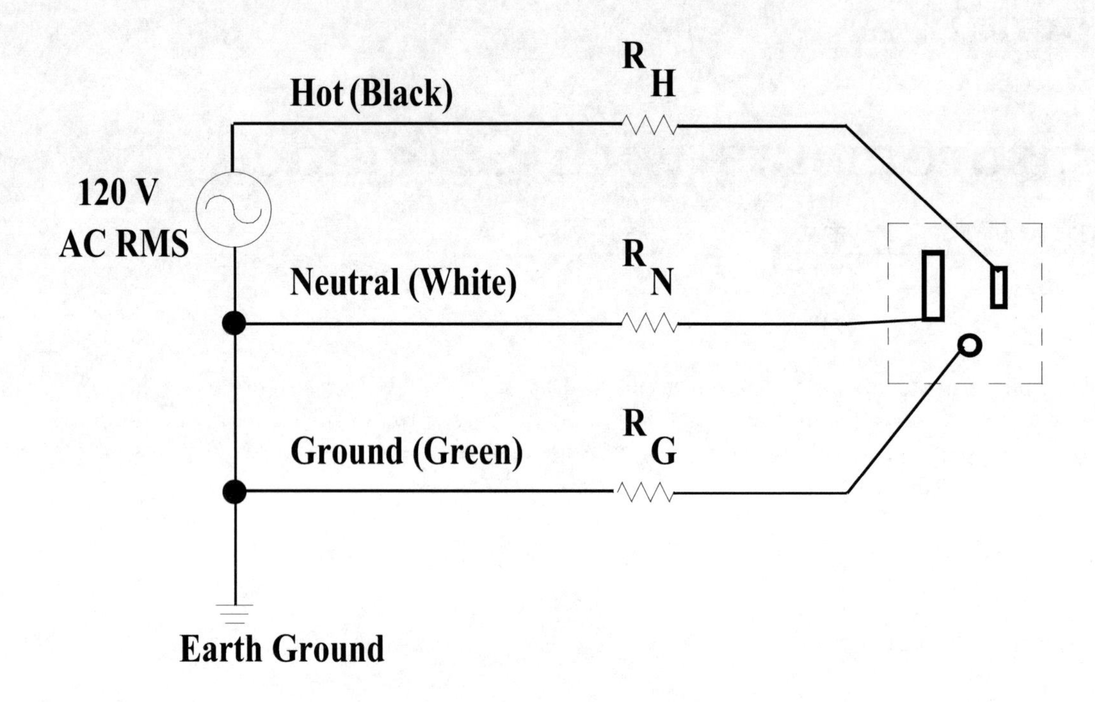

Figure A.2: 120 V ac Power System.

A conceptual representation of a 120 V ac power system is shown in Fig. A.2. The source voltage is 120 Volts ac *rms* which means that the instantaneous voltage is $120\sqrt{2}\cos\omega t$. Although only one ac receptacle is shown, there would normally be many such receptacles connected in parallel in a typical circuit.

The ac source would normally be the step down transformer servicing the building containing the circuit. The distribution system servicing the neighborhood containing the building would use a voltage such as 2,400 V which would be reduced by this step down transformer to a value such as 240 V. Typically this step down transformer's secondary would be a delta connected three phase transformer with each phase in the delta 240 V ac. Unless the customer has a use for three phase power, only one phase would be connected to a building. This phase would be centered tapped with the ac power ground connected to the neutral. This results in two 120 V ac sources that are 180° out of phase. This supplies 120 V ac circuits and 240 V ac circuits.

Three wires known as the hot, neutral, and ground connect the ac receptacle to the ac voltage source. The resistance of each of these three wires is R_H, R_N, and R_G and since these are normally copper wires of moderate length the value of the resistance is very small. These three wires are normally placed physically together in a conduit which means that they must be covered with insulation. Since all of the receptacles must be connected in parallel the National Electric Code mandates the following color code for the insulation: hot (black), neutral (white), and ground (green). If the conduit is itself an insulator, sometimes the insulation is left off the ground wire and it is a bare wire. For a 240 V circuit there are two 120 V ac hot wires that are 180° out of phase and the additional hot wire is covered with red insulation. (To needlessly complicate life, a different color code is used in the dc electrical systems of cars.)

The neutral and ground wire are connected together at the source and to a wire going down inside the earth (third planet from the sun). (The earth ground symbol is the three parallel horizontal lines of decreasing width below the voltage source.) These two wires are normally not connected together at any other point. The hot wire at the source end is 120 V above the earth ground voltage.

In a properly functioning 120 V ac circuit, current flows in only two wires: the hot and neutral. The ground wire is present primarily for safety reasons. Current flows up through the 120 V ac source out through the hot wire of the ac circuit, out the receptacle into the hot wire of the ac power cord, through whatever the load on the ac system is at the end of the power cord, back through the neutral wire of the power cord, through the neutral wire of the ac circuit to the ac voltage source. (Obviously in an ac circuit the direction of flow of the instantaneous current changes every half cycle, but, conceptually, it is easier to think of the hot wire supplying current and the neutral wire returning it.) Normally, no current whatsoever flows in the ground wire. When current flows in the ground wire a fault (an unintentional short circuit or other egregious event) has occurred in the ac circuit.

At the ac receptacle the voltage of the hot wire may be somewhat less than 120 V with respect to the earth ground if the current being drawn by the load is substantial. This is due to the voltage drop across the resistor R_H. Similarly, the voltage of the neutral may be a few volts above earth ground due to the voltage drop in the resistor R_N.

A.3 Grounding for Safety

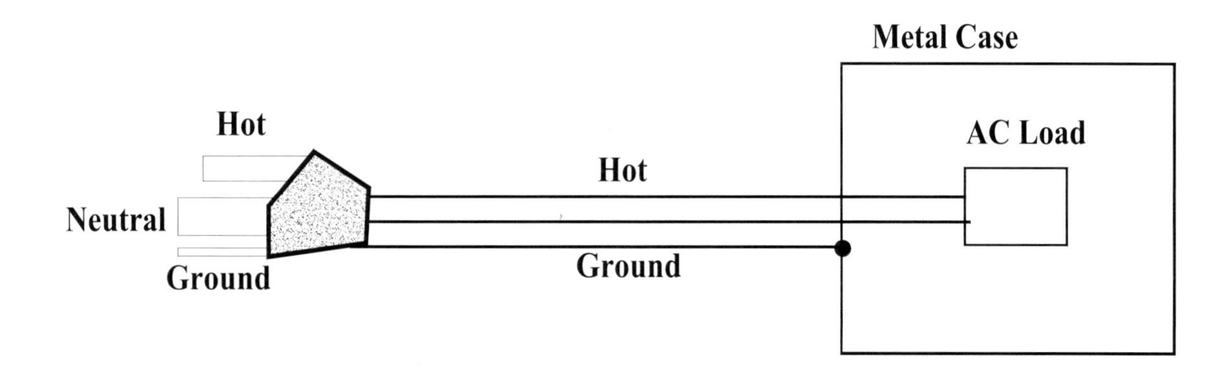

Figure A.3: Grounded instrument.

A typical load for an ac receptacle is shown in Fig. A.3. The ac load is what consumes ac power. It could be the dc power supply section of an electronic instrument or a motor of an appliance. This ac load is housed or placed inside a metal case or chassis. For safety reasons the metal case is connected to the ground wire of a three wire ac power cord. This means that, as long as the integrity of the ground wire is maintained, the metal case is always at the earth ground potential. This is usually desirable since it means that a human touching the case would remain at the earth ground potential.

With a two wire power cord there is no ground wire to ground the case containing the ac load. This has a potential shock hazard as shown in Fig. A.4. An ac load is situated in the metal case. The load is being supplied by a two wire power cord. A circuit protection device known as a fuse is placed in series in the hot wire to protect the ac load from too large a current. When the current flowing in the circuit reaches the value that the fuse is designed to blow for the fuse will melt and become an open circuit breaking the electrical circuit, stopping the flow of current, and protecting the ac load. Unfortunately, the fuse supplies little or no

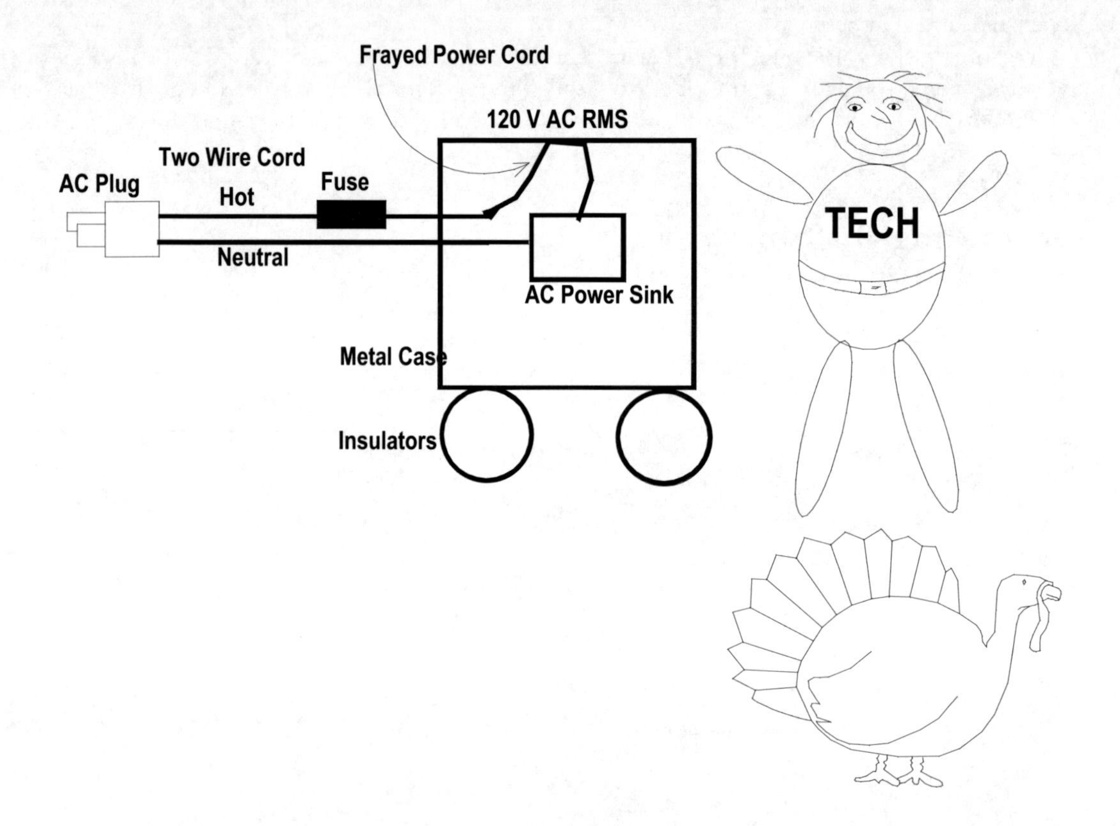

Figure A.4: Two wire shock hazard.

protection to a person touching the metal case who could receive an electrical shock (unintentional flow of electrical current though a portion of the body).

Since only two wires, the hot and neutral, are needed for an electric circuit, a two wire power cord can be more economical that the three wire cord. There is no shock hazard if the wires remain covered with insulation. However, if the insulation becomes frayed, broken down by chemical corrosion, eaten by rodents, or by some other means removed from one of these two wires, a bare wire may touch the metal frame and change the potential of the case to that of the conductor. If the neutral wire touches the frame, then the potential of the frame may only be a few volts with respect to earth ground. However, if the bare hot wire should touch the frame it would reach a potential of 120 V when, in the parlance of electrical and computer engineering, it becomes hot.

For the situation shown in Fig. A.4 the metal frame is sitting on an electrical insulator such as rubber, plastic, or wooden feet. Even if the frame reaches a voltage of 120 V ac the electrical circuit still functions normally with respect to the ac load. Current is still supplied to the load through the hot wire and back though the neutral. Since the frame is insulated nothing has happened to modify the flow of current in this circuit.

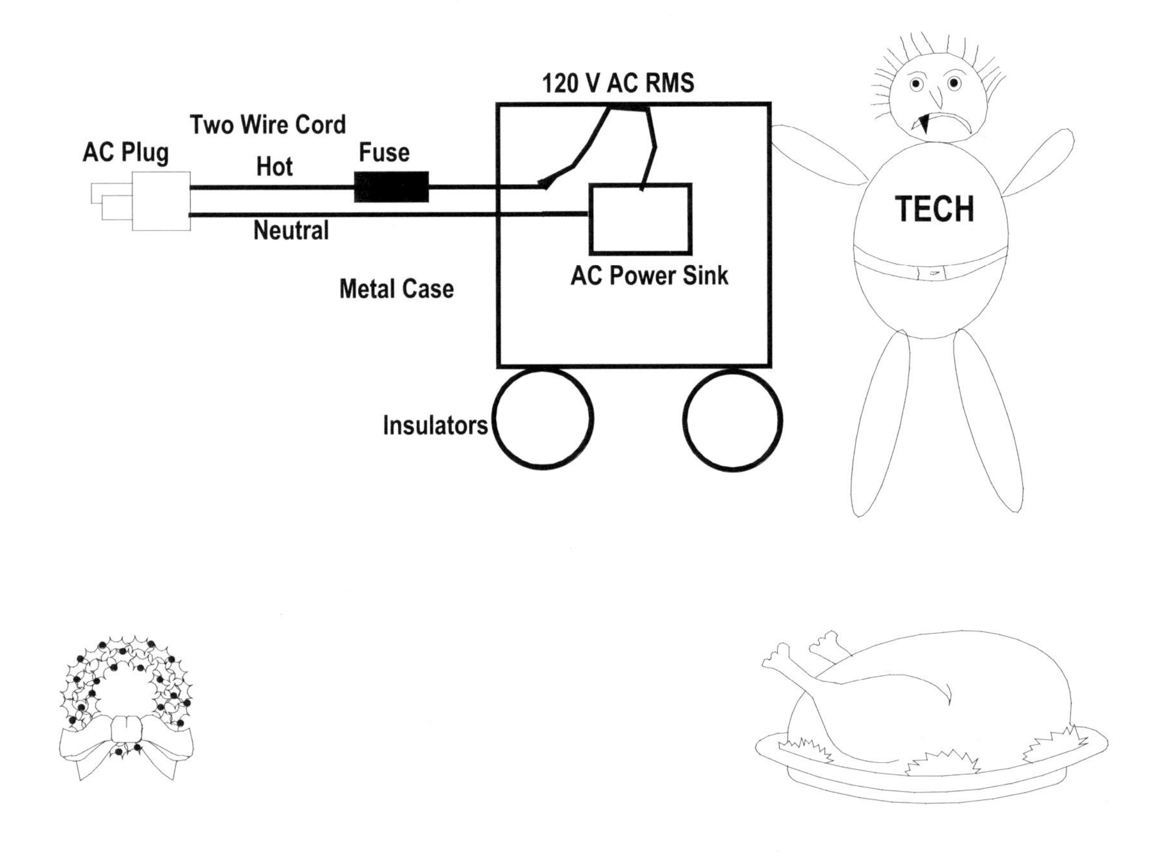

Figure A.5: Shock.

The situation could change if a biological organism such as a human touches the metal frame as shown in Fig. A.5. The person's hand would be at a potential of 120 V ac with respect to the earth ground and the feet would be at another potential. This causes a current to flow in the person which produces a sensation known as electrical shock. The severity of the shock is determined by the amount of current. If the current is large enough the fuse could blow but, by this time, the person would probably have received a severe shock. The fuse is selected to protect the ac load and not someone touching a hot case.

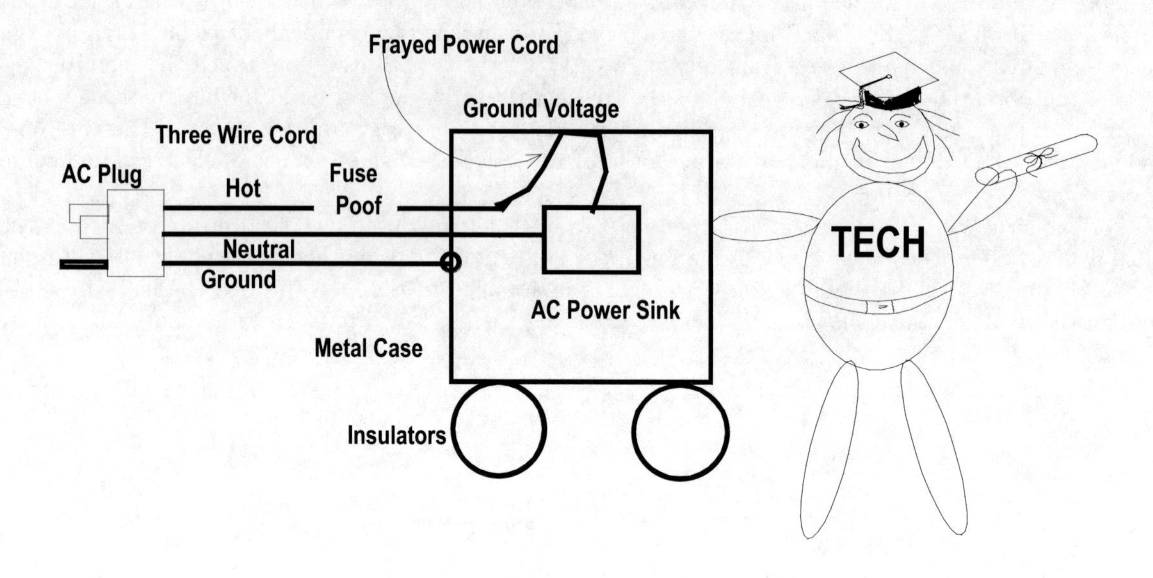

Figure A.6: Three wire protection.

The protection afforded by a three wire power cord is illustrated in Fig. A.6. The ground wire is connected to the metal case. The instant the insulation comes off the hot wire and touches the case a short circuit exists from the hot wire to the case to the ground wire. This causes a surge of current to flow in the ground wire which causes the fuse to instantaneously blow (become an open circuit and thus break the flow of current). Thus the case always remains at the earth ground potential (unless the ground wire breaks). This prevents most but not all electrical shocks in 120 V ac power systems.

Some appliances use a two wire polarized power cord. This can be inserted into a receptacle with one orientation. The case is normally connected to the neutral. This means that the case is always at a slight potential with respect to the earth ground. If the currents drawn by the load are slight, this provides almost the same safety as the three wire system. However, if there is even the slightest chance of an electrical shock hazard, the three wire power cord is preferable to either of the two wire versions.

A.4 Circuit Protection Devices

A circuit protection device is a device that monitors the current flowing in a circuit and breaks the circuit if the current exceeds a predetermined value. The purpose of the circuit protection device is to protect the ac load from excessive current which might damage it. Once an excessive current has been detected the circuit protection device switches from an short circuit to an open circuit and halts the flow of current. Three common circuit protection devices are fuses, circuit breakers, and ground fault interrupters.

A.4.1 Fuses

Shown in Fig. A.7 are some typical instrumentation fuses. A glass cylinder has a wire constructed of a special metal alloy in the center supported on either end by metal caps. The fuse on the left is known as a fast blow fuse and the one on the right is known as a slow blow fuse.

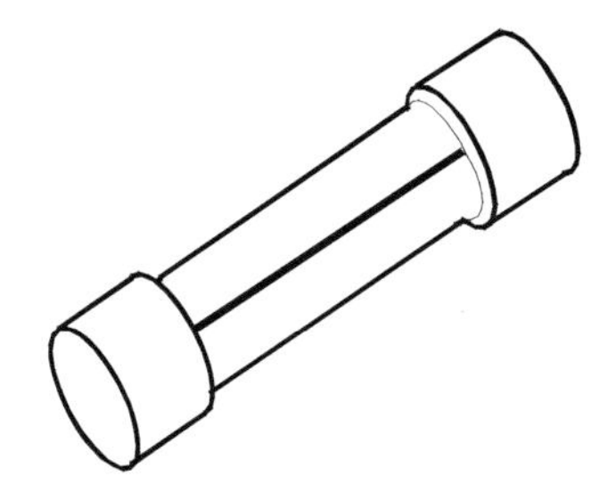

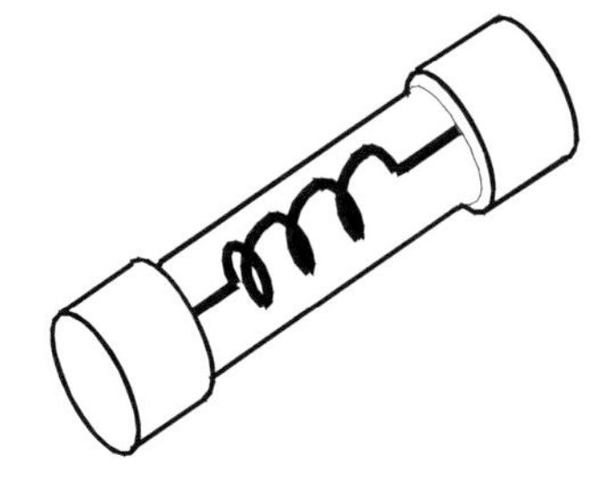

Figure A.7: Fuses.

For currents less than the value that the fuse is designed to blow for the fuse is just a section of wire or a short circuit. When the current flowing through the fast blow fuse exceeds the value that the fuse is designed to blow for the metal alloy melts and the fuse becomes an open circuit and halts the flow of current. Slow blow fuses will tolerate momentary current surges over the value of current that the fuse is designed to blow for. Once a fuse has blown it must be discarded and replaced which can be expensive. Circuit breakers and ground fault interrupters can be used perpetually since they are just switches which can be closed after they have been opened.

A.4.2 Circuit Breakers

An electromagnetic and bimetallic circuit breakers are shown in Fig. A.8. Numerous turns of the wire conducting the current are wrapped around the sensing element. Once the current in the electromagnetic circuit breaker reaches the trip value, the force exerted on the metal plunger is such that it mechanically opens a switch which breaks the electrical circuit. After the fault has been determined and removed this circuit breaker must mechanically closed.

The bimetallic circuit breakers uses a bimetallic strip as the sensing element. Two different metals are joined which have different temperature coefficients of expansion. When current flows through the wire wrapped around the sensing element the strip heats up which cause the two metals to expand by different amounts. This cause the bimetallic strip to bend and, if it is part of the conducting path, break the circuit. Once the strip cools the strip will bend back and automatically close.

Often both a bimetallic strip and electromagnetic circuit breaker are used in the same circuit. By placing the two in series protection would be afforded for an excessive current which could not be tolerated for even short instances with the electromagnetic circuit breaker and the bimetallic strip would provide protection for smaller current surges.

A.4.3 Ground Fault Interrupters

A ground fault interrupter is shown in Fig. A.9. Equal turns of wire from both the hot and neutral wire are wrapped around a metal toroid. The orientation of the turns of wire is such that the magnetic field set up in the toroid by the current flowing in the hot wire has an opposite orientation from that set up by the neutral. Thus, in a normally function ac circuit, there is no net magnetic flux in the toroid since the currents in the

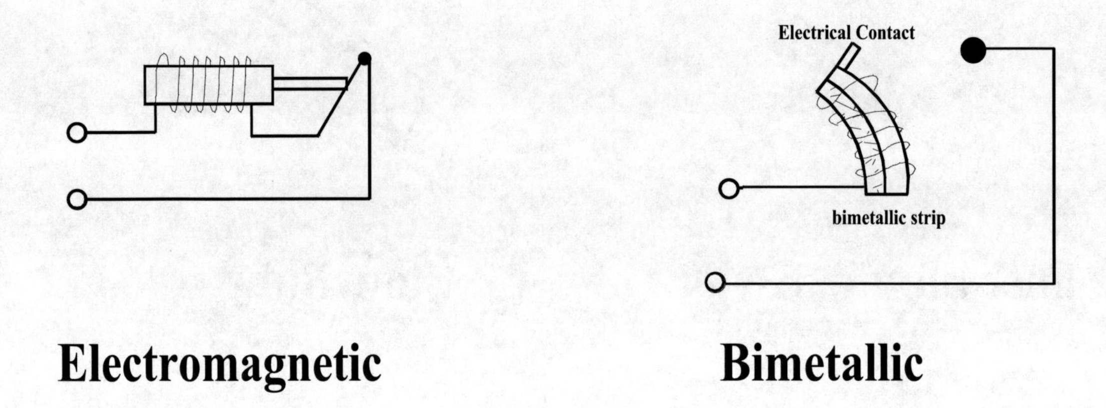

Figure A.8: Circuit breakers.

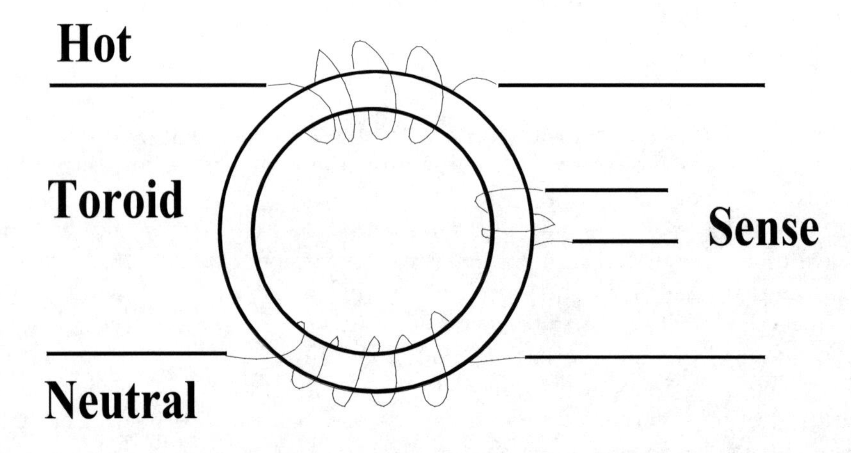

Figure A.9: Ground fault interruptor.

hot and neutral wire are equal. When a fault occurs there is an imbalance between the current in the hot and neutral wire which results in a net magnetic flux in the toroid. This time-varying flux induces a voltage in the sense coil which is then used to actuate a relay that breaks the circuit in both the hot and neutral wire. This affords the maximum protection since it breaks circuit in both current carrying conductors.

A.5 Electric Shock

Electric shock occurs when electrical currents flow through a biological organism (in the following the organism will be assumed to be a humanoid). The effect of the flow of current though the human body depends on a number of factors such as the type and frequency of current as well as the person's age, weight, sex, health, etc. The statistics found herein are averaged over a large number of test subjects.

It is current rather than voltage that produces damage to biological organisms. One can think of it as the i^2R losses in the organism. Walking across a carpet on a day when the humidity is low may induce a voltage of several thousand volts on an individual. When the person comes close enough to a large metal surface the electric field between the portion of the person's body nearest the metal surface and this metal object may reach the breakdown electric field intensity for air; this causes an arc of current to flow from the individual to the metal object in a manner analogous to a lightning stroke. Although the voltage is immense, the current is small and so the shock is merely an irritation rather than a life threatening situation. The effect of this type of shock is also mitigated in that it is a dc rather than an ac current.

The damage produced by the flow of electric current though the body is a function of the frequency of the current. Unfortunately, the susceptibility of the human body to damage by ac current reaches a maximum near the common ac power line frequency of 60 Hz. This power line frequency of 60 Hz was selected because of its importance in horology.

Current Level	**Effect**
1 mA	Threshold of Perception. Mild Tingle.
5 mA	Ouch. Slight shock. No Significant Damage.
9-30 mA (men) 6-25 mA (women)	Painful Shock. Maximum " Can't Let go" Current.
50-150 mA	Extreme Pain. Can't "Let go". Possible Death
1-4.3 A	Ventricular Fibrillation. Probable Death.
$\geq$ 10 A	Cardiac Arrest. Severe Burns. Certain Death.

Table A.1: Effects of 60 Hz Current Passing through Body

The effects of 60 Hz ac current flowing from hand to foot for 1 second are summarized in Table A.1. These statistics reflect the effects on relatively young to middle age people in good physical health. For the very young or old or the infirm lower levels of current may produce these effects.

The threshold of perception for current is about 1 mA. Below this level current cannot be perceived.

At 5 mA the shock is not painful and does no direct damage. The individual can certainly feel the current and will be disturbed by it but it does no direct damage. It may do indirect damage if the person is startled and inadvertently jumps into the path of a racing locomotive, etc.

The current range 9-30 mA (men), 6-25 mA (women) produces a painful shock. This is a particularly insidious range in that the individual cannot let go of the conductor due to muscular contractions, i.e. involuntary muscle spasms prevent the unfortunate to break contact with the electrical circuit. Unless contact is broken this may produce severe injury.

Current in the range 50-150 mA produces extreme pain. The individual still cannot let go of the conductor and muscular contractions become severe and respiratory arrest occurs. Death may occur.

For currents in the range 1-4.3 At this current level the individual will be "knocked off" the conductor. Nerve damage occurs and the muscular spasms are catastrophic. Ventricular fibrillation occurs which is interference in the rhythmic beating of the heart; the heart stops pumping blood and death will occur within minutes.

At current of 10 A or more cardiac arrest occurs, the burns are exceedingly severe, and death will almost inevitably occur. Of course there are exceptions.

Anent the frequency of the ac current. For currents in the kHz range the current is switching directions so fast that most body components can't follow it. Therefore, much higher current ranges can be tolerated for high frequency ac currents.

The amount of current that flows through the body once contact is made with an electrical circuit depends of the potential difference across the parts of the body in contact with the circuit and the impedance of the body. Body impedance depends on a number of factors such age, sex, etc. but the dominant factor is the moisture content of the skin. Dry skin may present an impedance as high as 500 kΩ whereas wet skin may be as low as 1 kΩ. Once current enters the body the impedance drops considerably since the outer layer of skin is burnt off and the blood stream appears to be a salty electrolyte of relatively high conductivity. Thus the effects of a 120 V ac, 60 Hz shock can vary from below the perception level to lethal. Therefore, it is imperative when working with electrical equipment where there may be a shock hazard to have dry hands and feet and to wear shoes with leather or rubber soles. Rings, bracelets, and other metal objects should not be worn if there is a shock hazard.

A.6 Grounded Instruments

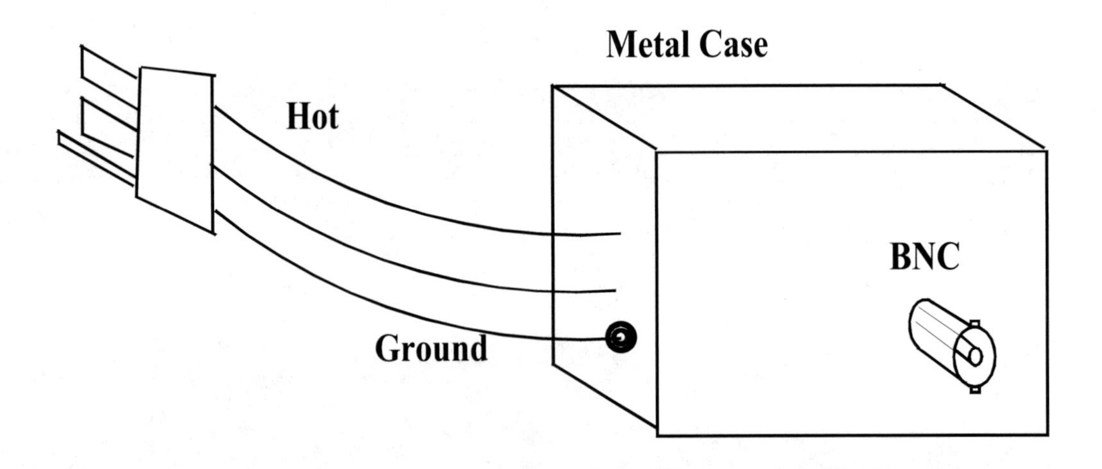

Figure A.10: Grounded instrument.

Certain laboratory instruments have one or more connectors internally connected inside the instrument to the earth ground in the ac power cord. These are known as grounded instruments. An instruments which does not have any of its connectors internally connected to the earth ground is said to "float" or the be a "floating" instrument or "to float" with respect to ground. All instruments which use 3 wire power cords have the case grounded for safety reasons but the connectors may be either grounded or floating.

Grounded instruments invariably use BNC connectors as shown in Fig. A.10. The BNC is a cylindrical conductor system for which the outer conductor, known as the shield, is grounded. (Certainly specialized instruments have the facility to float the shield but with elementary laboratory instruments a BNC connector means that the shield is grounded.) A cylindrical conductor system is much less vulnerable to pickup of noise and other interfering signals than two parallel wires.

The list of floating and grounded instruments that are found in ECE 3041 is given in Table A.2.

Floating	**Grounded**
Hewlett-Packard 3311A Function Generator	Tektronix 3012B Oscilloscope
Philips/Fluke 6303 RCL Meter	Tektronix 3022B Function Generator
Fluke 73 Hand Held Digital Multimeter	
Simpson Model 260-7 Analog Multimeter	
Agilent 34401A Digital Multimeter	
Agilent 33220A Function Generator	
Agilent 3630A Triple DC Power Supply	

Table A.2: Floating and Grounded Instruments

When two grounded instruments are used in the same circuit the possibility of a ground loop exits. A conducting path exits through the ground wire common to the two instruments. Because the resistance of this path is small large currents can be induced through time varying magnetic fields and a potential difference along the ground path. These ground loop currents may interfere with the measurement being made. Therefore, it may be desirable to float one of the instruments by using a 3 wire to 2 wire adapter. One needs to be careful because this also floats the case and a possible shock hazard then exits.

A.7 Measurement with Grounded Instruments

Grounded instruments must be used with care in making measurements in circuits to assure that an unintentional path is not introduced through the ground wire. In the circuit shown in Fig. A.11 a circuit is being energized by a grounded function generator and a grounded oscilloscope is being used to measure the voltage across circuit element Z_2. However, circuit element Z_3 is being shorted out due to the grounds for the function generator and oscilloscope being placed on either side of it. Only one node in a circuit can be used as the ground node and attempts to make more than one node ground node will introduce unintentional short circuits.

To make the desired measurement either the function generator or oscilloscope could be floated by placing a 2 to 3 wire adaptor in the AC plug for the instrument. It could also be floated by using an isolation transformer which is a transformer with a 1 to 1 turns ratio so that the primary and secondary voltages are the same but the secondary has no ground connection.

Another way of correctly measuring the voltage across circuit element Z_2 would be to make a differential measurement as shown in Fig. A.12. This requires an oscilloscope with two input channels. The grounds for the function generator and both oscilloscope channels are connected to the bottom of the function generator and the connections for the center conductor for the oscilloscope leads and function generator are as shown. The ADD feature is used and the $Ch\ 2\ INVERT$ function is activated. If the $VOLTS/DIV$ setting for both channels are the same this results in

$$Ch1 - Ch2$$

being displayed on the screen which is the desired voltage. Most digital oscilloscopes have a math functions menu and can directly display $Ch1 - Ch2$.

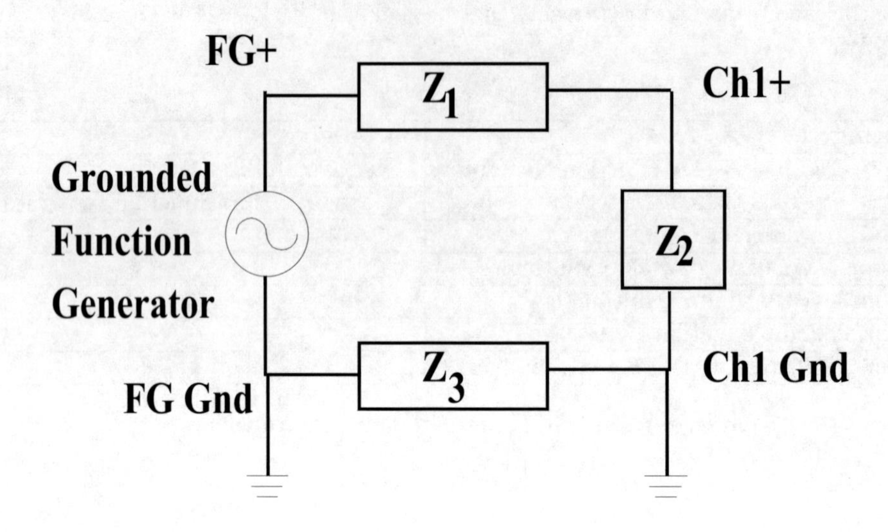

Figure A.11: Grounded function generator and grounded oscillocope.

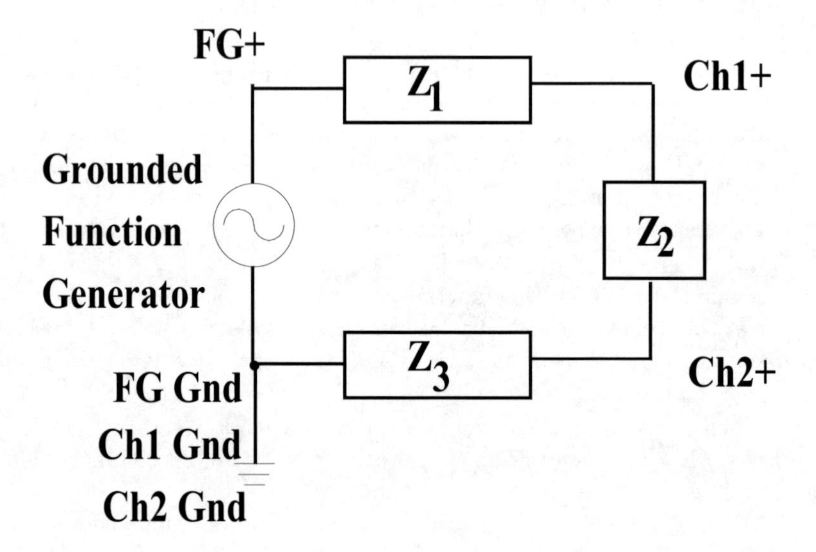

Figure A.12: Differential measurement.

A.8 References

1. *Guidelines for Incorporating Safety and Health into Engineering Curricula, Volume 1 Laboratory Safety*, Joint Council for Health, Safety, and Environmental Education of Professionals, Savoy, Illinois, 1994.
2. S. Wolf, *Guide to Electronic Measurements and Laboratory Practice*, 2nd edition, Prentice-Hall, 1983.
3. S. Wolf and R. F. M. Smith, *Student Reference Manual for Electronic Instrumentation Laboratories*, Prentice Hall, 1990.

Appendix B

Guidelines for ECE 3041

This is intended to be an introductory laboratory course in electrical and computer engineering. It is an instrumentation laboratory in which familiarity with basic electrical and electronic laboratory instruments is obtained by performing a set of elementary experiments. Each of these experiments has been selected, in part, to illustrate or feature the proper use of laboratory instruments and/or to highlight some practical problems encountered in making electrical measurements—there are no gedanken experiments.

B.1 Organization

This course consists of both a recitation and a laboratory session. These weekly sessions should be complementary and well coordinated.

The recitation session is a one-hour lecture in which the experiment for the week will be discussed. Some facets of this discussion should include: any relevant theory needed for the experiment, a description of any new equipment to be used, the format for the laboratory report, and, when applicable, the homework assignments. The experiment should be read before the recitation session so that any pertinent questions can be directed to the recitation instructor.

In the laboratory session, which occurs after the recitation session, the experiment is performed. There is normally one student at each laboratory station or bench.

A clear understanding of the purpose of the experiment and the data to be taken will result in a smoothly proceeding experiment which can be performed with a minimum of difficulty. A lack of proper preparation will be a hindrance in the performance of the experiment and may result in chaos. Proper preparation will also aid in the recognition of erroneous data and the determination of the cause of this faulty data. It is important that this be determined while performing the experiment, since it will, normally, not be possible to repeat the experiment after leaving the laboratory.

B.2 Conduct of the Experiment

The procedure section of each experiment details the steps to be taken. If a circuit diagram is given, the circuit should be connected will all energy source [both AC and DC] turned off. The circuit should be checked to make certain that it is correct prior to the application of power.

Some of the experiments in this course use an analog multimeter. It should be set to the correct function (volts or amps) prior to the application of power to the circuit. This analog meter should also be set to its maximum range. Failure to do so may result in a blown fuse or damage to the meter.

Reasonable efforts are made to maintain equipment in proper working order and calibration. If the equipment is malfunctioning, inform the laboratory instructor so that the equipment can be repaired or a replaced. Never remove equipment from another table without the permission of the laboratory instructor.

Occasionally, an instrument's fuse may blow during an experiment. If this happens, it is best for the laboratory instructor to check the fuse. When the fuse is checked, the instrument should first be unplugged from the AC power line before removing the fuse from the fuse holder. Failure to do so may result in a painful electrical shock.

B.3 Safety Precautions

Electrical or computer engineering laboratories contain equipment that can be hazardous. Lasers, microwave equipment, and rotating machinery require some rather obvious safety precautions. The most common hazard to be found is electric shock. Although most students never receive even a mild shock, it is best to insure that this does not happen by understanding something about the nature of electric shock and following some common sense safety precautions.

The case or chassis of all the AC instruments used in electrical or computer engineering laboratories are connected to the earth ground via the ground wire in the AC power cord to prevent the possibility of electric shock. Thus, the case of all AC instruments (those which have a power cord which must be plugged into an AC receptacle to operate the unit) are "grounded." One of the input or output connectors of an instrument may be internally connected to the ground wire; if this is the case, this instrument is said to have a "grounded" connector. If none of an instrument's input or output connectors are internally connected to the AC ground wire, the instrument is said to have "floating" connectors.

The following instrument that is used in ECE 3041 has a "grounded" connector: Hewlett-Packard 54602B Oscilloscope. The connector that is grounded on this instrument is the cylindrical shield on the BNC connector. All of the other instruments that are used in this course have "floating" connectors.

Electric shock is caused by current passing through the body. It is related to voltage, only in that the current is given by the voltage divided by the body impedance. Thus, a voltage of $1,000\ Volts$ may produce only a tingling sensation, while a voltage of $115\ Volts$ may be fatal. This is caused by the wide variation in the impedance of the body to the flow of electric current. For instance, dry skin may have an impedance as high as $500\ k\Omega$, while wet skin may have an impedance as low as $1\ k\Omega$. Also, electric shock depends on the type of current. The power line frequency $60\ Hz$ produces a more serious shock than either DC or high frequency currents.

The effects of electric shock vary widely from person to person and are also functions of age, sex, and physical condition. However, some statistical averages have been determined. The threshold of perception is about $1\ mA$. Currents from $1\ mA$ to $5\ mA$ are harmless. At about $10\ mA$, involuntary muscular contractions occur that make it difficult to remove one's hand from the source of current. Current levels from 100 to $300\ mA$ produce ventricular fibrillation which may be fatal within minutes. For currents above $300\ mA$, the muscular contractions of the heart are so severe that ventricular fibrillation (an interference with the rhythmic beating of the heart) does not occur. Above 6 $Amps$, cell destruction will occur. (All of the above current levels are for $60\ Hz$ AC.)

Electric shock can usually be prevented by following these common sense safety rules:

1. Never touch an uninsulated, energized conductor. First turn the energy sources off.

2. Never touch an electric circuit with wet hands. The resistance of wet skin is much lower than dry skin.

3. Always wear shoes in an electrical or computer engineering laboratory. Dry shoes act as a resistor in series with the feet and can, therefore, greatly decrease the current that results from touching an energized conductor and standing on a grounded conductor. Safety is more important than making a fashion statement.

With the exception of the AC power line, all of the current levels encountered in ECE 3041 laboratory are nonlethal. It is unusual for anyone to receive even a mild shock. However, it is best to develop good safety procedures early in one's career.

B.4 Laboratory Exam

The laboratory exam is a ten-minute exam administered at the beginning of each laboratory session. Anyone arriving late will receive a grade of zero. The subject of the exam is either the experiment for the current week or the previous week's experiment. Whether the exam is open or closed book will have been announced by the laboratory instructor the previous week. Should the laboratory instructor fail to make this announcement it should be assumed that the exam is closed book.

B.5 Laboratory Report

There are two types of laboratory reports used in this course: formal and informal. The formal laboratory reports are technical papers that must be produced on a word processor and meet composition, style, and grammatical standards expected in any technical report or paper that might be submitted to a technical journal or government agency. The informal reports are simple hand-written summaries of the experiment. Since they involve considerably more effort and are exercises in technical writing, the formal reports are weighted more heavily than the informal reports.

B.5.1 Informal Laboratory Report

An informal laboratory report is performed at the conclusion of the experiment. It is a hand-written summary of the data taken, any plots that are requested, and the answers to any and all questions found at the end of the experiment as well as those that may have been posed by the laboratory instructor. It is to be submitted to the instructor prior to leaving the laboratory.

B.5.2 Formal Laboratory Report

This is a technical paper and must meet the standards for such a document. It is both a report of the results of the experiment and an exercise in technical writing. It will be graded for both content (20 %) and style (80%).

The communication of the results of an experiment is as important as the care taken in performing the experimental work. No matter how important the findings of an engineer may be, they are useless unless he/she can communicate them to others. Consequently, much of an engineer's time is spent in writing reports on his/her own work or the work of the technicians or junior engineers working under him/her. Therefore, it is necessary to cultivate the ability to write a clear technical report.

The purpose of writing laboratory reports in instructional laboratories is twofold: first, to develop a report-writing ability; second, to assemble the information gathered in the laboratory into a form in which it can be easily remembered by the writer and presented to the reader. It should be emphasized to students taking an instructional laboratory course that they are being evaluated on both their experimental techniques and their ability to write acceptable technical reports.

There is no "best" way to write a laboratory report; rather, there are many good ways. A report should reflect careful organization, proper English, neatness, and attention to detail. The 5 C's of excellent technical communication are the following:

1. Clarity
2. Conciseness
3. Correctness
4. Consistency
5. Comprehensiveness

Indeed, excellent technical communication should be clear, concise, correct, consistent, and comprehensive. A guiding principle to keep in mind throughout report writing is that any outsider with a technical background similar to the report writer's should be able to duplicate the entire experiment, data, and conclusions by reading the report. Do not trust memory to fill in details—after a few days the experiment may be a mystery even to the report writer.

A good report should be carefully and concisely written and should be free from grammatical errors and misspelled words. A suggested outline for a laboratory report for ECE 3041 (which should be strictly adhered to unless the laboratory instructor directs otherwise) is as follows:

1. Each experiment must have a cover sheet or title page which contains the name of the Institute, the College, the School, the course, the section, the day of the week and the hours for the laboratory section, the name of the experimenter, the semester, the date the experiment was performed, and the date the report was submitted. This cover sheet or title page is the page in the report that appears first.

2. The Pledge of Academic Honesty is placed immediately after the cover sheet page. It must be signed with the student's name and number, or the report will not be graded.

3. Table of Contents. This lists the major sections of the laboratory report. Indicate the page number for each major section. It is not necessary to begin each section of the report on a new page, but a section heading should not be the last line on a page. The Table of Contents Page follows the Pledge of Academic Honest page.

4. The first part of the laboratory report is to be an overall Objective, and the final part is to be the General Conclusions or Summary. (Any appendixes follow the General Conclusions. Each laboratory report has a least one appendix, which is the rough data sheet signed and dated by the laboratory instructor.) This is so that a reader can tell what the experiment was about by a glance at the Objective section of the laboratory report and what degree of success the experimenter achieved by a glance at the General Conclusions section of the laboratory report. The Objective, Introduction, Purpose of the Experiment, Abstract, or whatever name is chosen for this initial section should appear as the first item on page 1 (it is not necessary to number page 1). The Conclusions or Summary should appear on the last numbered pages prior to the appendices.

5. Most of the experiments in the ECE 3041 laboratory manual end with a section called "LABORATORY REPORT." The portion of the report between the Objective and General Conclusions must be written in the order indicated under "LABORATORY REPORT" in the laboratory manual. These steps generally follow the order of the steps in the procedure.

6. Any numerical results must be placed in a box. If there are many numerical results, then they must appear in a table. This rule must also be followed with the homework assignments. The gist of these guidelines and/or rules are the following simple instructions: The report must be written in the order indicated in the laboratory manual, and numerical results must appear in a box or a table. Each time these instructions are violated, a 1/2 point penalty will result for the grade for the laboratory report (the maximum grade for a laboratory report is 10 points). In short, the report should be a concise, intelligible, and readable account of what was done. It should never contain a personal pronoun. It should never contain statements such as: "I did...," "We did," "Lee Roy measured the voltage...," or "Do". Instead, the wording should be as follows: "The circuit was connected as shown in Fig. 3, and the following data was taken," etc. Anthropomorphisms such as "the oscilloscope didn't feel well" are not acceptable—don't attribute emotions or feeling to inanimate objects. The report should be prepared on a word processor. Curves and diagrams must be drawn with a CAD system. Computer simulations such as SPICE should be included as an appendix. Results from simulations should be placed in the main body of the report when appropriate. The following are some common mistakes made when writing laboratory reports:

- Cover page missing information. The cover page should contain all the information requested. This is the first thing that a reader sees; thus, it should also be free of formatting errors and misspellings. Make certain that it includes the "College of Engineering" and the "School of Electrical and Computer Engineering". Chose font sizes and spacing that make the cover attractive, impressive, and informative.

- Pledge of Academic Honesty not included or not signed. The report will not be graded. The student must resubmit the report and receive a penalty for a late report.

- Answers not boxed. All final answers or results should appear in a box. This also applies to homework assignments.

- Circuit diagram not included with SPICE simulations.

- Inadequate conclusions. Pertinent conclusions should appear at the end of each section of the report, as well as those at the end of the report. The word "pertinent" means that some sections of the report may not be important enough to require a conclusion.

- Improper use of tenses. The Objective and Procedure sections are written in the past tense. The experiment was performed in the past and, therefore, the description of the procedure used for the accumulation of data, etc. should be in the past tense. The theory, data, simulation, and conclusion sections are written in the present tense. Stating fundamental scientific and engineering principles in the present tense does not constitute a tense change.

- Improper use of pronouns. Personal pronouns are not acceptable in a laboratory report and are not used. Besides, it is not necessary to say "I" or "we" performed anything; the names of the experimenters are on the title page, which removes any ambiguity about who performed the steps in the procedure. Correct usage is, "the function generator was connected to the oscilloscope" and not, "we connected the function generator to the oscilloscope."

- Casual and Imprecise Language. "This HP meter worked pretty good," or "We couldn't get the plots to print right, so we sketched them by hand," or "It looked to us like," or "the percent error was reasonable," are not acceptable. Be specific and exact.

- No labels for tables, graphs, circuits, and/or calculations. All tables, graphs, circuit, calculations, and other non-text figures should be clearly and logically labeled. The label should not only include a sequential number, but also a description of what is being labeled. Examples Circuit 3, Nonelectronic Voltmeter; Calculation 7, Example of Percent Error Calculation; Table 2, Theoretical and Measured Voltage with Percentage Error; and Graph 6, Voltage vs. Frequency for RCL Circuit of Circuit 5. Labeling something with the step in the procedure which uses it is not acceptable—Example: Step 6 would not be a appropriate title for a graph.

- Improper table format. Tables must have lines clearly separating rows and columns. Furthermore, decimal points should be aligned if the word processor being used is capable of this. All circuit diagrams, tables, sample calculations, figures, etc. should appear in the body of the report with the procedure step with which they are first associated.

- Graph format incorrect or inadequate. All graphs and/or figures should be computer generated. Axes should be properly labelled, as should multiple plots per graph. The graph should have a logical title that describes what is being graphed, not just the procedure step with which the graph is associated.

- Raw data just stapled to report. Raw data should be included as an "Appendix" to the report, not just stapled on at the end without a title or page numbers. Each report should contain at least one appendix (the raw data with the signature or initials of the laboratory instructor). The pages of an appendix should be numbered with a letter prefix indicating the appendix. If there are multiple pages in an appendix, they should be numbered with Arabic numerals following the appropriate letter prefix.

- Number pages. All pages of the report should be numbered including any appendices except for the first page in the main body of the report. The sections prior to the Objective should be numbered with lower case Roman numerals. The Objective should begin a new page—although this is page 1 of the report, no number appears on this page.

The laboratory report should be turned in the week after the experiment was performed along with any homework assignments made by the recitation instructor. Severe penalties will be imposed for late reports. Students' questions about the grading of an assignment should occur in the laboratory instructor's office and not during the laboratory session.

Appendix C

SPICE

SPICE is a computer program for the simulation of electronic or electrical circuits. The version of SPICE that is used in the ECE instructional labs at The Georgia Institute of Technology is the student version (aka evaluation version) of PSpice which is a version of SPICE that was developed by the MicroSim Corporation for use on personal computers using either the MS DOS or the Windows operating systems. The rights to the evaluation version of PSpice are currently owned by Cadence.

Two entry modes are available to use PSpice: the Text Editor and the Schematic Editor. The Text editor is used to create DOS code for the PSpice compiler. The Schematic Editor is used to make a drawing of the circuit and the code is generated from the drawing. Only the Text Editor will be described in this appendix. The Schematic Editor is discussed in the text *Herniter, M. E., Schematic Capture Using Cadence PSpice.* The text by Herniter also has a CD ROM with an improved version of the evaluation version of PSpice.

PSpice performs three types of analysis: 1) `DC`, 2)`AC`, and 3)`TRANS` (transient). Only the `AC` and `TRAN` analyses will be examined. When an `AC` analysis is performed the voltage at each node in the circuit is calculated at the frequency or frequencies specified by the user. The `TRAN` analysis requires that the user specify the time interval to be examined and the pertinent initial conditions. In addition to the node voltages, PSpice has the ability to calculate mathematical functions of these node voltages.

The first step in the use of PSpice is to label each node in the circuit. There must be a least one circuit element between each pair of nodes. There must be a node labeled node "0" which is used as the datum or reference node, i.e. each of the other node voltages are calculated with respect to this node. There must be a dc path to ground for each node (If two capacitors are in series there is no dc path to ground for the node between the capacitors. In this case one would place a large resistor in parallel with one of the capacitors.)

The text editor is used to create a source file; this source file must be of the form `MYFILE.CIR`. The first line in the source file must be a title line. Following the title line there are lines of source code specifying each circuit element, the two nodes at the endpoints of the circuit element, the type of circuit element, and its value. A distinct symbol or name must be used for each circuit element; this symbol may have a length of 7 letters or symbols with the first letter specifying the type of circuit element. The lines of source code for common circuit elements have the form:

VXXXXXX N+ N- AC MAG (Independent AC Voltage Source)
IXXXXXXNT NA AC MAG (Independent AC Current Source)
RXXXXXXN1 N2 RES(Resistor)
CXXXXXXN1 N2 CAP(Capacitor)
LXXXXXX N1 N2 IND (Inductor)

The first letter specifies the type of circuit element, the XXXXX's can be either letters or numbers, blanks are usually used as delimiters, the N′s simply mean node numbers (Note that the order for the nodes is important only for voltage and current sources. For a voltage source, the first node is the node of the plus terminal while for a current source the first node is the node of the tail of the arrow.)

After the specification of the circuit elements control statements are required to specify the type of analysis to be performed, to load the graphics post processor (a fancy name for the part of the software package that lets the user draw the desired plot on the screen of the PC and then send a hard copy to the screen). For an AC analysis, the control line is

.AC DEC NPTS START STOP

where the leading "period" specifies that it is a control line, AC states that it is an AC analysis, DEC specifies that the frequency will be plotted on a log axis, NPTS is the number of points per decade (10 is plenty), START is the start frequency, and STOP is the stop frequency. The next control line is

.PROBE

which loads the graphics post processor. The final control line is

.END

which, of course, tells the software package that this is the end of the source file.

After the source file, `MYFILE.CIR`, has been created using the text editor, PSpice is then run by the MS DOS control statement

PSPICE MY

which causes the execution of the program. If no error have been made, the graphics post processor will be automatically loaded after the computer program has been run; the graphics post processor is a menu driven program which is self explanatory. If errors have occurred, then use the text editor to examine a file called `MYFILE.OUT`.

Additional information on using PSpice with the Text Editor can be obtained from Banzhaf, W., *Computer-Aided Circuit Analysis Using SPICE* or Tuinenga, *SPICE: A Guide to Circuit Simulation and Analysis Using PSpice*.

For a transient analysis, the voltage and current sources must be specified as functions of time. The control line that is used is

.TRAN PS FT RD SC UIC

where PS is the print step size which sets the time interval at which the results are printed, FT is the final time or last value of time for which the results will be calculated, RD is the results delay which is amount of time that passes before the results are printed (usually 0 seconds), SC is the step ceiling which sets the time step size for the numerical solution of the circuit differential equation.